Luminous Clay

BELOW: *Unfurled Light II*, Angela Mellor, 2016. Bone china paperclay. 31 x 48 cm (h x w). *Photo: Stephen Bond.*

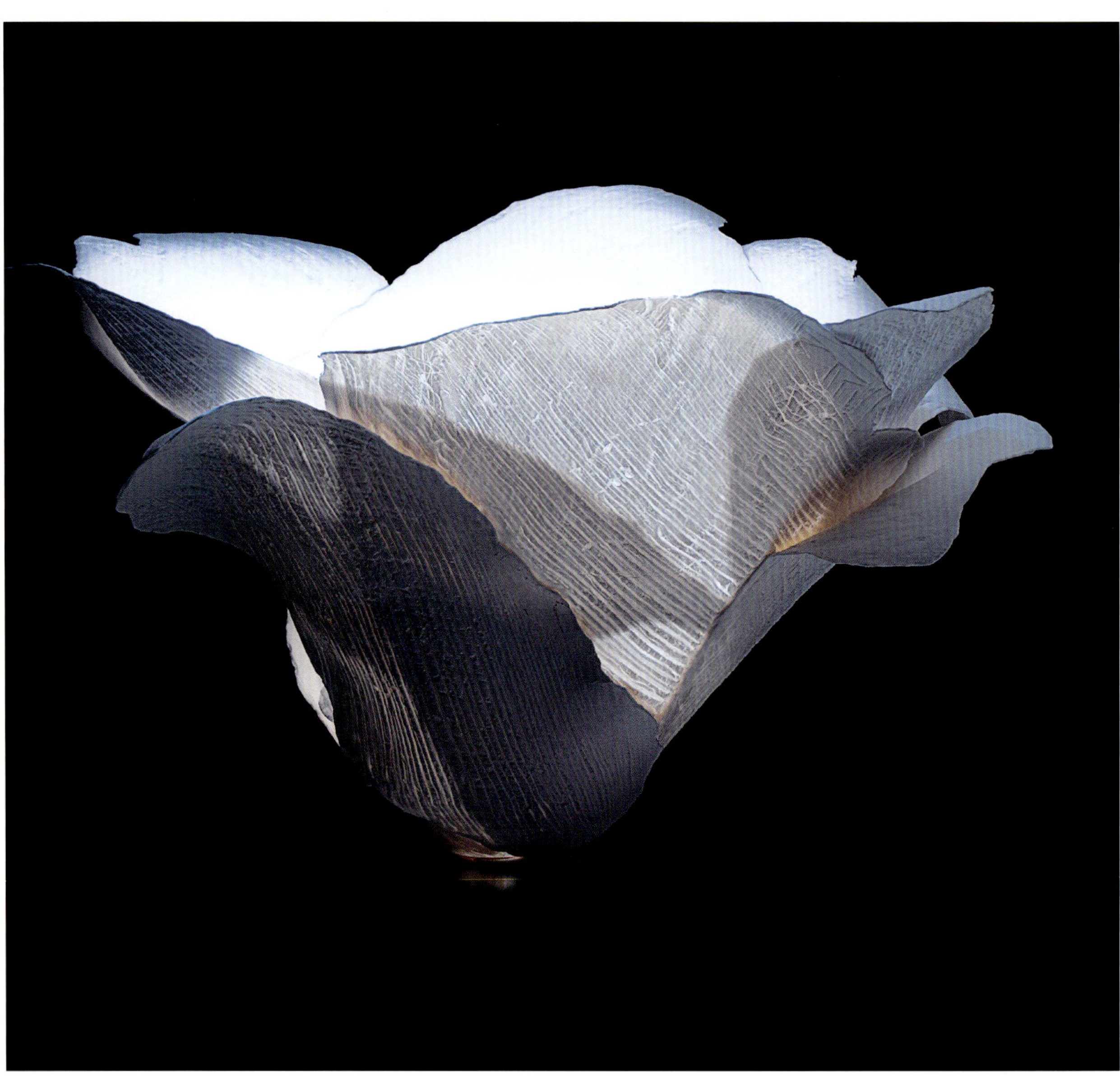

Luminous Clay

Working with Bone China and Porcelain Paperclay

Angela Mellor

HERBERT PRESS
LONDON • OXFORD • NEW YORK • NEW DELHI • SYDNEY

HERBERT PRESS
Bloomsbury Publishing Plc
50 Bedford Square, London, WC1B 3DP, UK
Bloomsbury Publishing Ireland Limited,
29 Earlsfort Terrace, Dublin 2, D02 AY28, Ireland

BLOOMSBURY, HERBERT PRESS and the Herbert Press logo are trademarks of Bloomsbury Publishing Plc

First published in Great Britain 2025

A catalogue record for this book is available from the British Library

Library of Congress Cataloguing-in-Publication data has been applied for

ISBN: 978-1-78994-389-4; eBook: 978-1-78994-390-0

2 4 6 8 10 9 7 5 3 1

Designed and typeset by Laura Woussen Design
Printed and bound in China by C&C Offset Printing Co., Ltd.

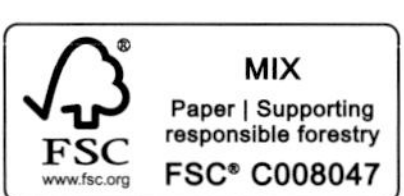

FRONT COVER:
(Top) *Coastal Light* bowl, Angela Mellor, 2012. Bone china with paperclay inlay. 13.5 x 22 cm (h x w). *Photo: Stephen Bond.*
(Bottom) *Coastal Light* (detail), Angela Mellor, 2012. Bone china paperclay. *Photo: Stephen Bond.*

BACK COVER:
(Top left) *Symphony*, Arnold Annen, 2020. Porcelain. 36 cm (h). Gas kiln reduction fire 1330°C. *Photo: Arnold Annen.*
(Bottom right) *Bowls*, Arnold Annen, 1993. Porcelain translucent. 27 x 20 cm (diam x h). Wheel thrown and turned, decorated with latex resist, brushed over and trailed with porcelain slip, unglazed. Gas kiln reduction fire 1260°C. *Photo: Reto Bernhardt.*

SPINE:
Hibiscus flower.

Contents

BELOW: Angela in her gallery, 2019. *Photo: Layton Thompson for* Ceramic Review, *Issue 298, July/August 2019.*

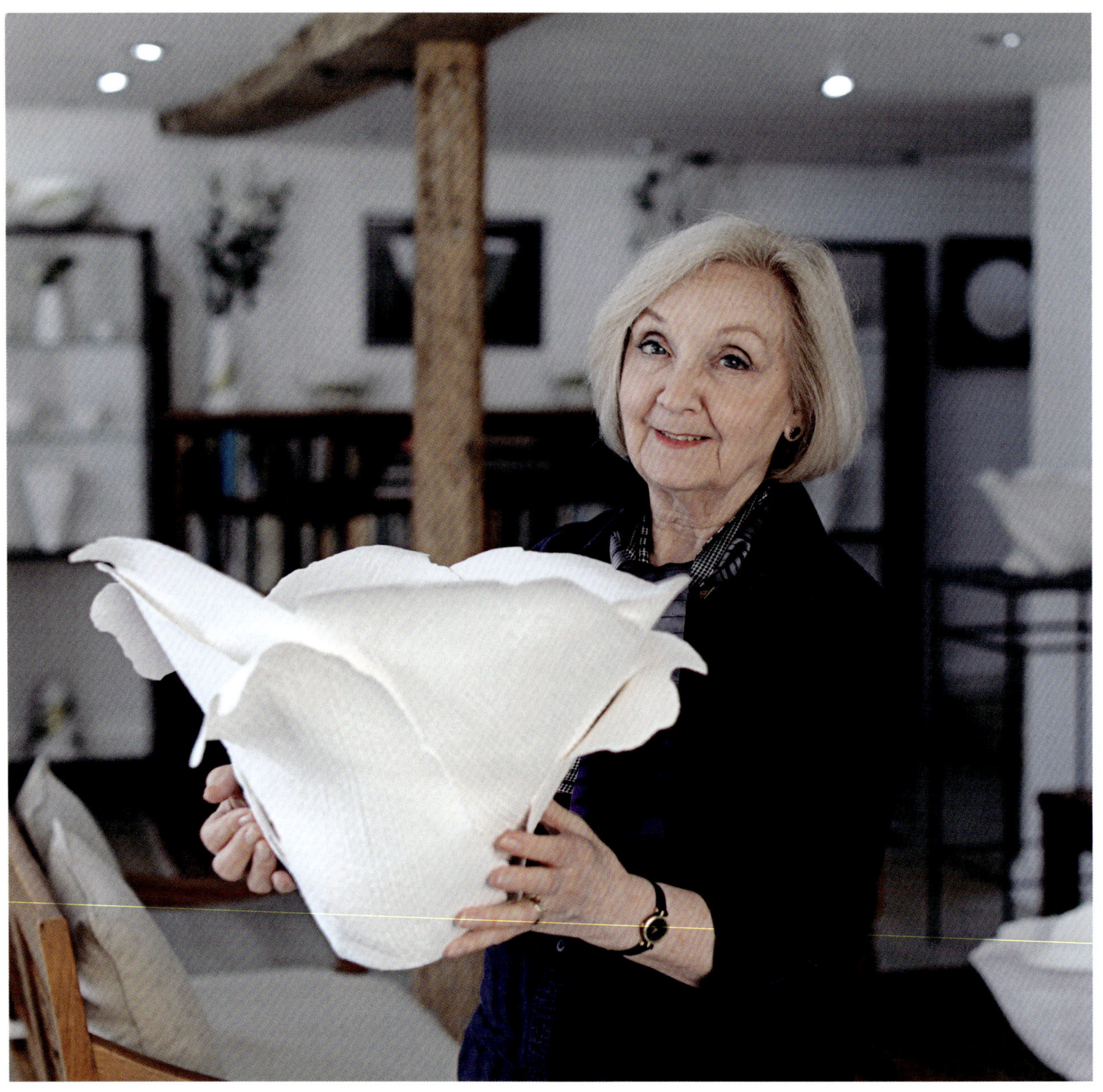

Foreword

I first met Angela Mellor when she arrived at the University of East Anglia (UEA) on an advanced teaching course. She had chosen Ceramics Education as her main subject of interest. While at UEA, porcelain became her chosen material, its potential for translucency encouraging her to research it further through a variety of exciting white vessel forms.

Moving to live and study in the bright light of Australia immediately appealed to Angela, who continued to search for the maximum transmission of light achievable in ceramics. She not only used porcelain, but also adapted bone china, which is more commonly used in industrial production. Her enthusiasm for it was further fuelled by producing some sets of 'light works' inspired by coastal images in collaboration with a lighting designer. She travelled widely around Australasia during her ongoing research. At this time, Angela pioneered the use of bone china paperclay, which allowed a previously impossible freedom of expression through hand-building with bone china. She has received honourable mentions for her work in Japan and South Korea.

Those potters who choose to work in porcelain or bone china do so because their very natures dictate it. They lean towards a perfection that is slowly and meticulously achieved. The road ahead is hedged by technical difficulties, and patience and respect are required owing to the limitations imposed by the material; there is little leeway for the spontaneous approach. If a degree of translucency is to be achieved, the wall of any porcelain or bone china vessel must be made as thin as possible.

This book reveals the increasing number of ceramists using porcelain and bone china paperclay, and I am certain that it will have a broad appeal to students, teachers, potters, and collectors alike.

Peter Lane
Potter, author, and expert in working with porcelain

BELOW: *Coastal Light* bowl, Angela Mellor, 2014. Bone china with paperclay inlay. 13.5 x 22 cm (h x w). *Photo: Stephen Bond.*

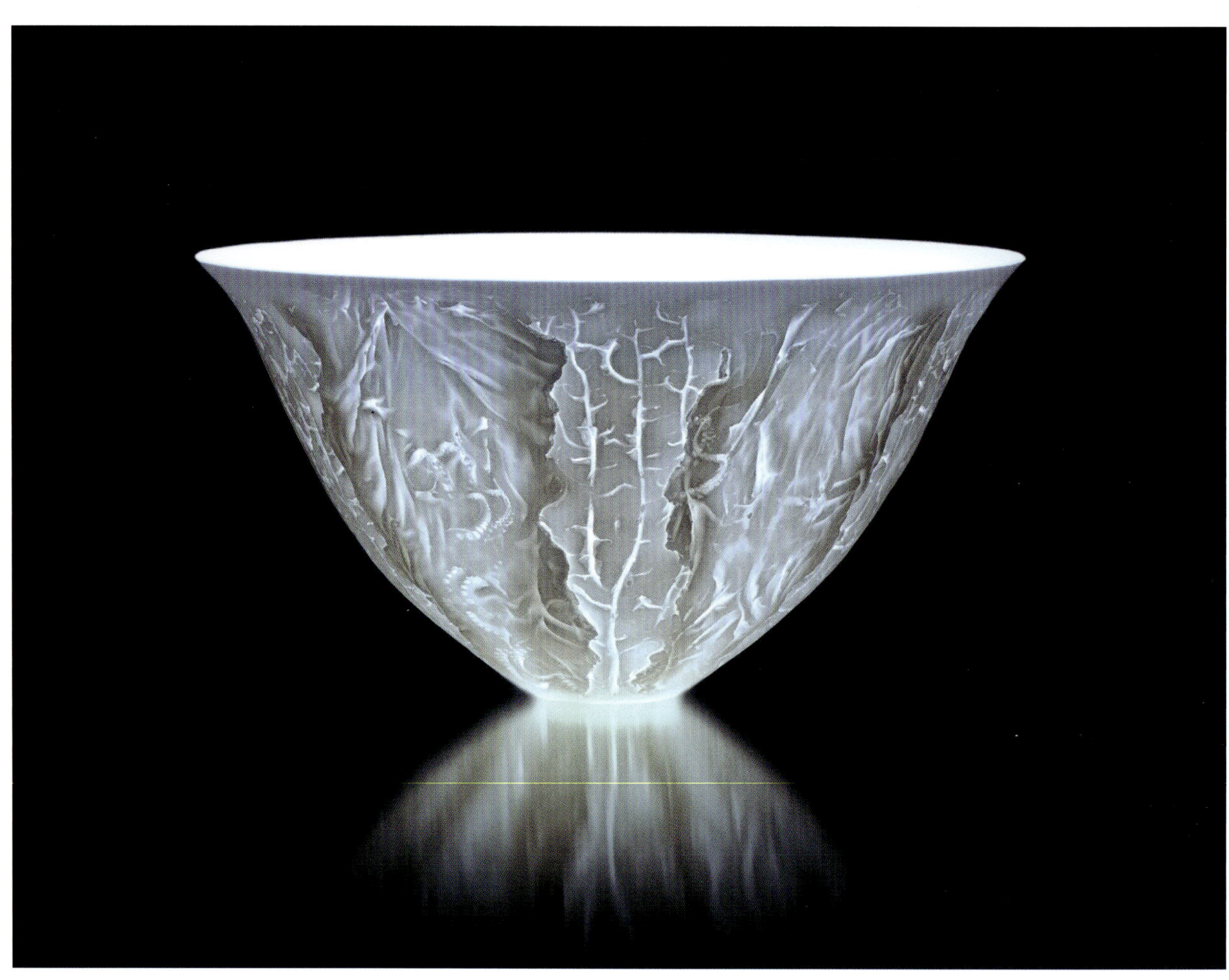

Introduction

My personal journey

My life completely changed when my sons, then in their early twenties, decided to leave the nest and see the world. They took off on their great adventure with their girlfriends in tow and rucksacks on their backs. Little did I know then what impact that would have upon my life! I was still teaching ceramics in Cambridge, feeling lost and very much alone. This was the first time in my life I had been without my boys, having brought them up single-handed since they were little.

As Christmas approached, I felt very depressed at the thought of spending it on my own. When my son Nick called and said, 'Why don't you join us in Bali for Christmas?' it came as quite a shock, and the thought of travelling all that way alone was a little intimidating. However, the nearer it got to Christmas, the more I realised that this is what I needed to do. Nick said that he would arrange the accommodation when I booked the flight.

Back in 1994 there were no mobile phones, and the only way to let him know I was coming was to write to him in Bali using the Poste Restante service. Although he never thought I would actually join him, he checked his post just in case and found my letter, and was – thankfully – at the airport to meet me. As I was unsure whether he would be there, I had booked accommodation at a beautiful complex near the airport, and we all went there.

After three glorious weeks exploring the island, its flora, fauna, marine life, and the wonderful hospitality of the Balinese people, it was time to return to the UK. I was feeling refreshed and ready for the start of the new term. I was unaware that a chance meeting on board the plane home would change my life! I started talking with the man I was seated next to; Manfred Dinse was returning from his hometown of Perth, Australia, to work in Jakarta. He had lost his wife, and had been home to see his four daughters for Christmas.

He proudly showed me photos of his girls, who were in their early twenties, and I reciprocated, showing him photos of my two boys of the same age. He said he had always wanted a boy and could we do a deal! This made me laugh, and we exchanged life stories over the next hour and a half. He was German by birth but had moved to Australia as a teenager. He handed me his business card and asked me if he could write to me. He also asked me to send him photos of my work. I was very impressed by his honesty, openness, and devotion to his family.

After sending a letter and photos of my work, I heard nothing back, which I thought quite strange given his keen interest. However, early on Easter Sunday morning, I received a phone call from him, wishing me a happy Easter. I was taken aback and asked why he had not replied to my letter; he said he had not received it. We reconnected, and he began faxing long letters, detailing his family life and work in Indonesia. He told me he was coming to London with one of his daughters, who had just graduated, and he would like to visit me. They spent a few days in London before coming to visit me in Cambridge. Manfred loved Cambridge and looking around the colleges. One day, as we sat on the banks of the Cam, he proposed! This all seemed to happen so quickly, but I had no doubt in my mind that this was the man I had been waiting for all those years.

I remember going into school to hand in my resignation before the autumn term started, so that I could leave at Christmas and fly to Perth, Australia. My sons were in Sydney at the time, and had just a few days left before they were to fly back to the UK. We arranged for them to come to Perth first so that they could give me away at our wedding, which was held in a beautiful garden on the Lesmurdie hillside.

Australia opened many doors for me and changed my life. It allowed me to follow my dream and become a full-time potter. I was very fortunate to find such a supportive partner in Manfred. I lived mainly in Jakarta for the first two years, and was inspired by the beautiful and unfamiliar plant life, which I photographed while travelling around Indonesia.

While accompanying Manfred on a short business trip to Surabaya, I visited an exhibition of work by two esteemed Australian lecturers, Les Blakebrough and Penny Smith, and their students from the University of Tasmania in Hobart.

A beautiful porcelain piece by master potter Les Blakebrough 'spoke to me'. It was a simple unglazed bowl with a band of blue watercolour decoration around its centre. I had never seen anything like it before, and wanted to learn more about this potter and his work. I later saw an advert in an Australian potters' magazine for a potters' conference in Canberra, to be chaired by Les.

I went to the conference and approached Les during a coffee break. I told him how much I admired his work and that I was new to Australia. I also asked him if it would be possible to study at the University of Tasmania for a few months to make some new moulds. I had read that a mould-maker from the Arabia porcelain factory in Finland was to be working there for a year. Ceramics lecturer Penny Smith had also spent time at the factory. After some discussion with Penny, he said that they didn't think three months would be long enough and suggested that I do a BA Honours for a year. How was I going to tell my new husband that I wished to study away for a year? However, Manfred was very supportive and encouraged me to do it.

My research began tentatively, as this was a big role change for me. Having taught for 23 years, suddenly becoming a student again was quite daunting. However, I loved having the time to research, making new moulds and working out how to make bone china paperclay. I also experimented with soluble colourants, which I found very exciting. I gained a first-class BA Honours, and I was offered an MA research scholarship at Monash University in Melbourne, combining ceramics and glass.

The aim of this book is to increase the understanding and appreciation of the translucent qualities of bone china and porcelain with special focus on paperclay, and to encourage more ceramists to explore this exciting medium.

This book deals with my own personal methods of working with bone china paperclay. Although paperclay is familiar to some, I have spent a great deal of time researching and developing the use of paperclay with bone china slip.

RIGHT: *Coastal Light* (detail), Angela Mellor, 2012. Bone china paperclay. *Photo: Stephen Bond.*

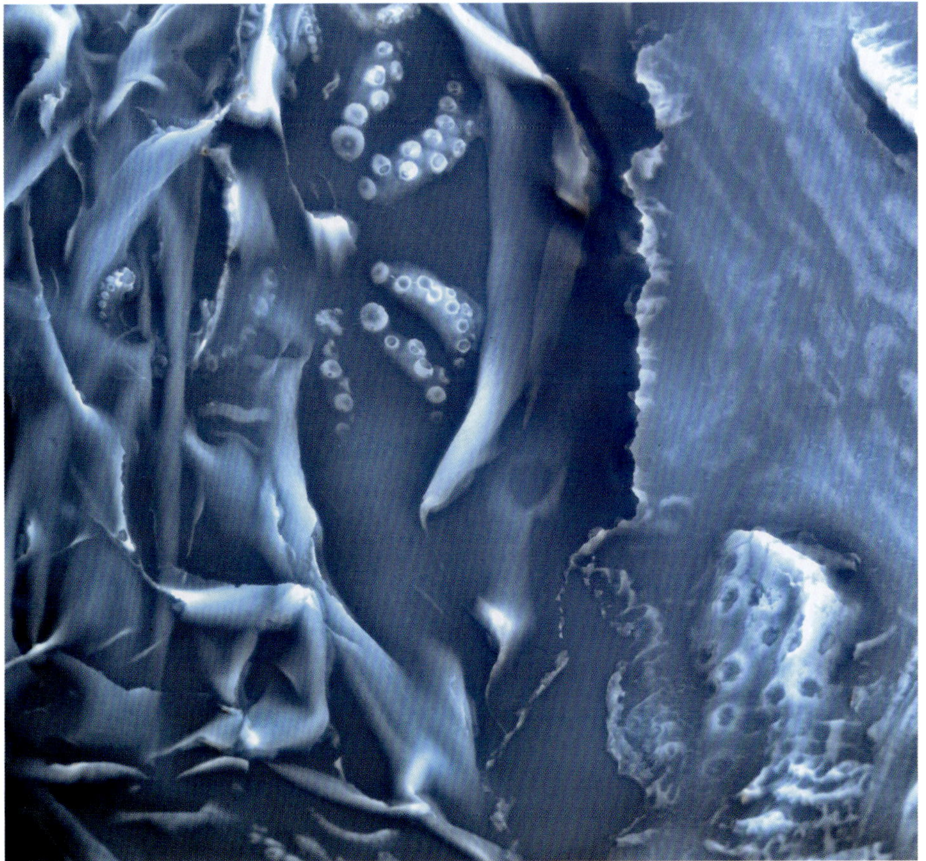

RIGHT: Inside the Manufacture Nationale de Sèvres, one of the principal European porcelain factories, October 1944. *Photo: AFP via Getty Images.*

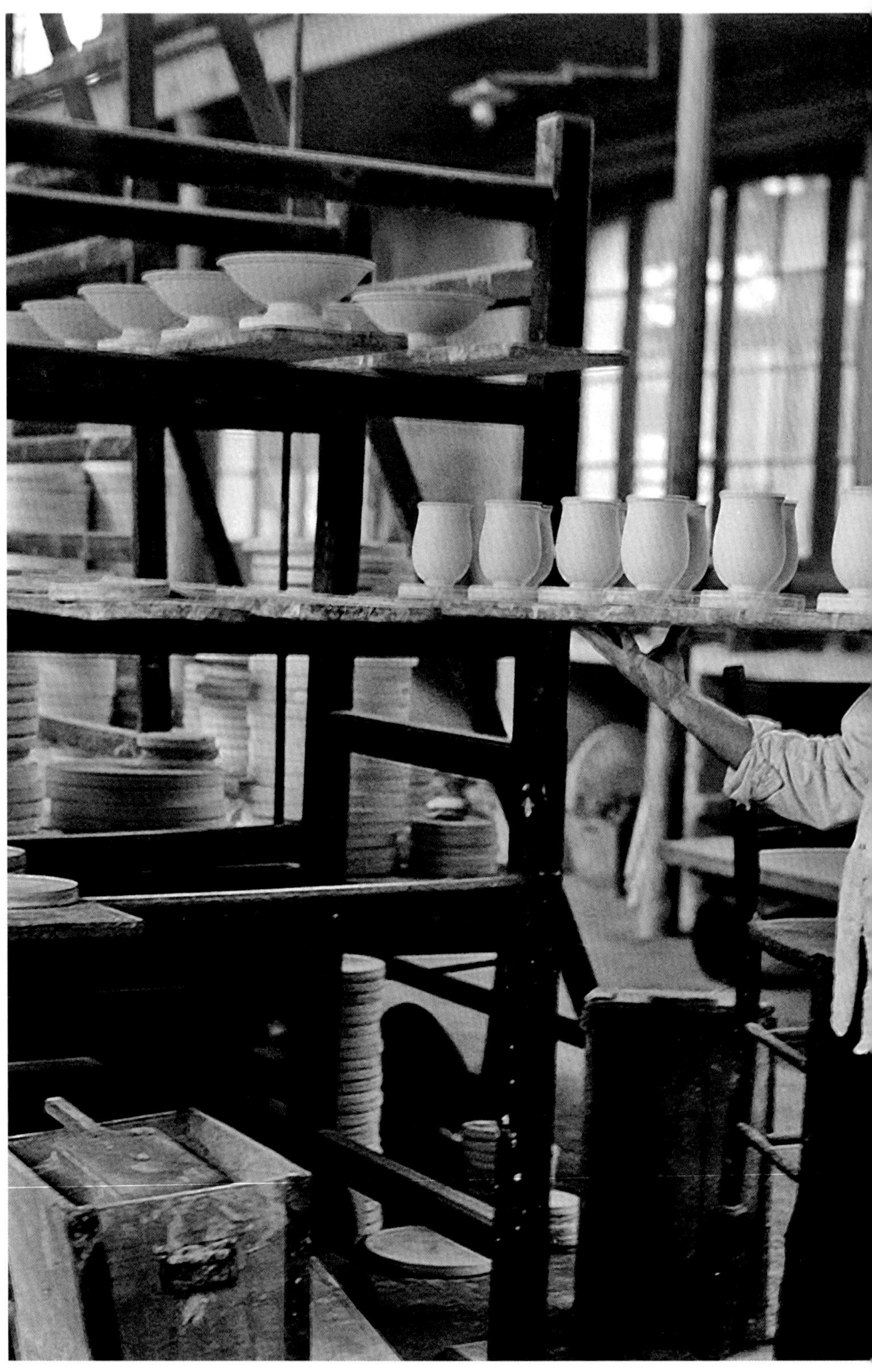

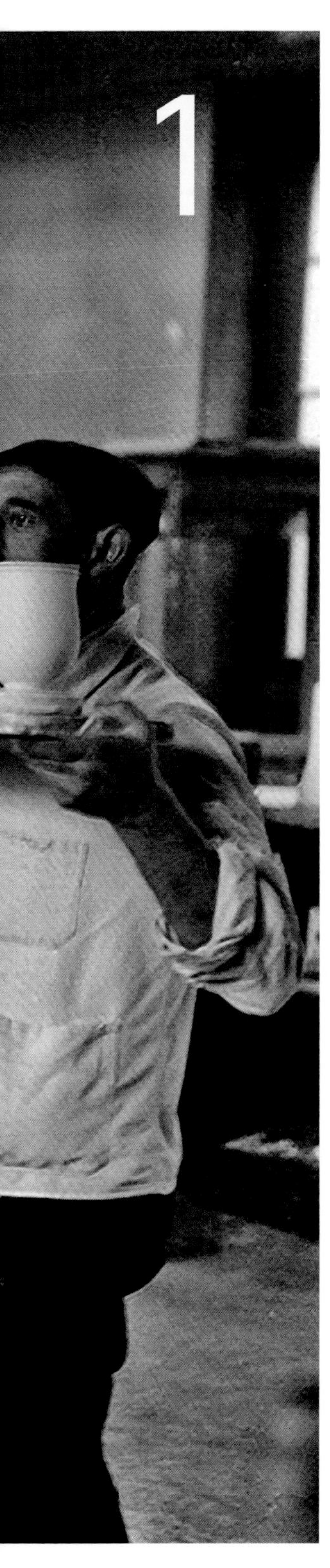

A brief history of porcelain

Much has been written about porcelain and bone china, so I will just give a brief account here. In *The History of Porcelain*, Paul Atterbury writes:

> *Porcelain is miraculous. A type of earth or clay, mixed with powdered rock and water, can be turned into a fluid material capable of being cast, moulded, or modelled into an endless variety of shapes. When fired at high temperatures it becomes hard, translucent, waterproof, and pure white in colour. It can be decorated in a variety of ways to create objects of lasting beauty or that are essentially functional.*[1]

Three hundred years ago, porcelain was available only to the very rich. Kings and princes fought over it, and paid large sums for the privilege of owning a piece. It was not made in Europe until the eighteenth century, though it was manufactured in East Asia for centuries before that, becoming an integral part of Chinese culture.

Porcelain's beginnings

Porcelain originated in China between 618 and 907 AD during the Tang Dynasty, and was further refined during the Ming Dynasty, between the fourteenth and fifteenth centuries. Marco Polo likened it to a cowrie shell, coining the term *'porcellana'*, and it became known in Europe as 'porcelain'. Chinese porcelain bodies were naturally occurring clays, which, when fired, made a ringing sound when struck. In contrast, the hard-paste (*pâte dure*) porcelain developed in 1709 by Johann Friedrich Böttger in Meissen, Germany, was a mixture of clays: kaolin, alabaster, and quartz. Hard paste, better known as 'true' porcelain, fired above 1350°C.

Early oriental porcelain was rarely seen in Europe. The few pieces that did survive the long and dangerous journey to Europe via the Arab trade routes were regarded as exotic treasures. It was looked on as a mysterious and mystic material, the secret of its manufacture unknown to many generations of alchemists and savants. The earliest recorded attempts to recreate the Chinese porcelains were carried out in Europe in 1570 by Francesco Mario de Medici in his alchemical laboratory in Florence.

[1] Paul Atterbury, *The History of Porcelain*, Orbis, London, 1982, p.7.

French factories developed soft-paste porcelain (*pâte tendre*) in the late seventeenth century as a substitute for porcelain. The early French manufacturing process used a glassy frit as a flux instead of feldspar, which distinguished it from other European porcelain.

The Sèvres porcelain factory was founded in 1738 and, in 1745, was granted a twenty-year monopoly for the production of 'porcelain in the style of the Saxon, that is to say, painted and gilded with human figures.'[2] Products of this early Vincennes period were made in soft paste (which was used exclusively until 1768, and occasionally until 1800) but were of uneven, generally inferior quality to those produced by other French factories.

In 1752, Louis XV became the principal shareholder, and in 1753, a royal edict granted Sèvres exclusivity over the production of porcelain; it soon became a luxury item. Decorative porcelain vessels and sculptures, and functional tableware were popular in the eighteenth century.

In 1745, several European factories were experimenting with the development of a porcelain-type body; centres including Pomona, Limehouse, Chelsea, Derby, and Bow in London all introduced bone as a flux. A glassy frit was used in these early types, which proved difficult to combine with the refractory and non-plastic nature of English china clay, inhibiting the manufacture of porcelain.

By 1747, the production of bone ash had begun in Europe. However, it was not until 1749, when Thomas Frye of the Bow works patented the use of bone ash for the second time, that any progress was made in the development of a relatively high-fired white clay or 'china' clay.

In 1794, Josiah Spode initiated the change to the mixture of bone ash, stone, and china clay that we use today. Originally termed 'Stoke China', and later 'bone china', its composition dispensed entirely with the use of frit, resulting in a body which, used in conjunction with slipcasting, paved the way to modern manufacturing methods in England.

Bone china is a uniquely English product, esteemed throughout the world for its delicacy, whiteness, translucence, and clean beauty. Best described as an English hybrid between soft-paste and hard-paste porcelain, it has stood the test of time and, as Josiah Spode had forecast, is a prized commodity.

Today, porcelain is widely used for tableware, sanitary ware, and the transmission of electricity in the home. Refined porcelains are used in aero engineering, hospitals, laboratories, and scientific research. On some levels, it is still an expensive luxury nowadays, but it can also be acquired cheaply and easily all around the world. The styles of porcelain tableware seen in china shops today reflect contemporary fashion.[3]

Porcelain as it was developed in China has little in common with bone china; likewise, bone china has little in common with the porcelain used in the aerospace and electrical industries, but they still share the name. The term has, over time, come to encompass many products; the materials and methods of production are fairly similar, but the end result is infinitely variable. The fluid and variable nature of porcelain is one of its greatest attractions.[4]

[2] John Fleming and Hugh Honour, *The Penguin Dictionary of Decorative Arts*, Allen Lane, London, 1977, p.723.

[3] Fleming and Honour, *The Penguin Dictionary of Decorative Arts* p.7.

[4] *Ibid.* p.8.

Contemporary studio porcelain

Porcelain was not used by studio potters until the mid twentieth century. Before that, pottery was mass-produced in factories like Meissen and Sèvres for high-end decorative pieces and tableware, sought after by the gentry. Porcelain was not considered plastic enough for throwing on the wheel and is still a challenge to studio potters to this day.

> *Modern studio ceramists do not wish to copy historic models but seek to explore the broader world of aesthetic concerns shared by painters and sculptors. They have found artists more interested in ideas than objects, for example, Conceptualism, Minimalism, performance art, who often disdain precepts such as beauty or decoration, previously associated with porcelain. Clay artists have now broken the rules of classical ceramics. This can be seen in the Funk movement of the 1960s, when San Francisco Bay area artists rejected the preciousness previously associated with ceramics by shunning the porcelain figurine tradition. This paved the way for a new use of porcelain in the studio.*[5]

> *Contemporary ceramics is more accessible than more established art disciplines where prices are usually much higher. Developments in aesthetic concerns, technical achievements and working methods have been rapid during the later part of the twentieth century. Porcelain offers a wide range of options and applications. The main qualities associated with high-fired porcelain such as delicacy, translucency, fineness, whiteness, density and purity are still available, but many potters choose to break with the traditional expectations of the medium. This usually comes with the exploration and realisation of the three-dimensional form, using porcelain as a suitable material to experiment with.*[6]

Bernard Leach was the father of modern studio pottery. He worked mainly in stoneware, but produced a porcelain body in 1950, although it was not very translucent.

It was his son, David Leach, with the help of Edward Burke, who developed a highly plastic translucent body, which became available commercially; the recipe became a reference for new porcelain bodies. David Leach was well known for his finely thrown porcelain teapots, fluted bowls, and decorated vases.

> *During the 1960s, interest in the orientally inspired stonewares of Leach and his followers waned, and the work of Lucie Rie gradually gained a new audience, with a more contemporary concern for 'modern design'. Her elegant and meticulously formed porcelain and stoneware pots, with their minimal decoration of finely scratched and inlaid lines, and her mastery of unusual and colourful glazes, marked her as one of the twentieth century's greatest potters, and increased the interest in porcelain among potters and the public.*[7]

[5] Jan Axel and Karen McCready, *Porcelain: Traditions and New Visions*, Watson-Guptill, New York, 1981, p.184.
[6] Atterbury, *The History of Porcelain*, p.8.
[7] Caroline Whyman, *The Complete Potter: Porcelain*, Batsford, London, 1994, p.15.

Porcelain has gained popularity where craftsmanship is encouraged and recognised, and this has resulted in many developments in technique and style. The new limit of porcelain is now defined not only by the artist's technical mastery, but also by the scope of their imagination.

Qualities and properties of bone china

The special characteristics of bone china are thinness, translucency, whiteness, and great strength. Most of these qualities are provided by one of its constituents, calcined bone, which acts as a flux to the body, making it fuse into a tough, glass-like substance at temperatures above 1240°C. The calcined bone makes up approximately half the body material.

Bone ash is produced from cattle bones. The bones are first boiled – the residue is sent to glue manufacturers – after which they are calcined (the bones are heated to a red heat, or more, to remove the water content) and ground with a mortar and pestle before being ball milled. Finally, the ash is introduced into the bone china body. Because of its poor mechanical strength before firing and a limited top firing range (after which it quickly loses shape and collapses), bone china is only used by potters with considerable experience in ceramics. Artist potters have only recently used bone china to any extent, and great skill is needed when handling it. It is only during the last 20–30 years that the translucency of bone china has been fully explored for artistic purposes.

Bone china lacks plasticity and is unsuitable for working by hand, and so it is usually slipcast. Due to the high loss rate, working in bone china can be expensive. The short firing range can cause considerable warping, because as soon as the body reaches the lowest melting point, large amounts of liquid rapidly form, which can cause the body to deform.

It is also notorious for its memory. Knocking or denting the ware in the damp stage and then reshaping it means that the dents or warps will return in the firing – the body will remember.

Bone china bodies tend to go 'off-colour', and this needs to be avoided if you want to retain its whiteness and translucency. Increasing the bone ash can help with this, but it can also make the body less plastic and more difficult to work with. To counter this, 1–2 per cent of ball clay can be added to the body to increase both its strength and plasticity.

Significant shrinkage of 15–20 per cent occurs in high firing (1220–60°C). When considering model and mould-making requirements, it is good practice to make the model between 15 and 20 per cent (an average of 17 per cent works well) larger than the desired finished size, to compensate for the high shrinkage.

Designing a drop-out mould avoids seam lines, and this is the type of mould I prefer. However, if the form requires a multi-part mould, then you need to design the form with a twist or edge where the seam can be placed when moulding. This will prevent seams reappearing during the firing.

Scale is important when using bone china; open forms tend to collapse during firing. Handling large, heavy moulds while draining the slip can also be a problem.

BELOW: *Tea Bowls*, Angela Mellor 2014. Bone china with paperclay inlay. *Photo: Stephen Bond.*

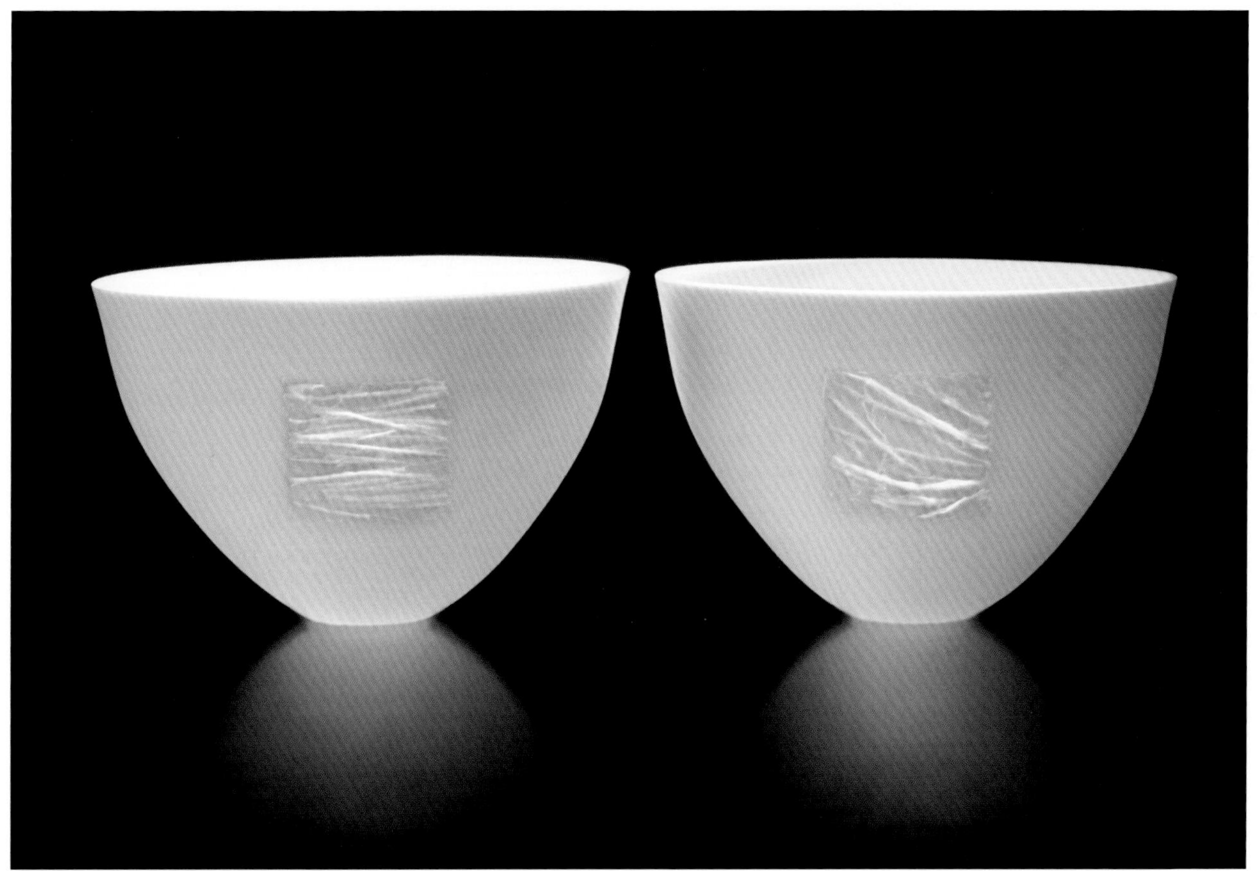

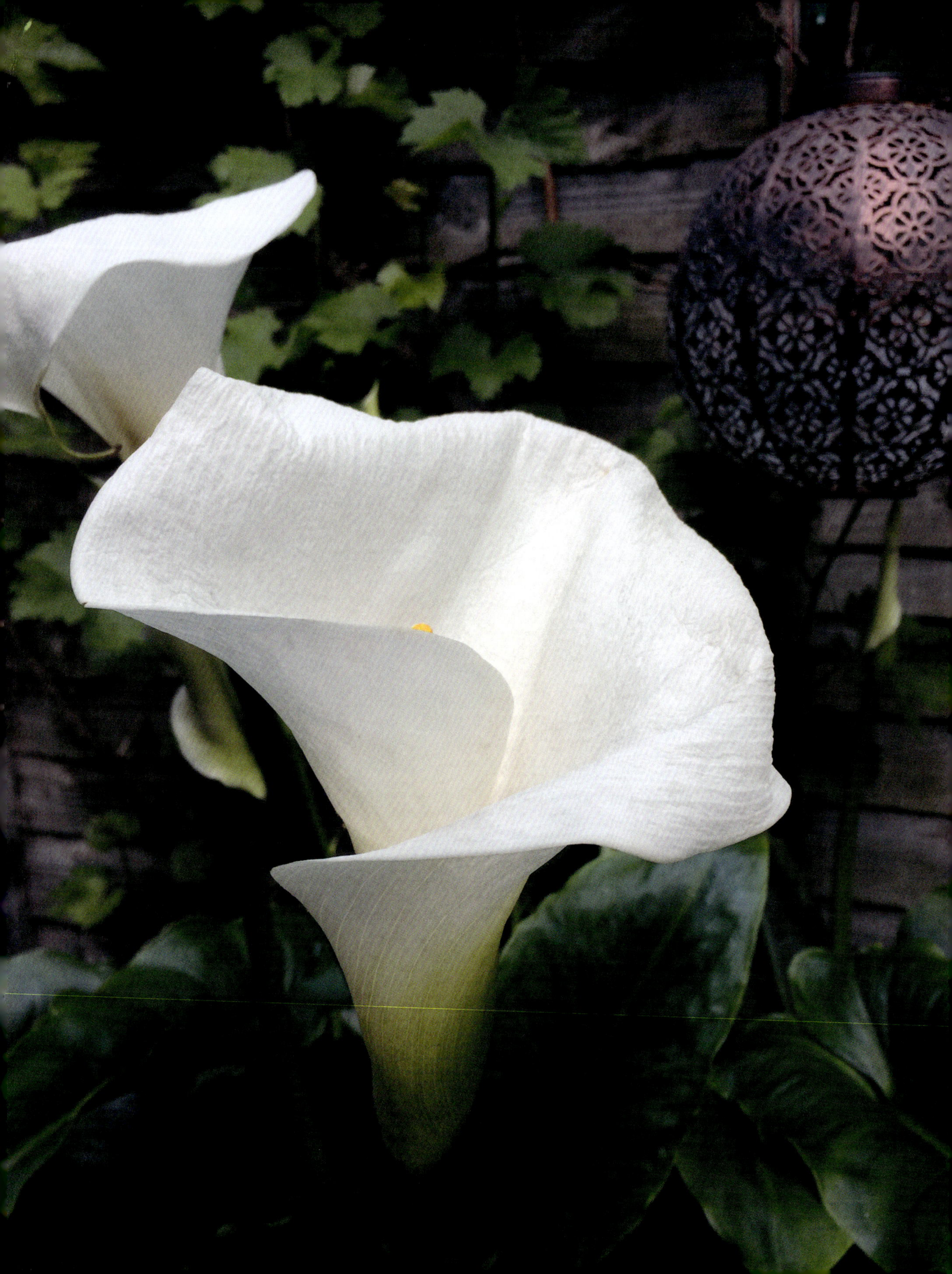

2

Early work and research

Why porcelain?

Riddle me re, riddle me re,
What rings like a bell, but pours out tea?
What is stronger than bone and smoother than silk,
Made from the earth, but the colour of milk?
Yet through it you can see.

This riddle, taken from Caroline Whyman's book, *The Complete Potter: Porcelain* (p.101) highlights porcelain's many attractive qualities.

Their colours and their forms, were then to me
An appetite

William Wordsworth, *Tintern Abbey.*

My early training was in textile design. I then gained a distinction in my Certificate in Education in Art and Design at the University of Manchester.

During my first teaching post, the headmaster asked me if I would like to attend a pottery course for art teachers so that we could introduce pottery into the curriculum. I was quite excited, as I had never worked with clay before. I became hooked during my very first class, and was eager to pass on my knowledge to the students.

When teaching art in the 1970s, I saw Mary Rogers give a talk at Sudbury Hall in Derbyshire. She held a small luminous object up to the light, and I wondered how this dense, hard material could encapsulate light and depict such details as one might see in the veins of a leaf or petal, at the same time reflecting and entrapping subtle colours found in nature.

I was immediately intrigued by the medium of porcelain and its sublime but mysterious qualities. I decided to learn more about it, and began to follow Mary Rogers' work. She was born in Derbyshire and studied ceramics at Loughborough School of Art, setting up her first studio nearby. In the 1960s, she had embraced the challenges of porcelain and was one of its leading exponents, using its natural translucency to experiment with effects of light and colour inspired by nature. She made small coiled and pinched forms, which she scraped when dry, sometimes carving the edges to make them paper-thin. She often used dots of metal oxides to decorate her work. Like hers, my method was hand-building, and my main source of inspiration natural forms, so I immediately felt an affinity with her work.

OPPOSITE: Arum lilies.

In the mid 1980s, I studied ceramics with Peter Lane at the University of East Anglia. Through Lane's teaching and his books on studio ceramics, I became aware of porcelain's physical properties, aesthetic qualities, and technical limitations. Porcelain can be an unforgiving and demanding material; it is prone to cracking and warping at high temperatures. As a consequence, working with it attracts a certain type of potter. David Leach comments:

> *Those potters who choose the medium of porcelain do so because their very natures dictate it. They lean towards a slowly and meticulously achieved perfection, in contrast to potters whose natures demand robustness, vigour and immediacy.*[1]

It has been a long and rewarding process and I hope that readers will learn something from this book and be inspired to try working with this magical medium.

Influences

Several artists influenced my development of translucent paperclay, as Peter Lane states:

> *There are many potters for whom those delicate and varying degrees of translucency, which can only be revealed after the final firing, will be the supreme attraction and ultimate goal. The interplay of light onto and through a finely made piece provides one of the most satisfying experiences in both seeing and touching.*[2]

Rudolf Staffel (USA)

Staffel began to use porcelain in the 1950s at the height of abstract expressionism. His hand-built vessels enhance the properties of porcelain, pre-eminently translucence, which became a focus and challenge for him. His approach was spontaneous, and he allowed his fingerprints and incised marks made on hand-built and thrown vessels to appear as 'ghost images' that let light emanate from the finished objects. He studied painting under Hans Hoffman between 1942 and 1943, and his surface and depth resonate with Hoffman's push–pull theory, which stated that paintings required contrast in colour and form to create depth and movement. Staffel translated this into his work with clay saying:

> *You are not aware of the push until you see the pull, you are not aware of the dark until you see the light. To be aware of translucency you must have opaqueness.*[3]

[1] Peter Lane, *Studio Porcelain*, Pitman House, 1980, p.6.
[2] –, *Contemporary Porcelain: Materials, Techniques and Expressions*, A & C Black, London, 1995, p.9.
[3] Rudolf Staffel, 'Light Gatherer', *Ceramics: Art and Perception*, 17, 1994, pp.34–8.

Studying Staffel's work, in which he pinched the clay to create thin and thick areas, made me aware of the varying degrees of translucency, and how these can be used to achieve dramatic effects.

Staffel is best known for his translucent porcelain pieces, which he called *Light Gatherers*, produced over three decades. These works can be seen the article 'Light Gatherer' in *Ceramics: Art and Perception*, no.17.

Translucency can captivate the observer, and many potters make use of this exceptional quality, although it does need direct light if it is to be fully exploited.

Jeroen Bechtold (Holland)

Bechtold explores slipcasting techniques, and carves away layers from the outside of his pieces. He then applies multiple glazes and fires in oxidation.

Arnold Annen (Switzerland)

Arnold Annen is renowned for his translucent, paper-thin porcelain bowls, which he developed in 1991. His bowls are thrown fairly thickly before being pared down with steel ribs. Annen paints latex resist onto the bowl and force dries the slip with a gas burner. The latex is then peeled away, and he uses a fine nozzle to trail lines of slip across the form, drying them with a burner flame. These bowls are extremely fragile and are fired upside down on a chuck made in the same porcelain body. Alumina mixed with water is painted around the rim of the chuck so that the bowl will release easily when removed from the kiln. These early works are fired to 1260°C with spectacular translucent results.

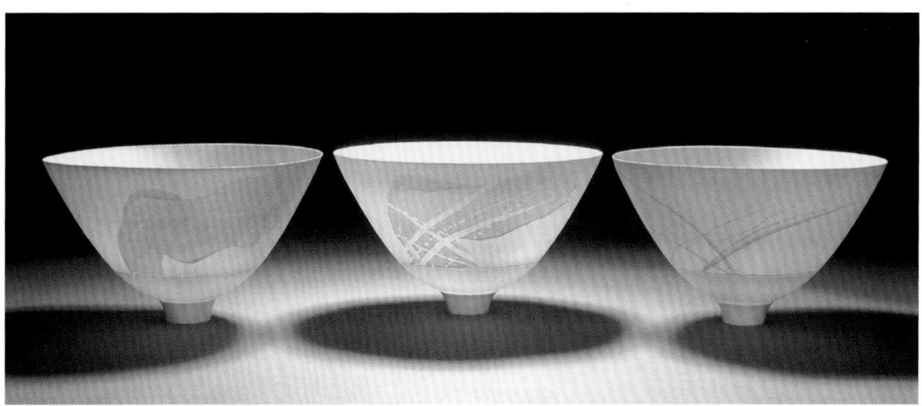

RIGHT: *Bowls*, Arnold Annen, 1993. Porcelain translucent. 27 x 20 cm (diam x h). Wheel thrown and turned, decorated with latex resist, brushed over and trailed with porcelain slip, unglazed. Gas kiln reduction fire 1260°C. *Photo: Reto Bernhardt.*

He later slipcasts his work using very large plaster moulds, pouring in slip while the mould is spun quickly on a wheel. Timing is essential to create a very thin cast. He works in Limoges porcelain and has perfected his techniques over the years, with a very disciplined approach and sophisticated methods.

When I asked Annen about his work, and specifically its translucency, he replied:

> *I work the material to its physical limit because I require maximum lightness and translucence. The atmosphere of light caught in the shell emerges in the relief. All works are made of Limoges porcelain from France, reduction-fired in a gas kiln at 1330°C. They are of pure, natural white porcelain, unglazed and very translucent. The parabola-like forms are made of slipcast porcelain. The slip is left in the plaster mould for between one and two minutes. The result is a wall of 2 mm thickness. The leather-hard bowls are then trimmed on the potter's wheel in many variations.*

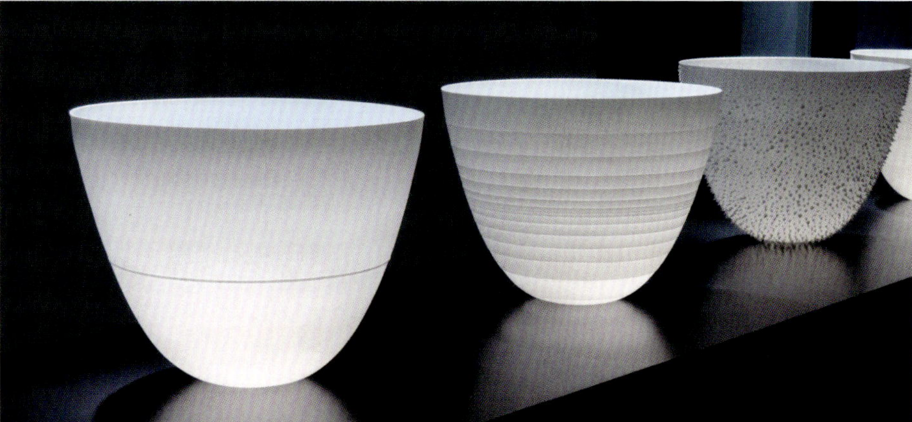

LEFT: *Bowls*, Arnold Annen, 2015. Porcelain translucent. 31 x 24 cm (diam x h). Gas kiln reduction fire 1330°C. *Photo: Arnold Annen.*

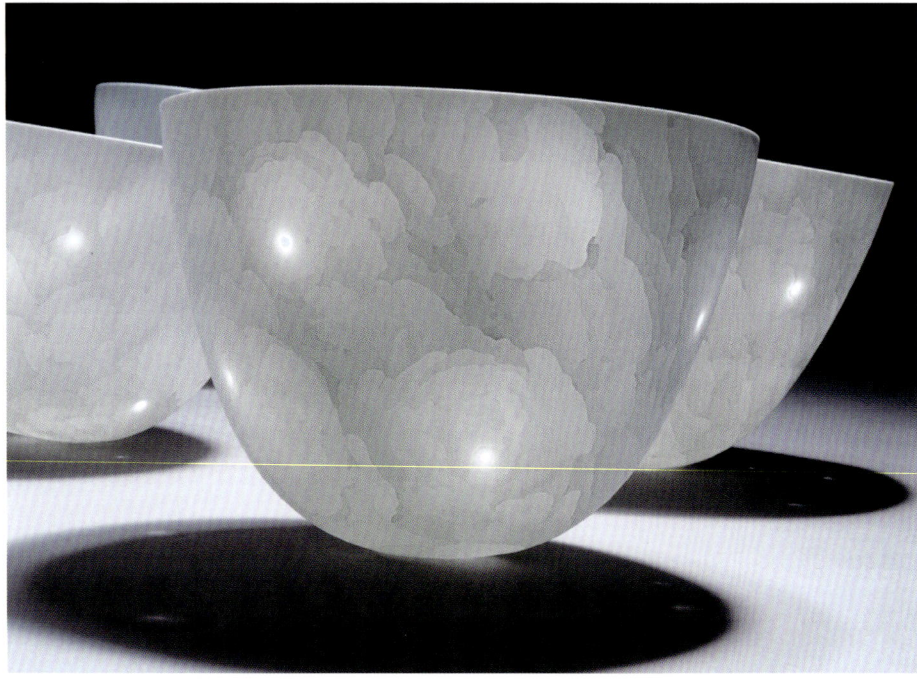

BOTTOM LEFT: *Floating*, Arnold Annen, 2017. Porcelain translucent. 28 x 21 cm (diam x h). Gas kiln reduction fire 1330°C. *Photo: Arnold Annen.*

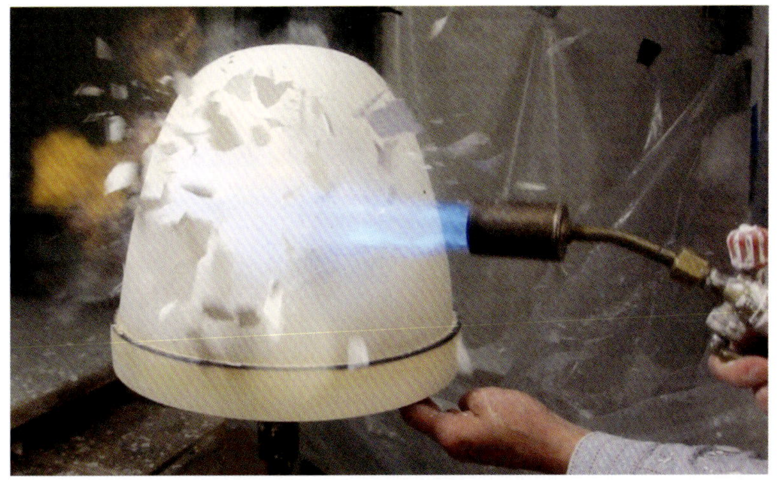

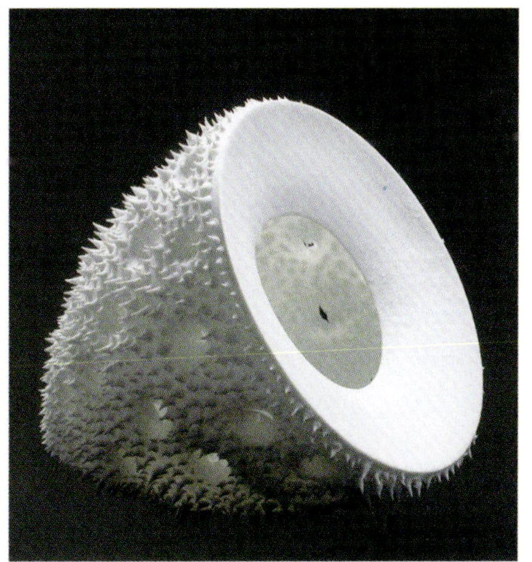

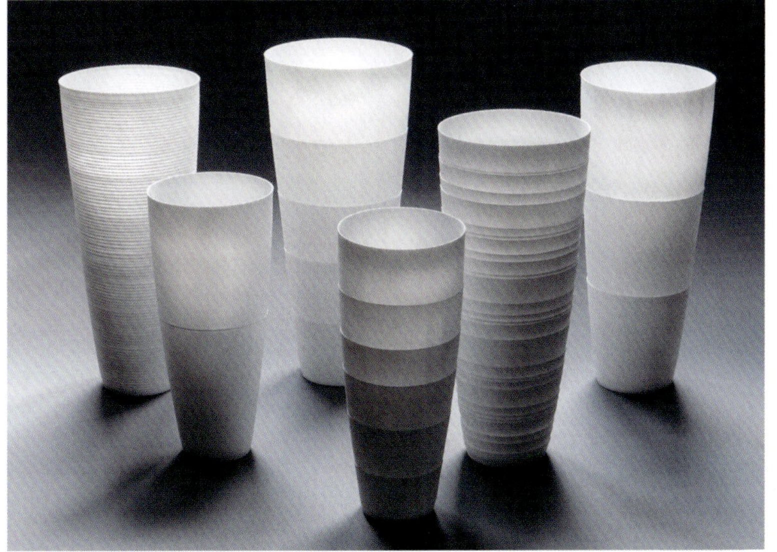

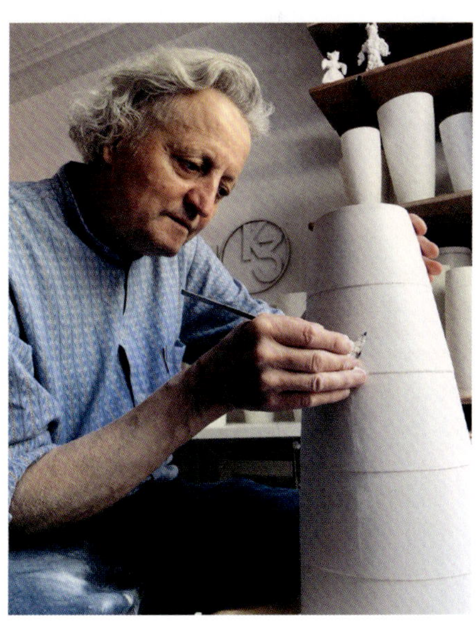

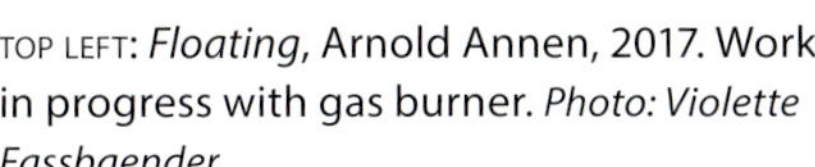

TOP LEFT: *Floating*, Arnold Annen, 2017. Work in progress with gas burner. *Photo: Violette Fassbaender.*

BOTTOM LEFT: *Symphony*, Arnold Annen, 2020. Porcelain. 36 cm (h). Gas kiln reduction fire 1330°C. *Photo: Arnold Annen.*

TOP RIGHT: *Sethocapsa*, Arnold Annen, 2007. Porcelain. 21 x 23 cm (diam x l). Gas kiln reduction fire 1330°C. *Photo: Arnold Annen.*

BOTTOM RIGHT: Arnold Annen trimming *Symphony*, 2020. *Photo: Violette Fassbaender.*

In 2011, I was honoured to receive an invitation from Laura Borghi to contribute to an exhibition called *Translucencies* at Officine Saffi in Milan, alongside Arnold Annen and Margaret O'Rorke. The gallery is beautiful and full of light. The photographer took some catalogue photos at night to show the full impact of the strikingly lit works by Margaret O'Rorke.

You can read more about her work and methods in the section on lighting later in this book, and in Margaret's own book, *Clay, Light & Water.*

Arne Åse (Norway)

Arne Åse is also interested in translucency, and to achieve it he paints shellac resist pattern on to his pots and then sponges away the unprotected areas. He creates a richly textured surface by layering sensitive brushstrokes of shellac, which allows the light to pass through the piece. He sometimes adds beautiful watercolour patterns with soluble colourants, which also enhance the delicacy of his work. He documents this fully in his books, *Water Colour on Porcelain* and *From Bowl to Art,* which I found very helpful when researching soluble colourants.

Les Blakebrough (Australia)

Les Blakebrough also used these techniques, but separately, either by applying soluble colourants as a band of colour around the pot, or by featuring only a small area of pattern with the use of shellac resist. Light played an important role in his work, and the evocative images of Antarctica, with their subtle variations of colour and the formation of ice patterns, are echoed in his decoration. The intrinsic quality of decoration is something that Blakebrough long sought in his work, the notion that if clay had a special enough quality of its own, there would be no need to cover it. Blakebrough states:

> *I was aware of the way in which glaze sat upon the surface of the work I always felt that the body was more beautiful, the nakedness of it has no flaws. Its pristine qualities exist between touch and look.*[4]

I was very fortunate to study under Blakebrough at the University of Tasmania in Hobart, where he introduced me to working with soluble salts. I experimented with soluble colourants, as they accentuate porcelain's translucence and delicacy without affecting its luminosity. However, I have taken these ideas a step further by using them on bone china.

[4] Les Blakebrough, catalogue entry in promotional leaflet for *Les Blakebrough: Ceramics* at FORM Gallery, Perth, Australia, 1997.

Working with bone china

Bone china is a clay body used mainly in industry. As a medium, it presents specific problems for the studio potter. The high proportion of bone ash in the bone china body means it lacks the plasticity of porcelain and is therefore mainly slipcast in plaster moulds. Models for the moulds can be made in clay, but plaster makes better original models, because solid plaster models can be turned on a lathe to achieve a glass-like surface. It is difficult to achieve the same degree of smoothness on a wet clay surface. I use plaster models as masters for my moulds, as they allow me to achieve a smooth, sensuous quality in the work.

My introduction of paper pulp allows for a freedom of expression and hand-building that had previously been impossible due to the clay body's lack of plasticity and its tendency to warp in the kiln. It also allows the potter to take full advantage of bone china's whiteness and to take the translucency and workability to new heights.

Bone china also differs from porcelain in its relationship with glaze. Whereas a porcelain body and glaze fuse together at one high temperature, bone china is fired to maturity in the bisque firing. Despite this tradition, I have glazed bone china at the bisque stage to 1000°C with no problem, as long as the glaze is applied thinly. I put a thin layer of glaze on the inside and let it dry before spraying it on the outside. I find it is best to warm the pot first.

Surprisingly, the results are quite different to those achieved on porcelain. For instance, cobalt produces a beautiful lilac on bone china, whereas it is blue on porcelain. I think this may be due to the bone ash in bone china.

This book investigates the transparency of bone china and its potential for the transmission of light. It is based on research undertaken as part of my BA and MA degrees, and through several series of work produced since then. Through an exploration of organic forms where light plays an integral part in the visual structure, I have developed work that accentuates the translucency and delicacy of these objects.

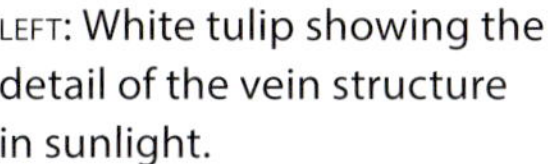

LEFT: White tulip showing the detail of the vein structure in sunlight.

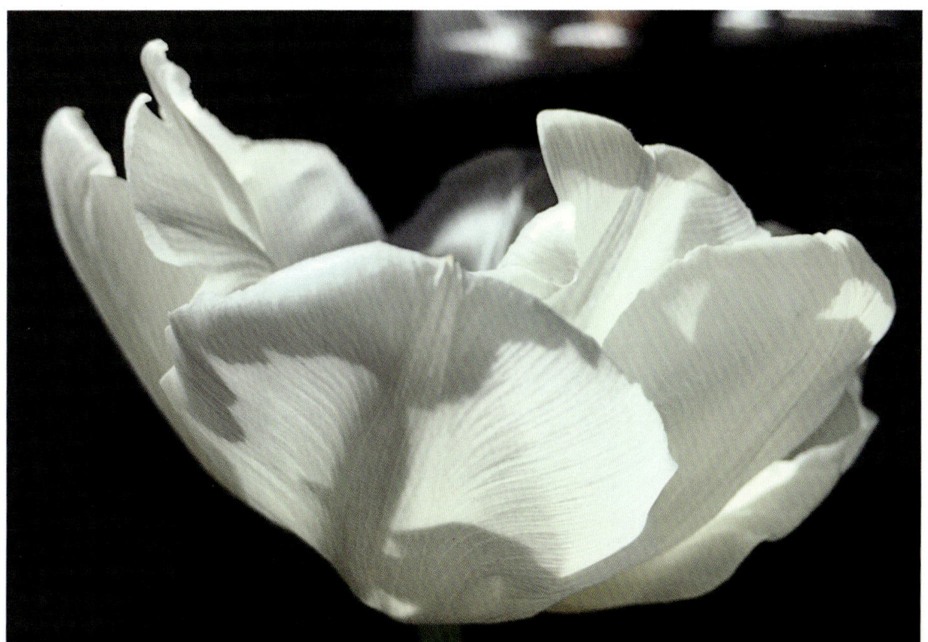

My journey with bone china

Before the 1980s, only a few potters in England – notably Glenys Barton, who worked for Wedgwood, and Jaqueline Poncelet – had attempted to explore the potential of slipcast bone china for individual work. Since then, others have been attracted to its response to light, experimenting and trying to solve its inherent practical challenges.

My first encounter with bone china was during the 1970s, when I visited Angela Verdon's studio at the Gladstone Pottery Museum in Stoke-on-Trent, UK. A skilful practitioner, her early work has an organic linear quality, with patterns flowing around simple bowl shapes. When illuminated from above, the light passes through the pieces, emphasising the sinuous lines, and giving them a quilted appearance.

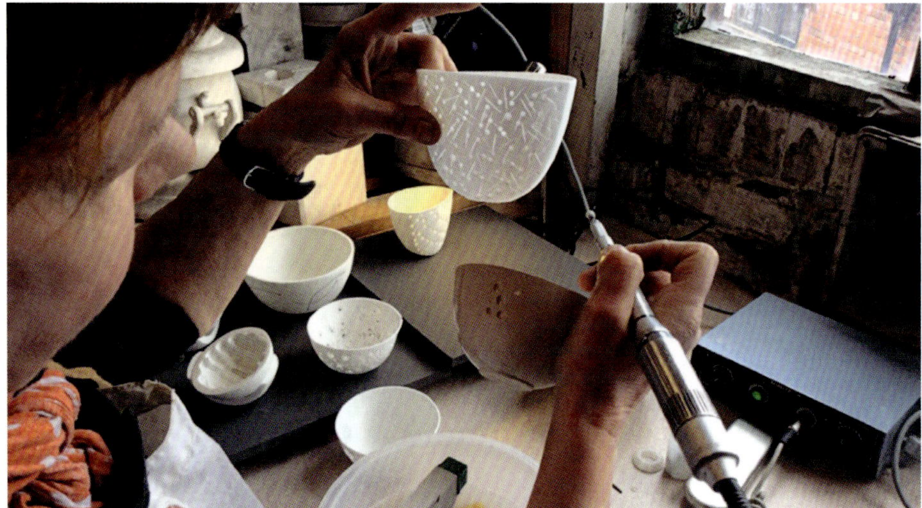

LEFT: Angela Verdon at work in her studio at the Gladstone Pottery, Stoke-on-Trent.

Verdon has been working in ceramics for over 50 years. Her first experience of clay was at an evening class, which led eventually to an MA at the Royal College of Art, London, where she was encouraged to explore other materials and working methods. Angela exploited this, as she has always been interested in both industrial production processes and sculptural techniques.

In 1976, she undertook a residency at the Gladstone Pottery Museum. Here, she discovered bone china, finding that when she carved into the surface, 'lines of light' appeared. She focused on this technique for the next decade, producing paper-thin pierced and carved vessel forms, which can be found in national and international collections.

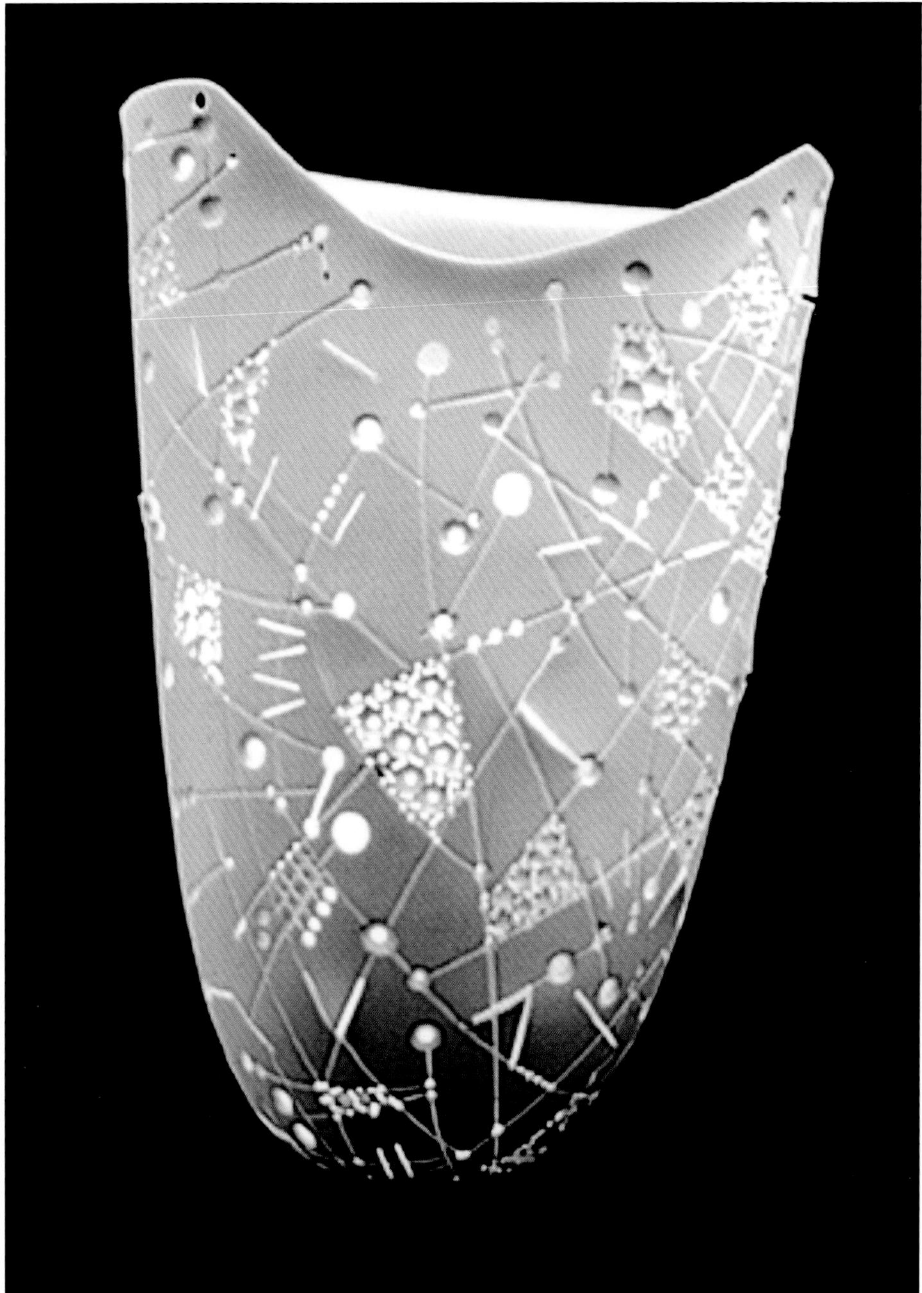

RIGHT: Untitled, Angela Verdon. Burnished bone china, pierced and incised. Asymmetrical form. 10 x 5 cm (h x w). Fired to 1260°C with two-hour soak. *Photo: Angela Verdon.*

Her recent work focuses on enclosed abstract sculptural forms. She investigates organic contours and contrasting hard edges. She is inspired by natural and architectural environments where line and form give way to shadows. Of this work, Verdon says:

> *Every aspect of the making process is exploited, through utilising the inherent qualities of the material, employing carving and mould-making techniques, to controlled multiple kiln firings. Working in this way results in surprising and unexpected forms that are reworked at various stages. The pristine surface finish is achieved through a laborious process of sanding and burnishing throughout the making process.*
>
> *The subtle play of light and shade gives definition to the sculptures with their oblique references to architectural and natural forms. The structure, balance and purity of form, combined with the strength of the material, produces their quiet but powerful emotional impact.*[5]

Even though these pieces are not translucent, their burnished surfaces have a soft luminosity that enhances the forms.

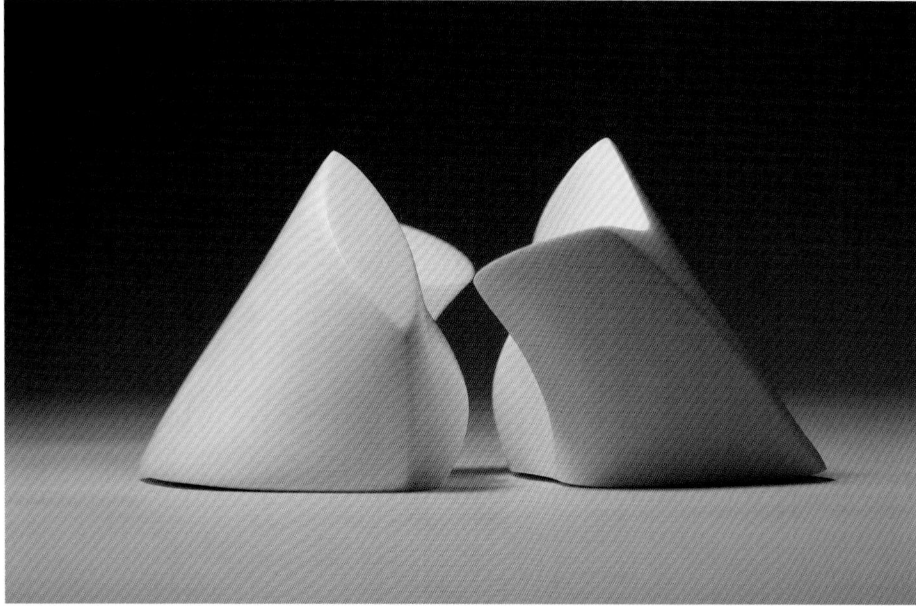

LEFT: Untitled, Angela Verdon. 17 x 25 x 15 cm (h x l x w). Fired to 1260°C with two-hour soak. *Photo: Point of View Photography.*

In 1984, I was granted a three-month sabbatical from teaching and attended an Advanced Art Teacher's Diploma course at the University of East Anglia. My aim was to further my knowledge of working in porcelain and to review my teaching methods. The course was organised and led by expert potter Peter Lane.

[5] Angela Verdon, www.angela-verdon.com (accessed 25 March 2025).

Peter was very generous with his wide knowledge of working with porcelain and glaze recipes. I learnt a great deal about working with this difficult medium, although I became frustrated with its lack of plasticity and its tendency to warp and crack at the seams, my work consisting of slab-built enclosed forms at the time. I was also disappointed by its lack of translucency. When I realised that I was not achieving the thinness, whiteness, and translucency that I was looking for, I felt the need and desire to move away from hand-built porcelain to slipcasting in bone china.

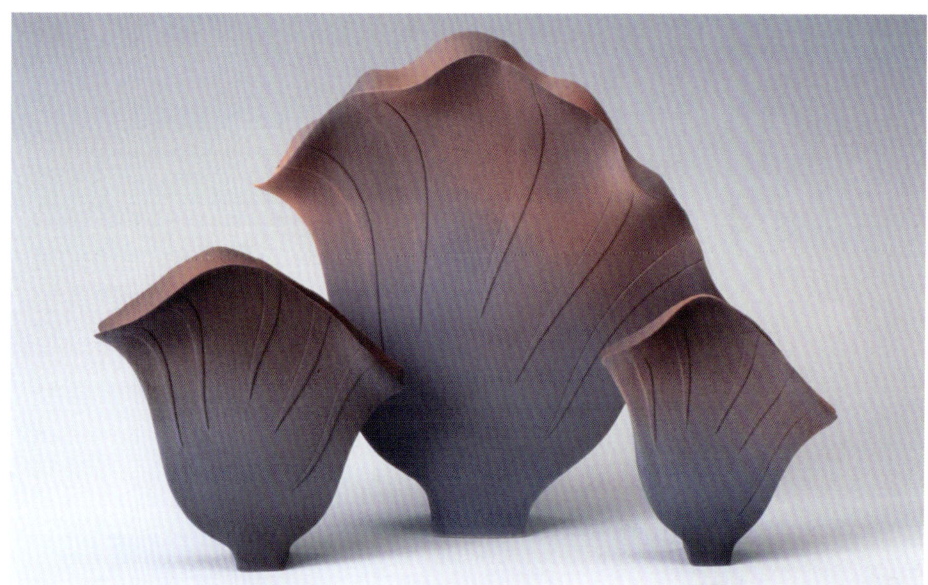

RIGHT: *Wing Forms*, Angela Mellor, 1984. Slab-built porcelain with impressed textured insert, sprayed with underglaze colours. 10–28 cm (h). Oxidised fired to 1240°C. *Photo: James Austin.*

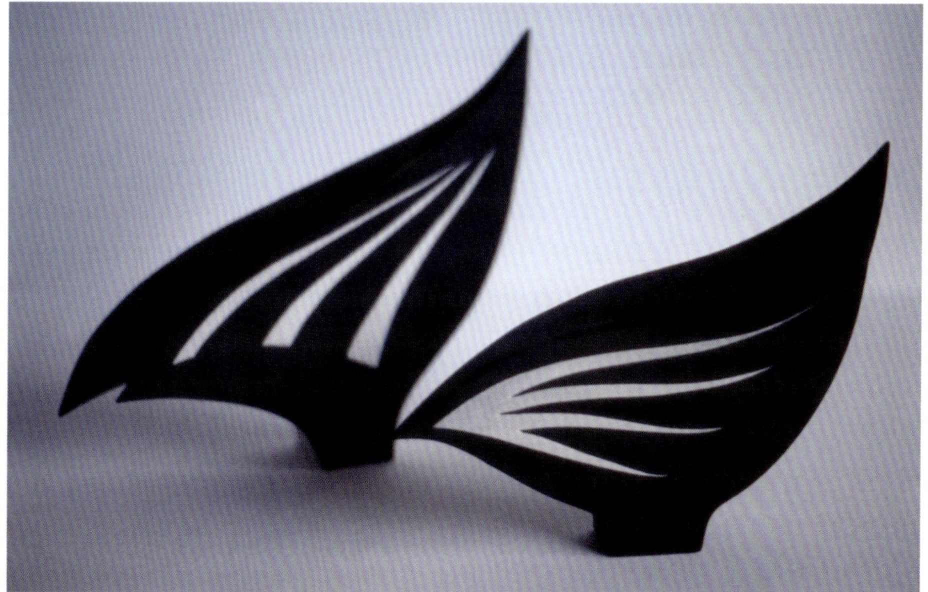

RIGHT: *Bird Forms*, Angela Mellor, 1992. Slab-built porcelain with impressed decoration, sprayed with ceramic stains. Oxidised fired to 1240°C. *Photo: James Austin.*

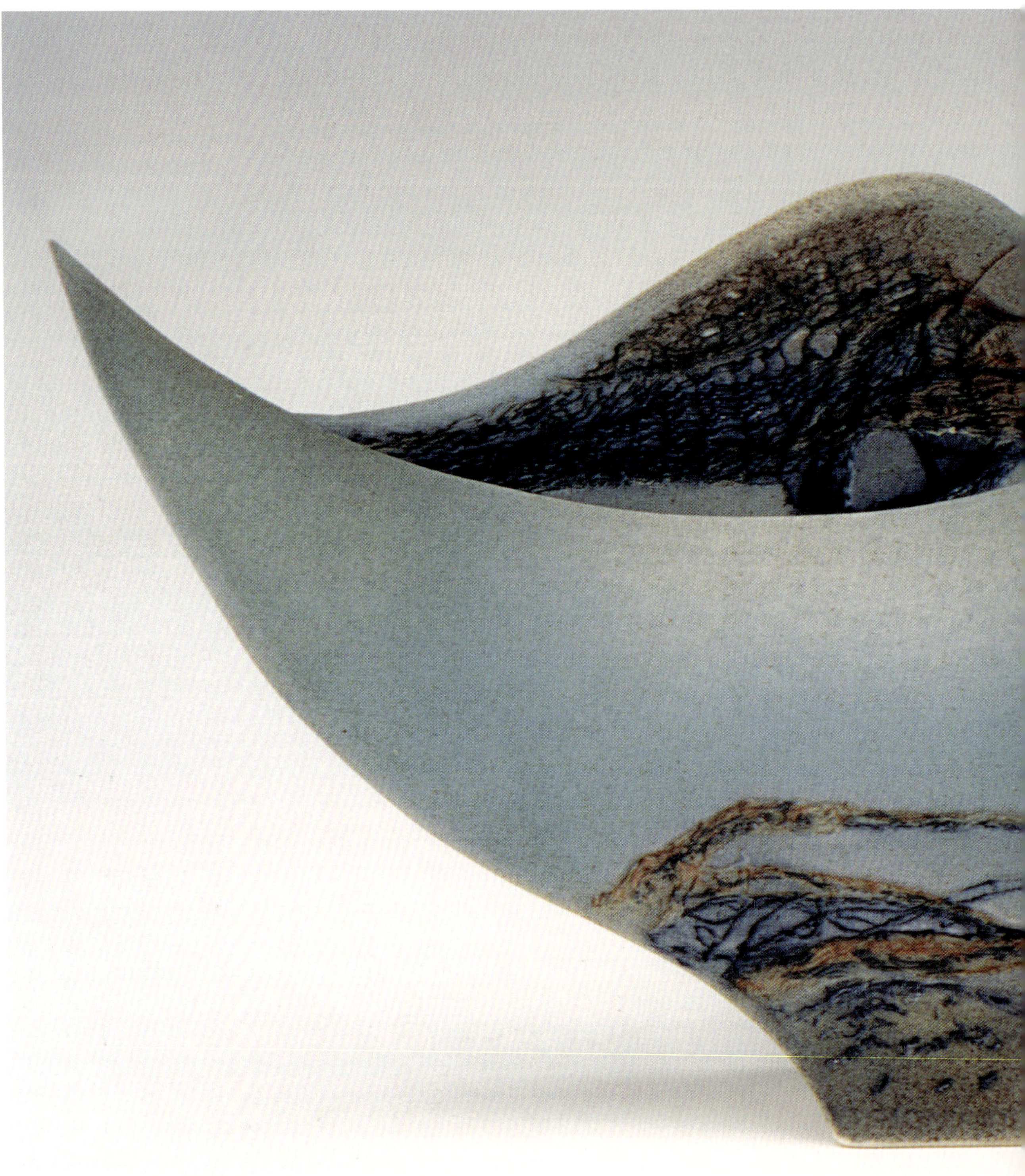

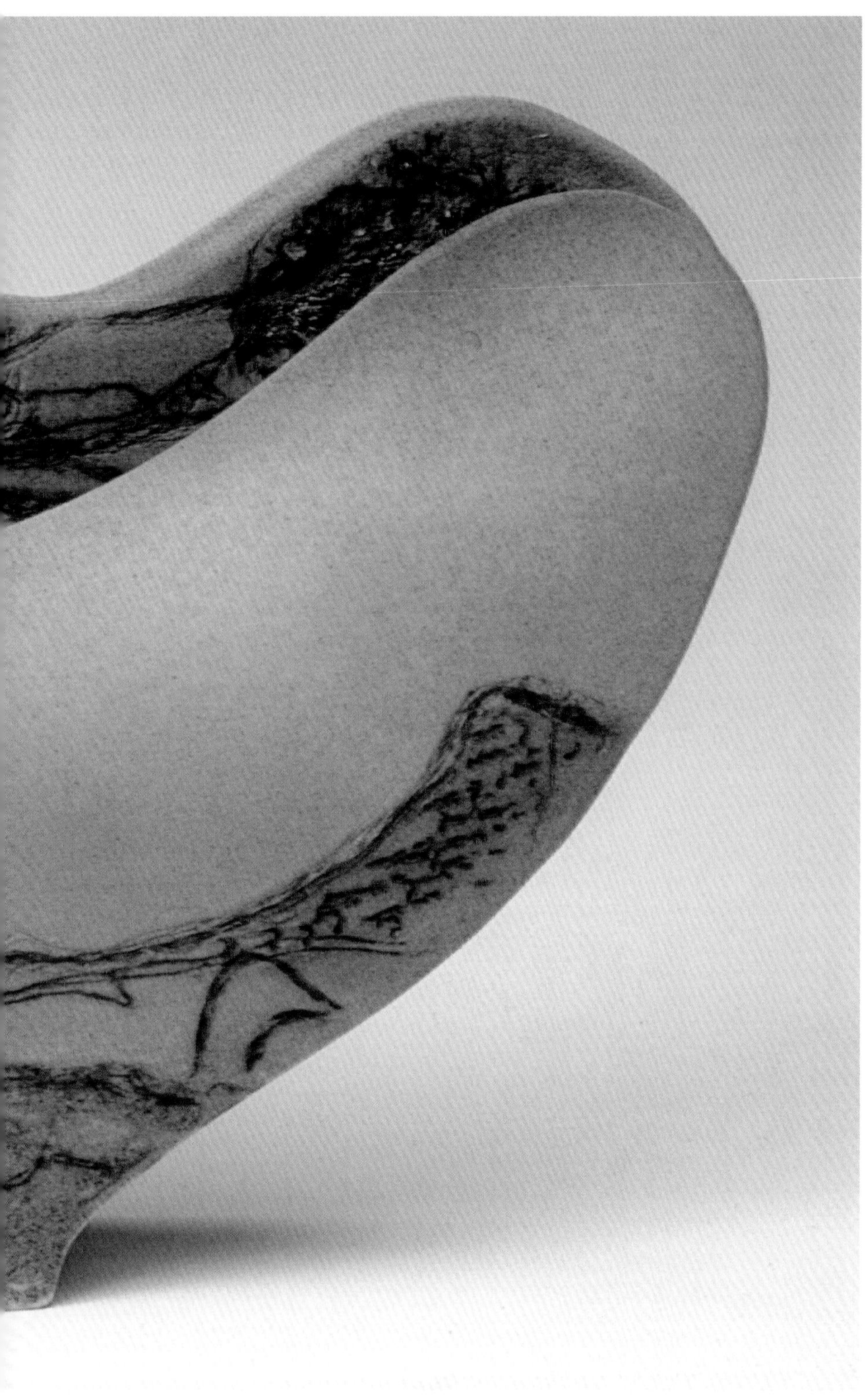

LEFT: *Wave Form*, Angela Mellor, 1992. Slab-built porcelain vase with impressed decoration and ceramic stains. 15 x 16 cm (h x w). Fired in an electric kiln to 1240°C. *Photo: James Austin.*

In 1991, I visited British ceramist Sasha Wardell at her studio in France. Sasha is renowned for her subtle yet delicate slipcast forms, which have won her international acclaim. She received excellent training in industrial techniques at the Limoges factory in France, which I visited while there. I learnt how to make a plaster model and mould from a small hand-built petal form I had made previously. This was my first slipcast object. It was so light, thin, translucent, and supremely white that it resembled a butterfly about to take flight. I was so excited ... at last, I had found the medium I had been searching for!

BELOW: *Natural Forms*, Angela Mellor, 1992. Slipcast and altered bone china. The form has been encouraged to deform gradually during the firing by reshaping part of the rim. Unglazed. 19 cm (h); 15 cm (h); 11 cm (h). *Photo: James Austin.*

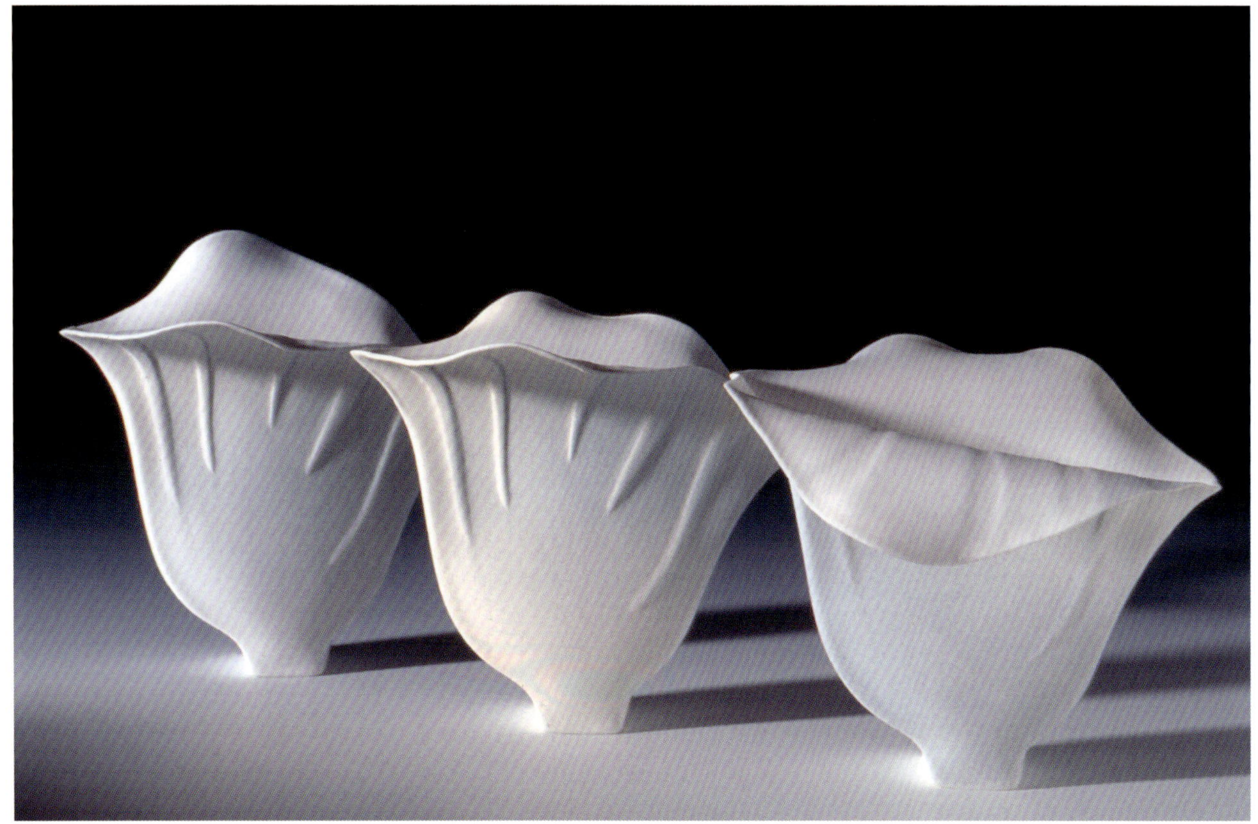

Like Sasha, I have found the idiosyncratic quality of bone china, especially its memory, both inspiring and frustrating. Coming to terms with its high shrinkage rate and its tendency to warp has been a central feature in my work through the years. I exploited the warping by not using setters in my early slipcast bone china, which allowed the pieces to move during the firing. This accentuated the organic forms I was making. I also worked on them in the damp stage, gently encouraging distortion of the rim, and packing them with alumina to retain the body shape during high firing.

RIGHT: *Natural Forms*, Angela Mellor, 1993. Slipcast and altered bone china. Smaller form sprayed with black underglaze colours. 19 cm (h); 16 cm (h). *Photo: James Austin.*

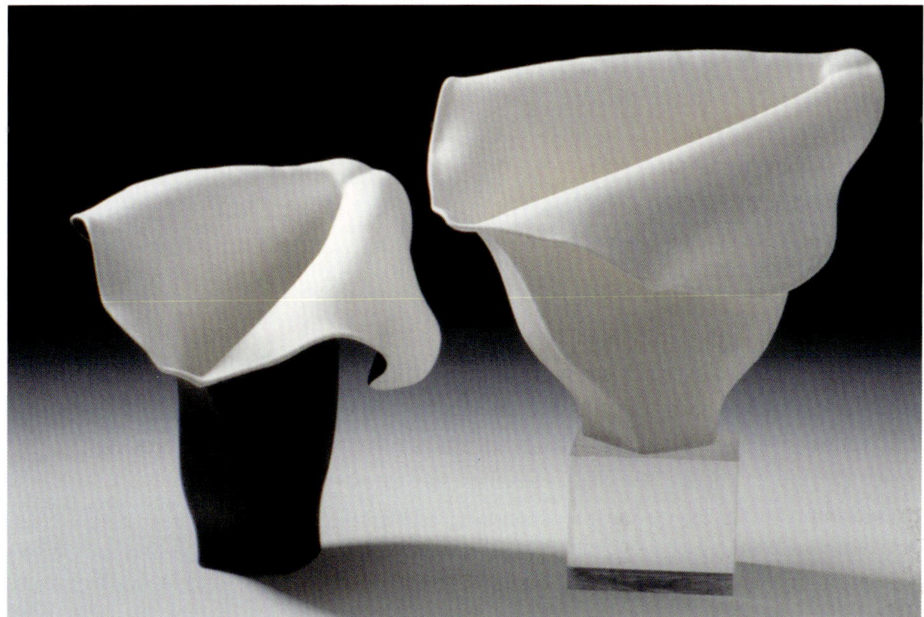

BELOW: *Natural Forms*, 1993. Slipcast and altered bone china. Sprayed with red and black underglaze colours. 19 cm (h); 15 cm (h); 11 cm (h). *Photo: James Austin.*

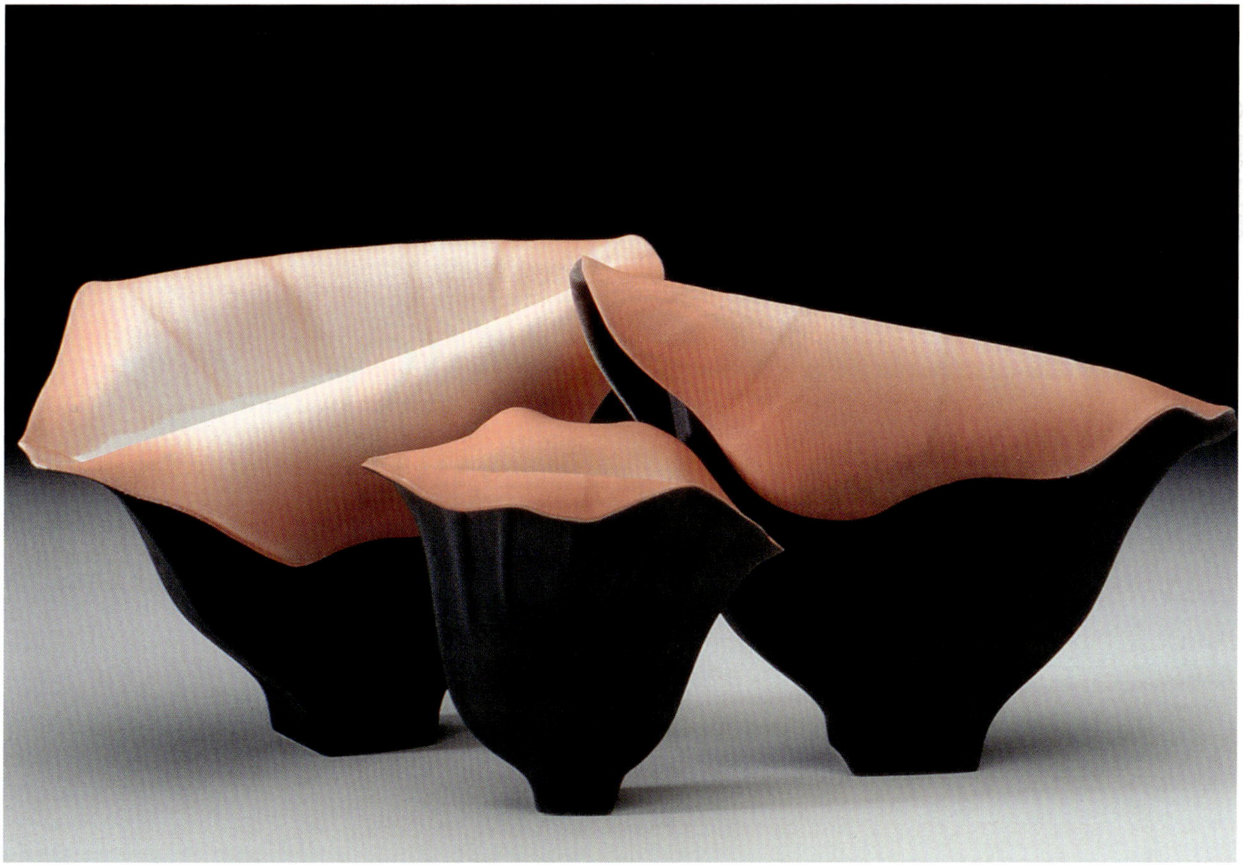

Studying in Australia

> *I find the tactile quality of bone china very appealing, and its marble-like surface has a warm sensuality about it. Slipcasting bone china allows me to make ultra-thin shapes, which are extremely translucent, pure and so white that ordinary porcelain seems almost grey in comparison. I find that the supreme whiteness of bone china makes the colours so much more vibrant.*[6]
> Angela Mellor

Since I first made this statement in the early 1990s, great strides have been made in researching clay bodies to achieve more whiteness. Les Blakebrough was a great exponent of this and, through his research at the University of Tasmania, he developed a very white and translucent porcelain body called Southern Ice, which is widely used today.

At the end of 1994, I left England and moved to Perth, Western Australia, to marry and set up a studio. As there was no supplier of bone china in Australia at that time, I found myself researching bone china casting bodies, eventually discovering a recipe that worked for me by Dr Owen Rye in Peter Lane's book *Contemporary Porcelain.*[7] This recipe is also included in the section on slipcasting (see Table 3, p.48).

Penny Smith, my tutor during my Honours year in Hobart, Tasmania, was a great support to me. She also explored the elusive quality of translucency in bone china during her three months' residency at the Arabia Tableware Factory in Finland in 1995. There, she created a series of fine vessels and light-emitting forms inspired by bulls' horns. Her surface decoration was achieved by layering latex into the moulds and allowing the resulting variations of thickness in the vessel walls to create different intensities of translucence.

Thanks to Penny's support in liaising with the wood and metal departments to turn a wooden base and fabricate a holder for my ceramic lamp, I developed a series of light forms based on the Angel's Trumpet (*Datura suaveolens*) flower, which I had seen in Indonesia. Manfred and I came across some growing by the side of the road. I was so struck by their beautiful forms that we stopped to take a photograph. I also picked one to take home to draw, not knowing of its toxicity! This elongated trumpet-shaped form became the basis of a design for a standard lamp.

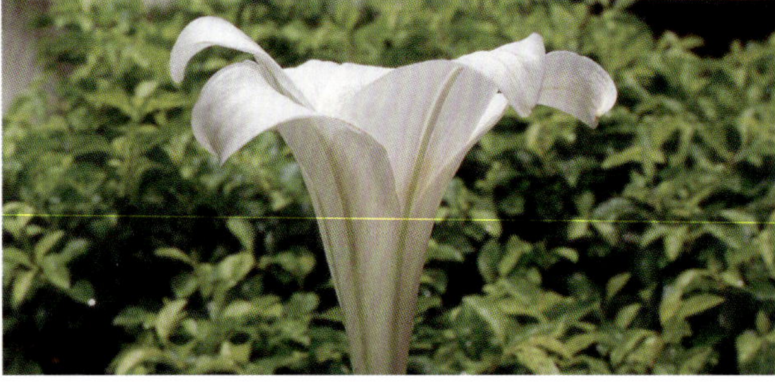

LEFT: *Datura* flower (Angel's Trumpet).

[6] Lane, *Contemporary Porcelain*, p.47.
[7] *Ibid.* p.44.

When light passes through porcelain, it creates a beautiful effect, making the walls luminous and seemingly lit from within. Even thick-walled pieces have a halo of light on thinner edges and portray the depth of alabaster. This effect is different to glass, which light passes through with a hard and glittering clarity, as through clear water. Light through porcelain is more like the soft milkiness of light passing through fine marble.

Initially, my main concern was with the passage of natural light through thin translucent bone china, but over time, the possibility of using electric light became of interest to me. In 1997, using the trumpet shape, I designed a set of three standard lights and a set of two wall lights. The surface design was inspired by the shadows of leaves and vines on the back of the fence outside my kitchen window. When decorating the piece, I created relief design with shellac resist, which allowed the light to penetrate in varying degrees. When fired, the effect of electric light on the uncoloured bone china gave a pale yellow glow, while under natural light the pristine white clay body was accentuated.

BELOW: *Angel's Trumpet Light Form*, Angela Mellor, 1997. Unglazed translucent bone china with shellac resist pattern, based on the shadows of leaves and vines. 20 cm (h).

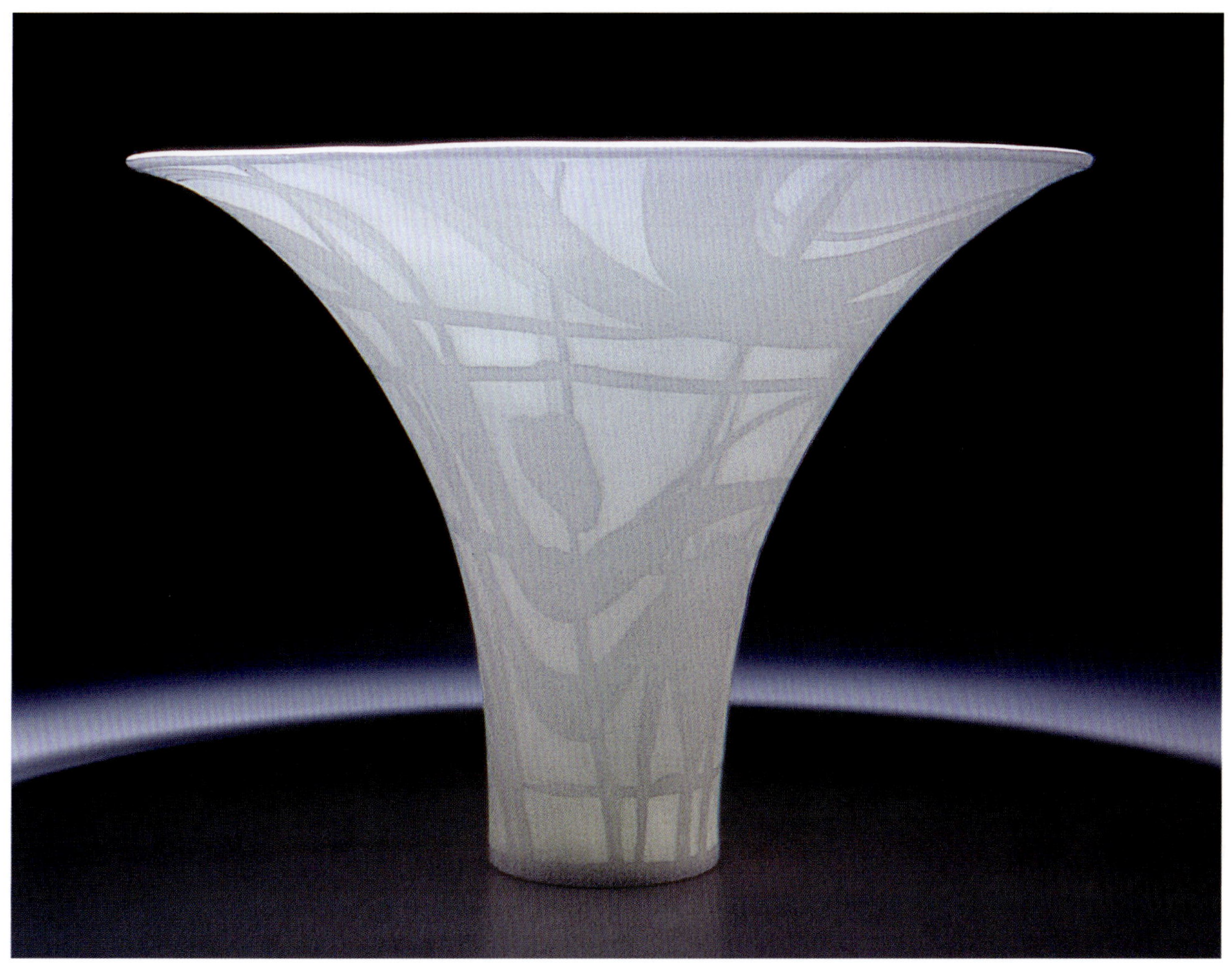

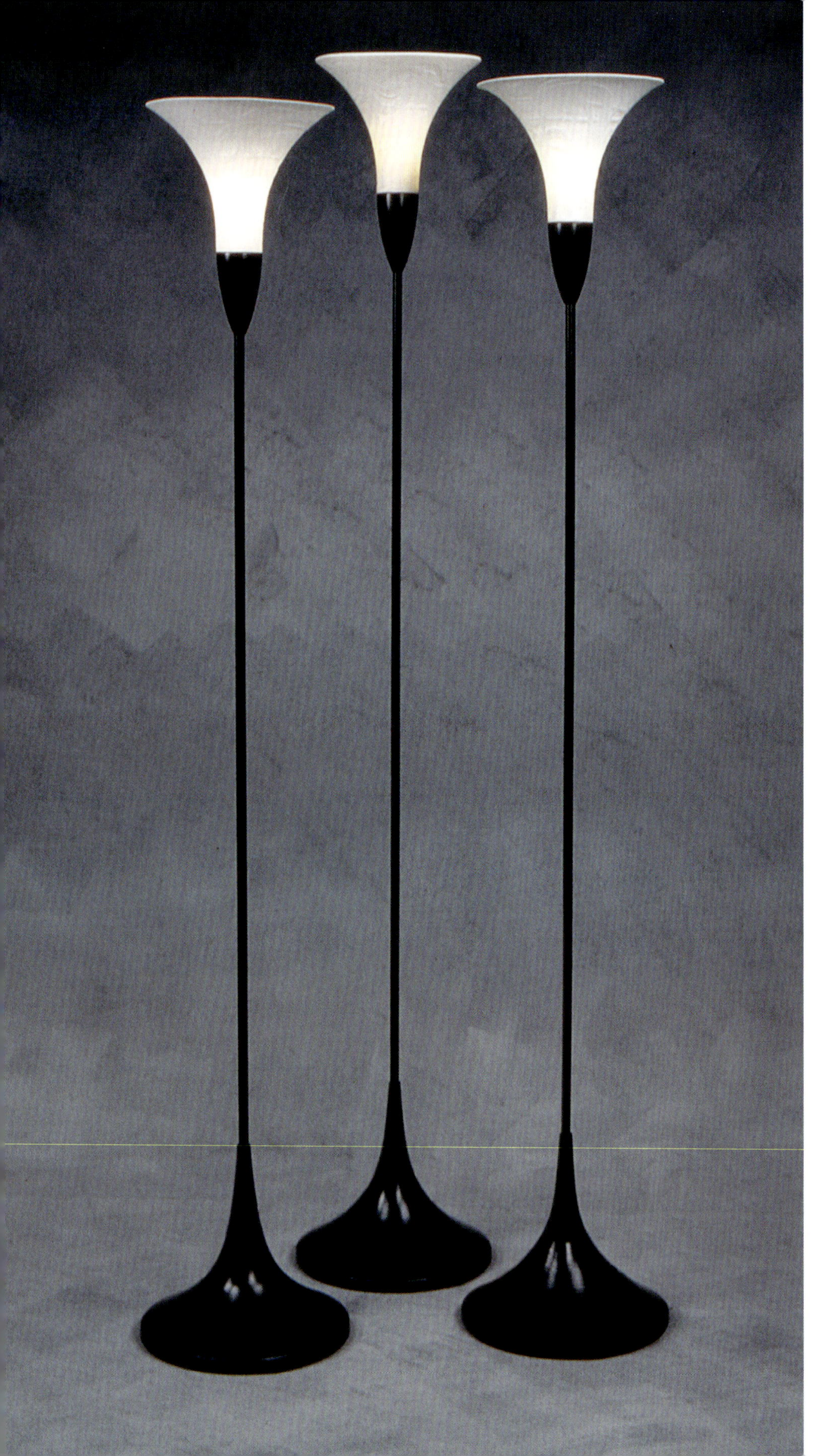

LEFT: *Datura* standard lamps, Angela Mellor, 1998. Bone china, metal rod, and blackwood base.

My photographs of contrasting natural textures on the beaches of Tasmania led me to look at the work of Adrian Saxe. Saxe is an American potter who was a resident at the Manufacture Nationale de Sèvres in France. He uses porcelain's white surface to explore glaze and lustre effects, using gold lustre to create aggressively opulent works. His choice of porcelain for the body communicates the wealth and status usually associated with the material. Both Sèvres and Saxe create display pieces and manipulate the effects of presentation rituals on how such objects are valued. His *Mortar Bowl with Stand* series shows the opposition between base and vessel. The works feature a beautifully rendered clay vessel positioned on a raku or stoneware base that resembles molten lava or rough stone. Saxe describes his hierarchal organisation in this way:

> *The bases represent the material stone, which in turn metaphorically indicate ceramic and aesthetic foundations. The vessel on the other hand, is more refined, culturally determined manifestation of the medium.*[8]

I liked the idea of using contrasting organic bases for my small delicate bone china vessels. I was inspired by the seashore, and since then have used coral, sand, stones, and even waveforms to create textured bases. In the translucent bone china vessels, I used iridescent lustres like mother-of-pearl to create shell-like effects, and they seem rather like little gems floating on the waves or caught upon the rocks. The vessels and bases contrast light against dark, refined translucency against coarse opacity.

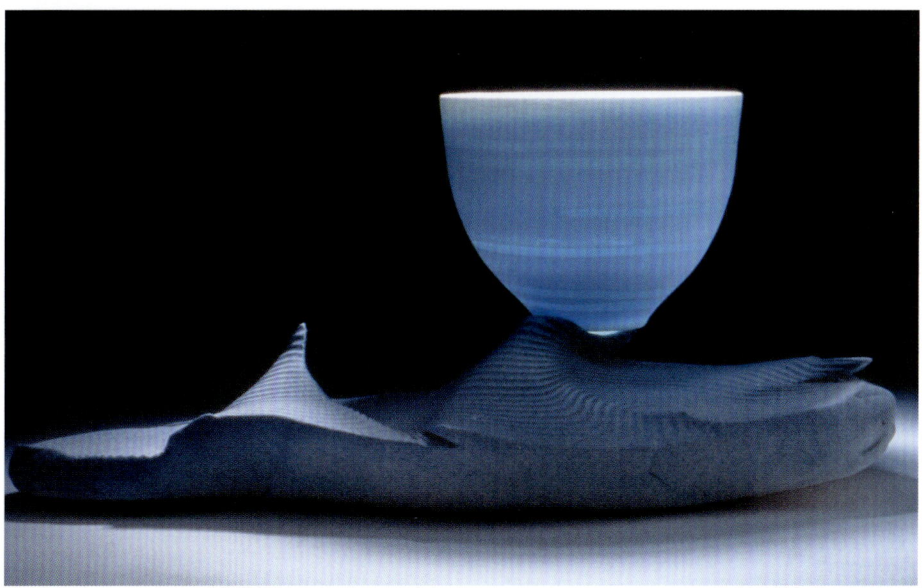

RIGHT: *Bowl on Wave,* Angela Mellor, 1997. Bone china with soluble colourants, on porcelain base sprayed with underglaze stains. 9 x 20 x 9 cm (h x l x w). *Photo: Uffe Schulze.*

[8] Martha Drexler Lynn, *The Clay Art of Adrian Saxe*, Thames & Hudson, New York, 1993, p.125.

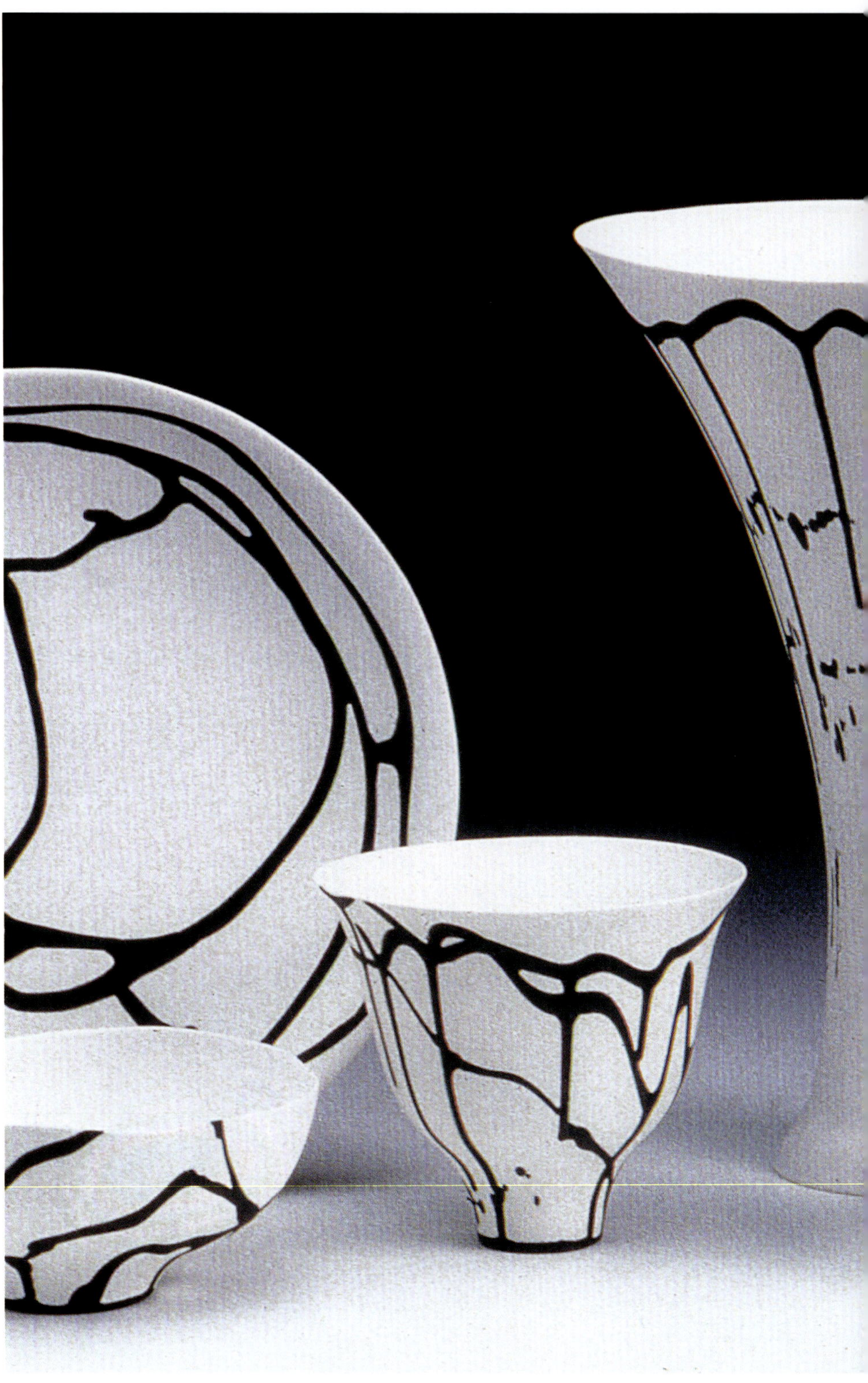

RIGHT: *Spider Web* vessels, Angela Mellor, 1996. Bone china with black slip trailed into the mould. *Photo: James Austin.*

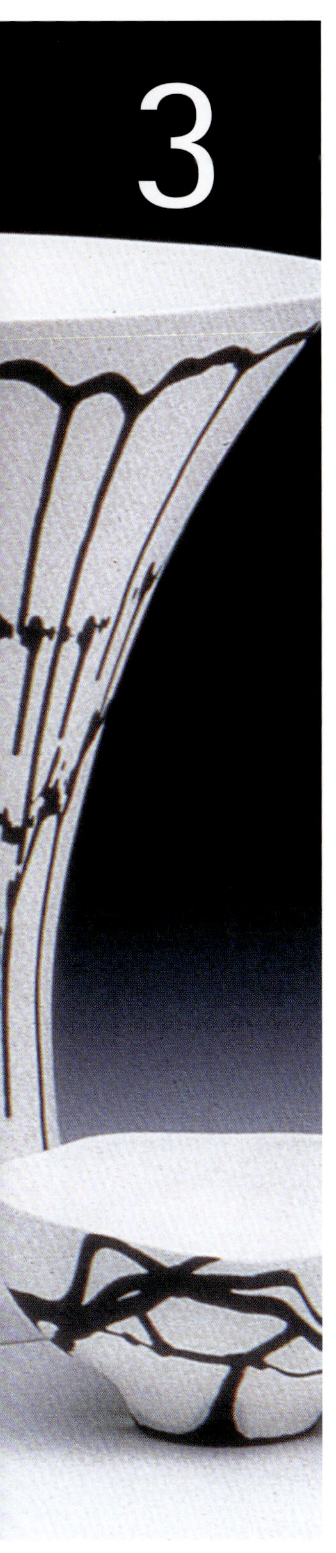

Methodology

Mould-making

I first attempted mould-making during a short course at North Staffordshire College in the early 1980s. I learnt how to make a couple of large models and moulds, and made a tall vase shape with a wide rim and narrow base. This happened to be a four-piece mould because the curve from the base turned in at the middle before flaring out at the top. During the short course, I met Jeff Beason, a tutor at the college, who specialised in bone china slipcasting using coloured slips. I invited him to come and give a demonstration to my students at a college I was teaching at in Staffordshire.

My teaching commitments meant that it wasn't until the early 1990s that I could further my knowledge of mould-making by attending one of Sasha Wardell's courses in France. I was able to carry out more in-depth study during my year at the University of Hobart and afterwards during my MA research scholarship at Monash University in Melbourne. As there were no facilities for mould-making at Monash, I attended an evening course in mould-making at the Holmesglen TAFE Institute.

This section is based on Mouldmaking Workshop notes, courtesy of the Ceramics department at the Holmesglen TAFE Institute.

Plaster moulds have been used for forming clay objects since the fifteenth century. Plaster is a material well suited to the purpose for a number of reasons:

- fine detail may be reproduced
- a chemically and physically stable mould is produced
- absorption may vary, depending upon the use (a dry mould will absorb moisture more quickly than a wet mould, the porosity allowing the release of the clay)
- a smooth, durable surface may be formed
- it is relatively cheap

When making a plaster mould for casting, you first need the original model of the form you wish to reproduce. This can be made of plaster, clay, wood, or metal. Next, you need to make the block mould – a negative of the original – also cast in plaster. After that, you can make the case mould: a mould, again in plaster, of the block mould. The case mould is necessary if you need a few moulds of the same shape. It is also good to have spares in case a mould breaks. I did not initially make case moulds, as most of my forms were for one-off pieces.

Life of moulds

There are two main factors limiting the life of a plaster mould: one is the change in dimensions due to mechanical wear; the other is the surface deterioration due to soluble materials in the body, such as deflocculants which, when casting slips, attack the surface of the mould to some extent.

Plaster

Plaster is homogeneous and, when properly mixed with water and allowed to crystallise, it becomes solid, without grain or hard or soft spots. It is an ideal material for making moulds.

The ratio of plaster to water will affect its setting time, its final strength, and its absorbency. The strongest mixture is 100 parts of water to 140 parts of plaster by weight. This gives a strong, heavy plaster with about 35 per cent absorbency. This is called a 140s mix, and it starts to set about four minutes from the start of mixing. It is used for cases and any original work that needs to be kept intact during the mould-parting process. A 130s mix is used for press moulds, casting moulds, and jolley moulds (hollow forms such as cups and bowls). A 120s mix is used for hump moulds and drying bats. Finally, a 110s mix is used for patching moulds. I use a 130s mix for all my casting moulds.

Chart of plaster mixes

The following is a guide for calculating the volume of variously sized cylinders that will be used as casting moulds, and the quantities of dry plaster and water needed to mix the correct amount of plaster to fill the mould. This uses the ratio 1:1.3 water to plaster.

First, use Table 1 to calculate the volume of the cottle cylinder (which can be either plastic or metal) that you will be filling with plaster. Measure the diameter of the cylinder in centimetres or inches and use the table to find the appropriate factor.

Next, use Table 2 to multiply the factor by the cylinder height to determine the volume in cubic centimetres or inches. The factors in the middle columns (water, plaster and weight) are the same for both centimetres and inches, so use whichever you prefer.

Table 1: Plaster mixes

FACTOR ←	DIAMETER (in.)	DIAMETER (cm)	FACTOR →
3.1	2	5	20
7.1	3	7.5	45
12.5	4	10	79
19.6	5	12.5	123
28.3	6	15	177
38.5	7	17.5	241
50.2	8	20	314
63.6	9	22.5	398
78.5	10	25	491
95	11	27.5	594
113	12	30	707
132.7	13	32.5	961
153.9	14	35	961
176.6	15	37.5	1104

Table 2: Volume calculation

VOLUME → (cu. in.)	WATER (kg or l)	PLASTER (kg)	TOTAL WEIGHT ← (kg)	VOLUME (cu. cm)
45	0.5	0.65	1.15	740
67.5	0.75	0.98	1.73	1110
90	1	1.3	2.3	1480
135	1.5	1.95	3.45	2220
180	2	2.6	4.6	2950
225	2.5	3.25	5.85	3690
270	3	3.9	6.9	4430
310	3.5	4.55	8.05	5080
355	4	5.2	9.2	5820
400	4.5	5.85	10.35	6560
445	5	6.5	11.5	7300
490	5.5	7.15	12.65	8030
535	6	7.8	13.8	8770

Plaster methodology

Mixing plaster should not be a problem, as long as you keep to certain rules:

- Measure the amount of plaster and sprinkle it through your fingers into the measured amount of water. Allow the mix to sit for two minutes to ensure that the plaster is saturated. This allows for air bubbles to reach the surface; an occasional tap with the foot on the side of the bucket will also help with this.
- Mix the plaster by hand with gentle flowing movements in one direction only. Keep your hand below the surface of the water, taking care not to incorporate air into the mixture, and breaking up any lumps that may form.
- If any air bubbles gather on top of the surface, they may be skimmed off. Alternatively, I use a quick spray of a 50:50 mix of water and methylated spirits to disperse them. It usually takes about 10–15 minutes to mix.
- A good indication that the mix is ready to pour is a slight visible trail when you draw a line over the surface with your finger.
- When you are ready, pour the plaster in a continuous, even stream into one spot near the inside of the cottle.
- Tap the sides of the cottle and gently vibrate the casting to force any air trapped within the plaster to rise to the surface.

Drying and storing moulds

Moulds should be dried slowly in a drying cabinet at a temperature not exceeding 40°C. It is not advisable to dry them quickly on top of a kiln, as the bases of the mould can become scorched and break up. When removing them from the drier, take care to avoid any thermal shock due to differences in air temperature inside and outside the drier. Moulds are best stored in a cool dry place.

Tools and equipment

The work area

It is best to keep the plaster and slipcasting areas separate. If this is not possible, try to divide the room into two different areas to avoid contaminating slip with plaster. Alternatively, do all your mould-making at once and clean up before commencing slipcasting. Your work area should have good light, a constant temperature, and be free of dust. For plasterwork, a bench with a non-porous surface such as Formica is an advantage. Ideally, the sink should have a trap and be deep enough to submerge buckets into. If possible, it is good to have two sinks: one for clean water and one for dirty water, in which you will clean buckets and tools.

Machinery

A plaster-turning wheel or lathe is essential to my practice. There are two types: vertical and horizontal. I have used both types, but I prefer to use a vertical wheel as you can see the profile of the model the correct way up.

For mixing small quantities of casting slip, I use a rapid glaze mixer. You can use a paint mixer on the end of a hand drill to do this.

Plaster mixing tools and equipment

- Weighing scales for measuring up to 7 kg
- A large plastic dustbin with a lid
- A scoop
- Plastic or rubber buckets with a lip to aid pouring
- Plastic bowls if mixing smaller quantities of plaster
- Jugs: a 1-litre plastic jug is good for measuring
- Spoons and whisks for mixing

Cottling material for forming retaining walls

- Flexible plastic sheeting 3 mm thick or thin vinyl flooring
- Aluminium sheet
- Plywood boards and wedges
- Plaster bats
- String and clothes pegs or an adjustable strap for securing the cottle

For soaping up models and moulds

- Natural sponges
- Soft-bristled brushes
- Soft soap or mould-maker's size

For forming wet plaster

- Plaster-turning or wood-turning tools
- Steel scrapers or metal kidneys
- Galvanised sheet metal for template-forming

For cutting hard plaster

- Surform files
- Various rasps, knives, files, and gouges
- Carpenter's saw
- Forged steel tools
- Hacksaw blades
- Coping saw
- Bow saw

Miscellaneous tools

- Callipers
- Flexible curve ruler
- Indelible pencil
- All-purpose adhesive glue
- Wet and dry sandpaper
- Compasses
- Straight-edged scrapers and steel square
- Assorted paint brushes
- Rubber mallet and steel hammer
- Spirit level
- Car inner tubes

Model-making

Throughout my career, I have used plaster moulds for slipcasting bone china. A wooden model can also be turned and then lacquered at least three times to make it fully waterproof. Where moulds are concerned, the creative stage is in the preliminary planning, the model-making which establishes the form, and the choice of colour and surfaces.

I keep a photographic record of my ideas, and I make drawings of the basic shapes before modelling. I use a metal or cardboard template of the profile as a guide in the forming process.

Although a variety of materials can be used to make a model, I use plaster, which is preferred by ceramists and industry for its many advantages. As well as being homogeneous and easy to smooth, it can be worked wet, damp, or dry without changing shape. A good, polished surface can be achieved and does not deteriorate with time. Because of its porosity, however, it does require sealing with layers of soft soap to prevent sticking when the mould is made from the original plaster model. A wide variety of forms can be made either by hand or using machinery.

A model can be made from plaster using a turning lathe to give a smooth, curved profile. The model is made 15–20 per cent larger than the final desired size to allow for shrinkage. A plaster case mould is made from the model so that the piece will drop out. I have avoided multi-part moulds for most of my work because the seams show up after firing, even though they can be disguised.

All open forms – vases, bowls, etc. – require kiln setters. These resemble a lid and help retain the rounded shape of the top edge of the cast. A separate setter model and mould is made for each piece. The mould and setter casts are dried and fired together. After the bisque firing, a thin layer of alumina wash is painted around the setter to prevent sticking during the high firing.

Machine-formed plaster models

Plaster wheel – vertical turning

I used a vertical plaster wheel to turn symmetrical forms throughout my research. The vertical wheel has variable speeds and is specially designed for turning plaster. This technique requires more speed and coordination than lathe work, as most of the shaping is carried out while the plaster is still soft.

Wheel preparation

The wheel (or whirler) head is made from plaster, usually about 6 cm deep and 30–45 cm in diameter, with grooves turned into the surface. These act as an anchor for the new block of plaster, which is cast on top. It is a good idea to have an extra layer of plaster on the wheel head to save having to replace the whole wheel head if it is constantly turned and grooved. It is easier to replace a new block than it is a new head.

Soft soap the plaster whirler three times so as to avoid a hard layer of plaster at the join between the two layers. This will also help release the top block when necessary. A plastic cottle is then placed around the head and secured tightly using nylon string.

The required amount of plaster to make a block 6 cm deep is poured carefully down the side of the cottle to avoid air bubbles being trapped. Turning must be commenced as soon as the cottle can be removed, as it is easier to turn soft plaster. Start turning with a triangular cutting tool, the long handle resting on the adjustable metal bar, the end of which is tightly tucked under the arm. It takes time and practice to perfect this technique. The block should be turned completely flat and checked with a spirit level. One or two grooves are then turned into the block plus a natch. With an indelible pencil, the extremity of the model needs to be marked. This line will remain during turning and act as a guide. Soft soap the plaster block as before and cottle up the diameter required. Once the plaster has been poured in and cottle removed, the turning can begin. Don't begin with more plaster than necessary because it is important to rough out the form quickly before the plaster has matured. Time is lost turning unnecessary excess away.[1]

Drop-out moulds with outside fitting spares

This is the type of mould I have used for most of my work, mainly to avoid seam lines, which are extremely difficult to remove from bone china. Also, my choice of forms – mainly bowls and cone-shaped vessels – are designed to suit this type of mould. The model usually tapers at the base, has no undercuts, and can be simply moulded in one piece, with a separate reservoir ring or 'spare'.

Before turning the plaster, the design is drawn on to draught paper and a metal or cardboard profile is made, to act as a guide when shaping the form.

When the model has been turned and smoothed, allow a collar of 3–4 cm around the top of the model to make the mould. Soap up the model and spacer (if using one) three times. I tend to work directly onto the wheel head. Then place the cottle on the line marked on the wheel head or around the spacer, making sure that it is high enough to clear the model. Secure it tightly with string. Then, mix up the plaster. It is better to overestimate than underestimate, as any excess can be used to make a plaster bat.

Pour the plaster mix steadily between the cottle and the model. After the plaster has been poured, tap the sides of the cottle to allow air bubbles to come to the surface. When the plaster has set, usually after about 25 minutes, remove the cottle and take off any sharp edges with a surform or metal scraper.

Lift the model and mould off the wheel head, keeping the model inside. Place a preformed disc of plaster on top of the mould and ensure that it covers the model by about 1 cm all the way around, leaving space to make the spare. Trace around the disc with an indelible pencil. You will then know where to put the natches. Remove the disc while you make the natches, which can be made by driving a coin into the top of the mould using a semicircular movement, which makes a half-sphere impression. Make two natches close together and two further apart around the rim of the mould. This makes it easy to line up the spare and the mould.

[1] Sasha Wardell, *Slipcasting*, A & C Black, London, 1997, p.89.

Soap up the mould and disc three times and place a smaller piece of cottling around the top mould. Mix up a small quantity of plaster and pour it carefully between the cottle and the disc.

When set, remove the cottle and carefully remove the disc from the spare or reservoir ring by tapping gently in the direction of taper. Place the ring upside down to avoid damaging the natches.

Release the model from the mould by turning the model upside down onto your hand, catching it when it drops out. Provided that the model has been properly soaped, there should not be a problem. However, if it is difficult to remove the model from the mould, this indicates that there has been a fundamental error as a result of bad soaping. Possible causes include:

- the model was too dry, and the soap was absorbed and did not create a seal
- the soap was too thin and was therefore absorbed into the model
- there is an undercut, which means that a multi-part mould was required

There are several ways you can try to release the model:

- place a metal plate on the base and tap sharply with a rubber mallet
- run water between the model and the sides
- apply compressed air around the join between the model and mould
- drive two screws into the top of the model, and prise the model out with a pair of pliers

Mould-making in itself is quite an arduous task, and many technical problems must be overcome if good working moulds are to be produced. Good models and moulds are essential when working with bone china.

I highly recommend Sasha Wardell's book, *Slipcasting,* as a resource for mould-making information.

Slipcasting

Slipcasting is a fairly simple process, widely used by industry for making objects from domestic tableware and delicate figurines to washbasins and lavatories. Complex slip moulds require many pieces. A craft potter will usually use between one and four pieces for a multi-part mould. Pottery of a uniform thickness can be made using plaster moulds and liquid clay (casting slip).

The slip is poured into a plaster mould and, as the water from the slip is absorbed, a layer of bone china is gradually built up on the surface of the mould. When the desired thickness is reached – my work usually takes between two and five minutes – the mould is inverted and the surplus slip poured out; the mould is wedged at an angle until the slip has stopped dripping. Because of bone china's tendency to warp, it is best to leave the cast piece in the mould until it is dry, preferably overnight. It can then be neatened with a sponge and fettled with a sharp knife or metal scraper.

Although the casting process is relatively simple, the design and making of the moulds and the formulation and mixing of casting slip are exacting. It takes time and research to obtain satisfactory, repeatable results.

Until recently, very few potters used bone china. Commercial suppliers are rare, and it is expensive. I found a bone china slip recipe that worked for me after much research and liaising with Walkers Ceramic Supplies in Melbourne, who made some up for me, relieving me of the tedious job of making my own. They have since become suppliers of bone china slip in Australia. My supplier in the UK is Valentine Clays, based in Stoke-on-Trent.

In addition to clay and water, casting slip contains certain alkalis, better known as deflocculants. Deflocculants help to disperse the particles in the slip, making it fluid enough to pour. This reduces the amount of water required, which also reduces the shrinkage. Deflocculants are not usually more than 0.5 per cent of the dry weight of the ingredients; more than this will make the cast difficult to trim.

Soda ash (sodium carbonate) and water glass (sodium silicate) are the most commonly used deflocculants. They can be used individually or together, in a ratio of 1:3. Water glass is produced by fusing sodium carbonate with silica sand by heating them under pressure. This process forms concentrated, syrupy solutions. These are measured using the Twaddell scale (°T), which is used to measure liquids denser than water. 75°T is a thin, free-flowing, gummy liquid suitable for use in china slips, where an increase in plasticity is required. The proportions of clay, water, and alkali are carefully balanced for each type of casting slip, and no more water should be added. Thoroughly blunge or mix it, and if it is still too thick, add a few drops of Dispex™. A good casting slip should not shrink too much from wet to leather hard and should have good dry strength.

Slip preparation

If making slip from dry ingredients, Caroline Whyman advises:

> *Always add the mixed dry ingredients to the measured amount of water, beginning with the most plastic and finishing with the least plastic ingredient. This will result in a very stiff mixture. Mix the deflocculant in a little hot water, taken from the total amount calculated for the slip, and add this to the mixed ingredients drop by drop while you blunge it.*[2]

I found it is best not to add all the measured water or dissolved deflocculant all at once as the slip may become fluid without using the full amount. Blunge thoroughly for 1–2 hours before use. If only mixing a small amount, a rapid gaze mixer will do. Allow to stand overnight and add a few drops of Dispex™ if the slip seems too thick. Scraps of casting slip may be recycled but do not add more than 20 per cent recycled slip otherwise there will be an imbalance of deflocculants, which can cause problems. Because of the high content of bone ash, it is necessary to ball mill the mixture for 4–8 hours.

Various recipes are available, but the most successful one that I have used is by Dr Owen Rye, a lecturer at Monash University for many years (Table 3).

Table 3

CASTING BODY B6	FIRING TEMPERATURE: CONE 8–9
Eckalite no. 1 kaolin	30%
Bone ash – natural	40%
Potash feldspar	22.8%
Silica	2.2%
+ 2 g Dispex™ per kg (dry weight) + 1 g sodium silicate per kg (dry weight) + 600 ml water per kg (dry weight)	

The slip must be mixed well and strained through a number 60 sieve before casting. Pour into the mould and leave for two to five minutes, depending on thickness and translucency required, and pour out. Table 4 shows a recipe I trialled during my Honours research.

[2] Whyman, *The Complete Potter: Porcelain*, p.49.

Table 4: Reconstituting bone china into casting slip

MIXING PLASTIC BONE CHINA	FIRING TEMPERATURE: 1220–60°C
Bone china plastic clay	50 kg
Sodium silicate	65 g
Water	4.5 l

Preparing plastic clay

Measure out the required amount of water, less 100 ml for mixing with the sodium silicate. Break up the plastic clay into small pieces and add to the required amount of water. Leave to slake down overnight. The following day, mix with a blunger and slowly add the sodium silicate mixture until you have the correct consistency.

Slipcasting procedures

Check the mould

The mould should be dry and sponged to remove loose bits of plaster and excess soap left from the mould-making process. Assemble parts or secure the spare (if using a drop-out mould) with inner tube bands to prevent the seams opening when the slip is poured.

Filling the mould

Avoid filling the mould too fast, otherwise casting spots, swirls, pinholes, and blips may occur. Filling the mould too slowly may result in filling lines and variations in thickness.

Positioning the flow of slip

Pour the slip through a small plastic tea strainer or kitchen sieve, and try to aim it into the centre of the base of the mould as you pour. If the slip drips down the side of the mould, a casting spot can occur, creating a hard spot that can be seen through a translucent body. It can also discolour and be resistant to glaze.

A large mould may be revolved on a banding wheel while filling, which helps to avoid filling rings. If a shallow mould is used, for example, for a plate or saucer, revolving the mould helps to prevent air bubbles becoming trapped while filling.

While casting using a large mould, the slip may need to be agitated by stirring the slip inside the mould with a soft brush before tipping, taking care not to touch the sides.[3]

[3] Wardell, *Slipcasting*, p.75.

Tipping out the slip

Tip out the slip steadily over a bucket. Some potters use a circular movement, rotating the mould while they pour, so that it will drain more evenly, but I have not found this to be necessary to achieve a smooth, even surface.

Try not to pour too slowly, otherwise draining marks will occur. Pouring too fast may result in sucking in, especially with narrow-necked forms. Avoid causing a glugging sound – this signals that casts will be flabby and distorted.

Draining the mould

After emptying, drain moulds at an angle of 30–40 degrees: if the angle is too steep, 'droppers' will occur on the inside of the base; if it is too shallow, wreathing will occur. Avoid reversing the mould too soon, or the slip will run back, leaving drips in the mould. If you are using a drop-out mould, leaving it too long means that it will fall out; if using a multi-part mould, the slip contracts around the plaster in the mould.

Trimming the cast in the mould

Trimming too soon can cause bits to fall inside and stick to the cast while it is still wet. If you leave it too long before you trim, small cracks appear in the rim. Keep your fettling knife sharp; to help prevent cracking of the rim, avoid cutting with a blunt knife. Sponging in the opposite direction of the cut helps to smooth over any cracks. If the object has a flat, round top, I find that drawing lightly over it with a large, thick, damp sponge gives a nice, clean-cut edge.

Removing the cast from the mould

The cast will distort if removed too soon from the mould. If you leave it too late, complex shapes such as teapot handles and spouts are likely to crack.

To help avoid distortion of bone china, leave the cast in the mould overnight. However, this may shorten the life of the mould, because deflocculants or soluble salts in the casting slip cause erosion, which opens up the pores inside the mould.

If casts do not release easily from the mould, dust the mould with talc before casting.

Fettling the cast

When the cast is leather hard, remove the seam lines with a sharp fettling knife or scalpel, and neaten with a damp sponge. Bone china casts should be treated with care to minimise warping.[4]

Casting faults and their causes and remedies can be found in Sasha Wardell's book, *Slipcasting.*

[4] *Ibid.* pp.74-7.

Slipcasting allows me to exploit the traditional methods used by industry for production ware and use it in a very personal way. Slipcasting in bone china is the only satisfactory method that I have found that allows me to make ultra-thin, extremely translucent objects. I have also exploited the warping nature of bone china to achieve organic forms inspired by nature. Slipcasting has also played an important part in the research for this book, in conjunction with paperclay slip, which has enabled me to replicate fine textures found in nature and extend the scope of my work into large-scale sculptural forms.

Surface treatments: resist

There are many different methods of decorating slipcast bone china and porcelain to achieve transparency and accentuate light and shadow. One of these is acrylic resist, which I have used in many of my slipcast pieces.

Acrylic/shellac resist

Acrylic or shellac resist is a decorative resist technique used to emphasise light, shadow, and translucence. A relief must be made on the surface of the piece, in the same way that a design is etched onto a printing plate. The raised part of the design is painted on to dry, unfired clay, using water-based acrylic varnish, and when the acrylic is dry, the non-lacquered surface is sponged back with water, thinning the wall and creating a relief design.

I use an acrylic matt varnish to which a vegetable dye is added so that the brush marks can be seen. Some artists use shellac, but this tends to shorten the life of your brushes and acrylic is so much easier to wash out. It is also much smoother to apply.

After the acrylic is applied, allow the resist to dry out completely. It is important to remember when working with thin ware that clay is stronger when dry. A thin layer of the unprotected surface can then be washed away with a soft, damp sponge. Always use pure water, and avoid excess water in the sponge, which might make the pieces too soft, causing distortion and cracking. Allow the piece to dry out completely every time a layer is washed away.

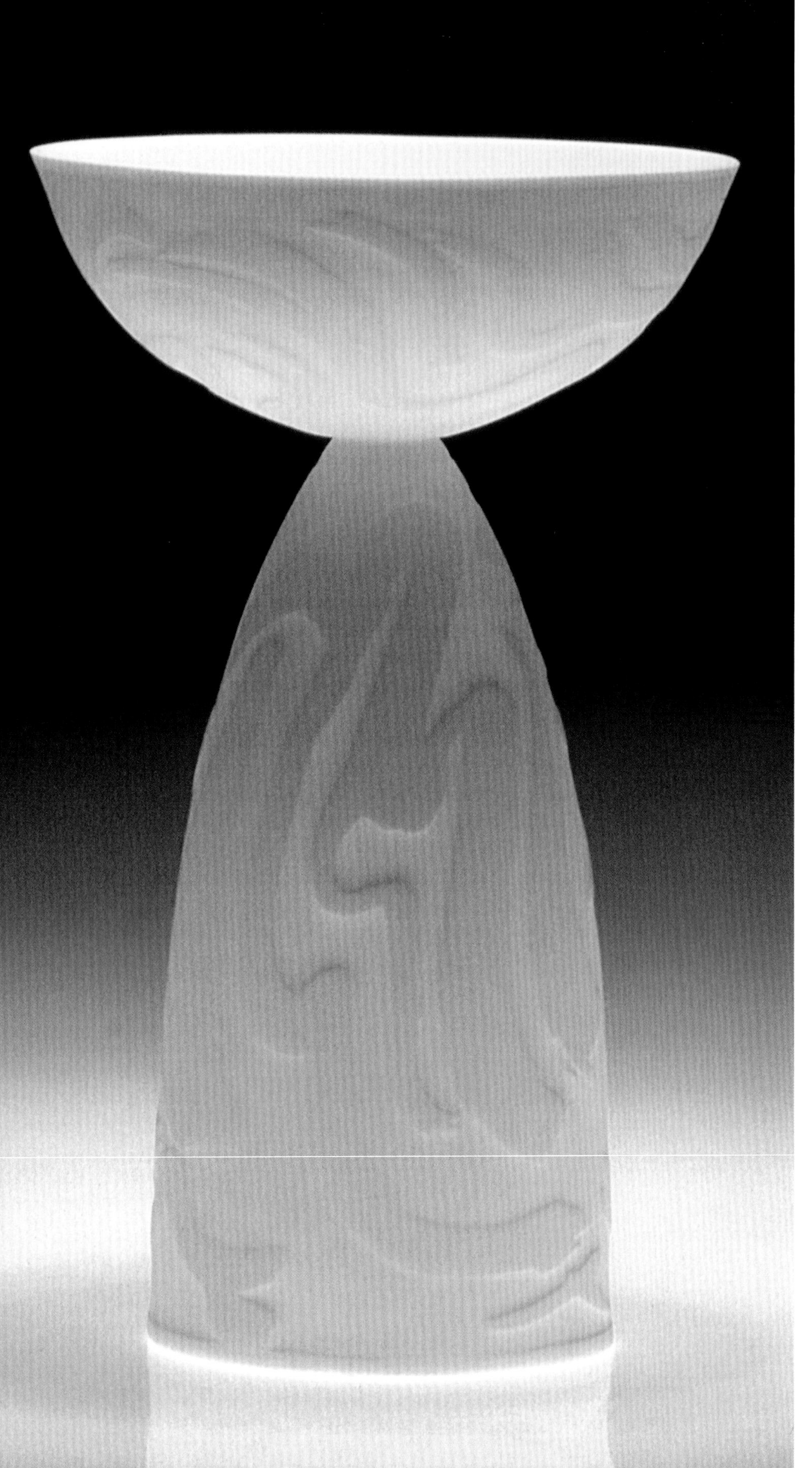

LEFT: *Chalice*, Angela Mellor, 1997. Bone china with water-etched relief pattern. *Photo: Uffe Schulze.*

Ceramists working with the relief technique

Sasha Wardell (UK)

When I asked her about her process, Sasha said:

> *Bone china is a very seductive material to work with, possessing qualities of intense whiteness, translucency, and strength. It is a very 'single-minded' clay, which forces the maker to work with clarity and precision. Its technical inflexibility and idiosyncratic making and firing characteristics might easily be a deterrent to investigation, however I consider these restrictions and limitations a challenge to creativity and working methods.*
>
> *I have adapted the industrial techniques associated with bone china production of mould-making and slipcasting to studio production. My particular interests lie in the translucent properties of this material and the differing degrees of luminosity possible. In order to enhance these qualities in my work, I have developed three variations of decoration which aim to exploit this characteristic.*

One of Sasha's processes is water erosion, which she explained to me:

> *The process of 'water erosion' focuses on the entire surface of the piece using a single cast and no colour.*
>
> *The variations of translucency are achieved by painting in acrylic medium onto the raw cast. This acts as a resist when the piece is 'washed' with a damp sponge until a relief pattern appears. Marks and patterns are derived from organic sources and fabric design.*

RIGHT: *Ripple Veil* vessels, Sasha Wardell, 2018. Slip cast bone china, using a water erosion technique. 18 cm (h); 14 cm (h); 11 cm (h). Fired at 1280°C. *Photo: Mark Lawrence.*

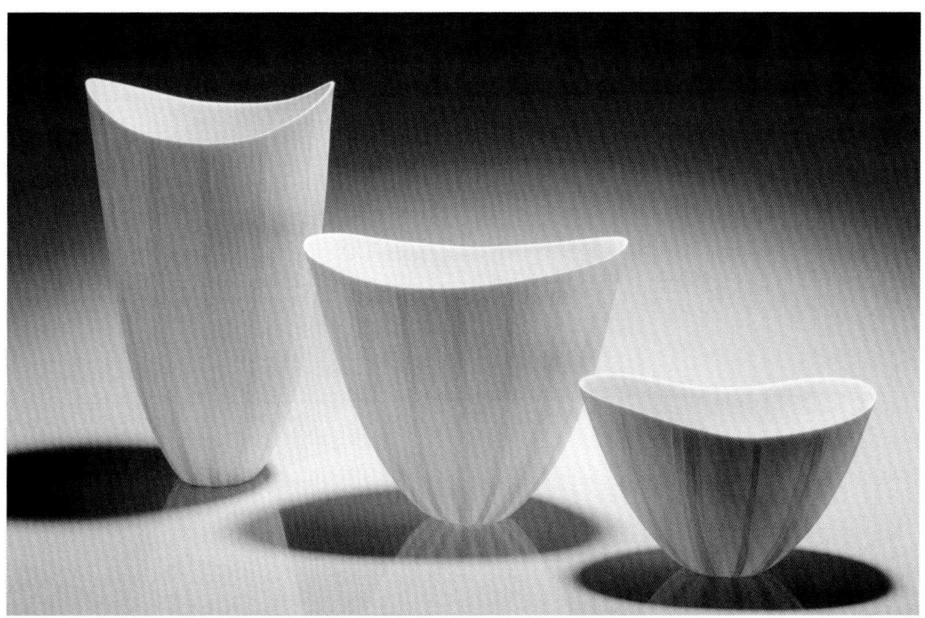

Arne Åse (Norway)

Norwegian potter Arne Åse uses shellac resist on porcelain, repeating the procedure many times; this creates a layered effect showing varying depths of translucency, and becomes more translucent after the firing.

Les Blakebrough (Australia)

Blakebrough also used the resist technique on porcelain, but instead of creating an all-over pattern, he might have only used it to define a single motif or pattern band.

Astrid Gerhartz (Germany)

Astrid Gerhartz favours the relief technique on the top halves of her thinly thrown cylinders to achieve translucency. After firing, she continues the design through to the lower half of the cylinder using soluble colourants. The inside wall is covered with glaze, while the exterior is left unglazed. If the colourants are applied to the unglazed exterior, they penetrate inwards as coloured sections. Different brushes are used to create different images. She uses the method to create an all-over embossed pattern on bone china. She says of her technique, 'It is a difficult method to use, as the clay is very fragile until after it has been fired.'

Latex resist

Latex can be painted or trailed on to bisque-fired ware and then painted over with nitrates. Layers of colour can be built up this way. Small drops of colour may stay on the latex areas, but these can either be wiped off or left in place to create textural effects. The latex burns off during the firing. Finally, the pieces are polished with a fine 400-grit wet and dry paper after the high firing.

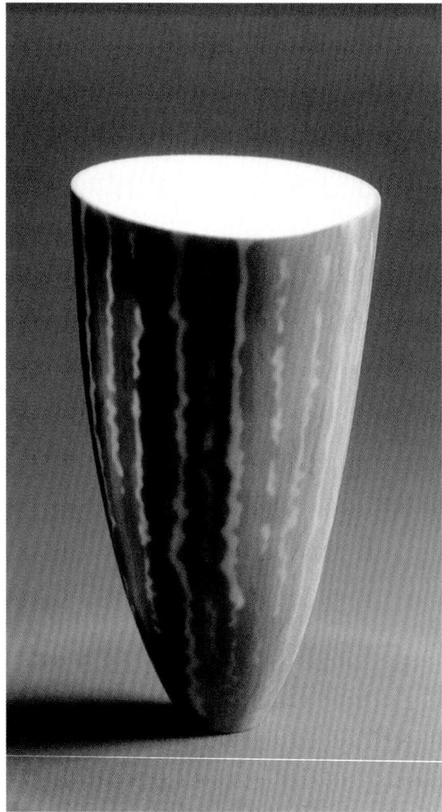

LEFT: *Coral Reef* (detail), Angela Mellor, 1998. Bone china, latex resist, soluble colourants. *Photo: Victor France.*

Transparency and soluble colourants

For many years, I was fascinated by the pure white body of bone china and was continually exploring its translucency. When I moved to Australia, light played an important part in my work, although I have always been inspired by the colours in landscapes and seascapes. It was the question of how to introduce colour while maintaining the translucency of the body that had drawn me to bone china.

Having seen Les Blakebrough's work, and studied the methods described by Arne Åse in his book *Water Colour on Porcelain*, I felt that soluble salts would be a very suitable technique to try on bone china. Åse says something in this book that felt very true for me:

> *Ceramics is a pictorial art, and artists relate to visible phenomena, to draw their inspiration from. This governs their modes of expression.*[5]

He also says:

> *As an artistic mode of expression, the use of water-soluble colourants (which may be classified as being the ceramic artist's watercolours) contains possibilities which are exceptional compared to the use of pigments which is common today. However, these colourants may be rather difficult to control. Many factors combine to develop the colours, and this, in turn, requires a high degree of craftsmanship, dexterity and precision.*
>
> *This [lack of control] is probably the reason why the ceramic industry has so far not found these colourants suited for their purpose. To the ceramic artist, on the other hand, [soluble colourants offer] an exciting challenge, demanding individual and personal solutions.*[6]

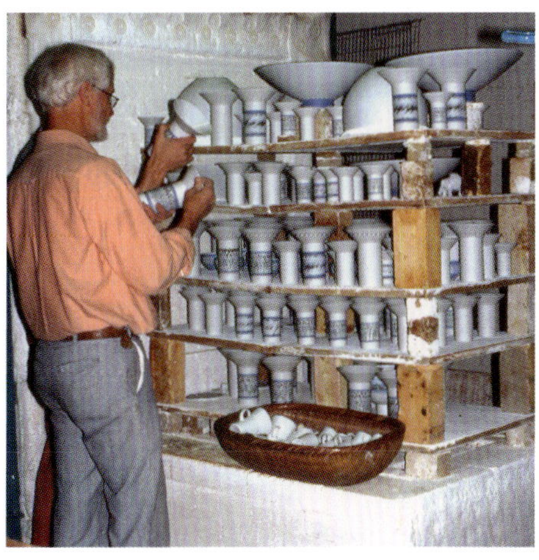

RIGHT: Les Blakebrough unpacking the kiln in the Research Centre in Hobart, Tasmania.

The late Les Blakebrough first introduced me to working with soluble salts during my year at the University of Tasmania in 1997. During that time, I carried out a few experiments and was very excited about the possibilities for using these colours on bone china. I was able to continue this research in more depth and to realise my conceptual aims while working towards my MA at Monash University.

[5] Arne Åse, *Water Colour on Porcelain: A Guide to the Use of Water Soluble Colourants*, Norwegian University Press, Oslo, 1998, p.21.
[6] *Ibid.* p.21.

Water-soluble colourants are chemicals in solutions that give colour to clays and glazes in the temperature ranges used in ceramic production. Comparing water-soluble colourants with ceramic pigments is rather like comparing oil paints with watercolours. Water-soluble colourants can be absorbed into the pores of the clay when they are applied to raw, dry, or bisque ware.

Water-soluble colourants were rarely used until the late 1980s, when Arne Åse wrote *Water Colour on Porcelain.* Before that, there was no literature on the subject; even now, many years later, very few artists work with them.

In 1999, at the 9th National Australian Ceramics Conference in Perth, Western Australia, I attended a workshop by Astrid Gerhartz, where she demonstrated her methods of applying colour with inorganic salt solutions. She glazes the inside of the pot with a clear glaze, leaving the outside of the pot unglazed, first applying cold wax to the rim to stop glaze from dripping down the outside of the pot. If glazing both inside and outside, there is no need to apply wax, but the pot needs to be dry before glazing the outside. Like Les Blakebrough, she enjoys the contrast between shiny and matt effects. She then covers the relief area with cold wax where no salts are required.

When painting over shellac with salts, it is best to wipe off the salt residue on the shellac with a cloth before firing. Gold tends to stick to shellac and needs to be removed with sandpaper after bisque firing. After applying shellac and one coat of colour, more shellac may be added and then a second layer of colour. Layers of colour may be built up this way. This technique of using acrylic resist and soluble colourants to build up layers of colour is one that I have used in my work.

Gerhartz normally uses Limoges porcelain, but during the conference she tried out a porcelain body developed by Les Blakebrough called Southern Ice, which is best fired to 1280°C. It seems to be an excellent body for throwing. Since then, a Limoges paperclay has also been developed, which is good for ceramists working in porcelain as it saves them having to make their own.

My own work with water-soluble colourants is restricted to dealing with their effects as colouring agents on unglazed bone china fired to 1250°C in an oxidised atmosphere.

RIGHT: *Sea Bowl*, Angela Mellor. Bone china with soluble colourants. 11 x 8 cm (h x w). *Photo: Victor France.*

BOTTOM RIGHT: *Cretaceous Bowl*, Angela Mellor, 2003. Bone china with soluble colourants. 11 x 8 cm (h x w). *Photo: Victor France.*

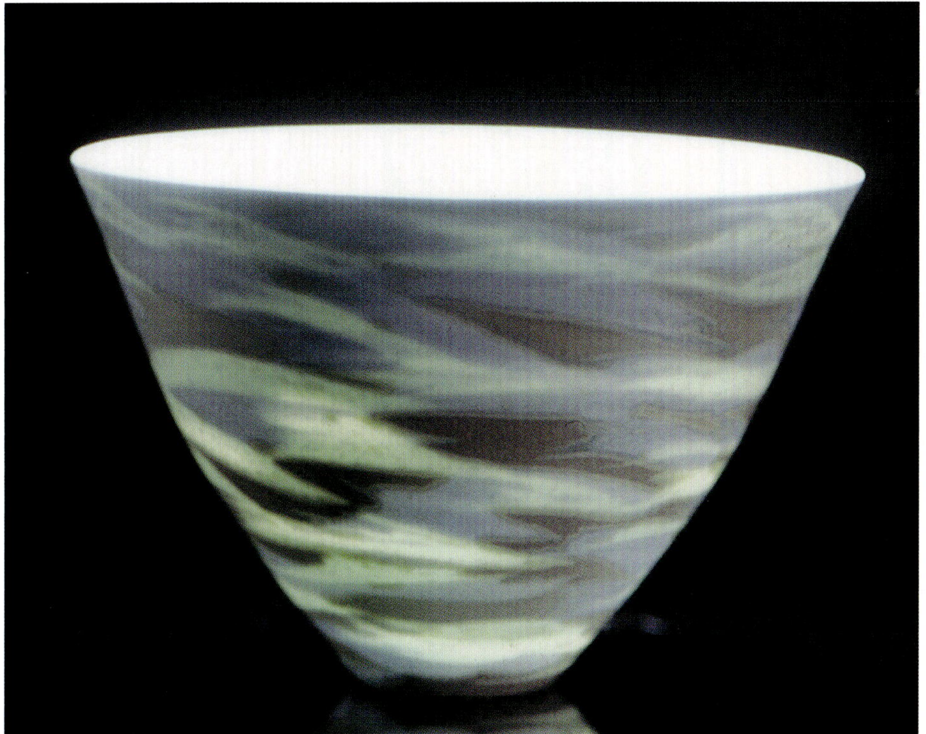

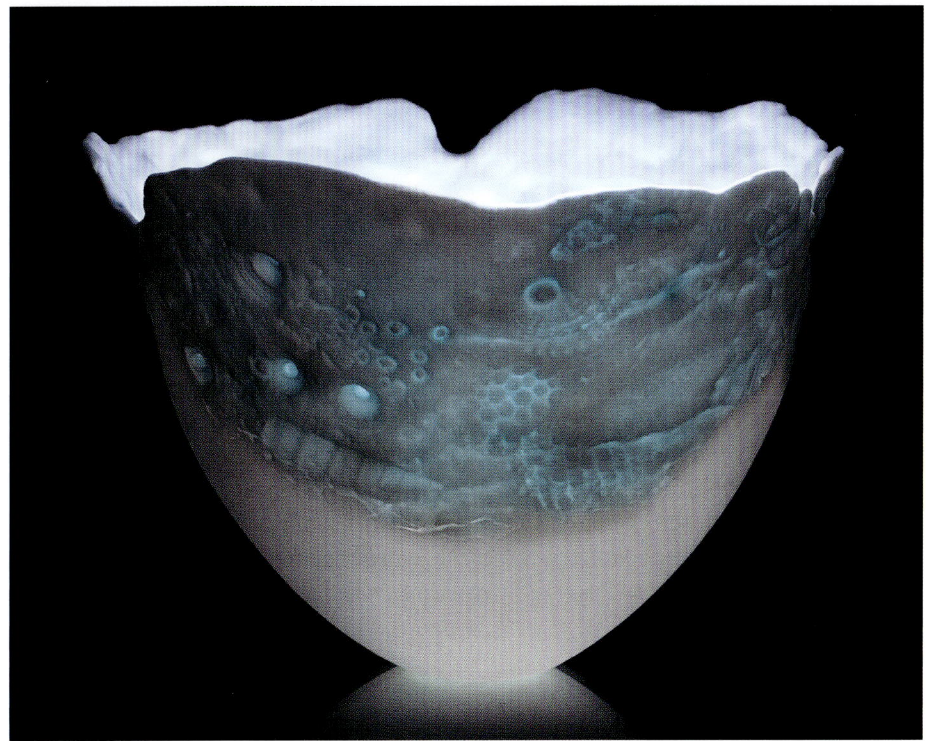

Artists working with soluble colourants

Mark Goudy (USA)

When asked about his work, Mark replied:

> *My current body of work centres on the interplay of light and shadow, translucency, and form-highlighting colour gradation. As a materials-oriented artist, I delve into the constraints and physical properties of my media: slipcast translucent porcelain and soluble metal salts.*
>
> *I have spent decades travelling, wilderness hiking, and exploring the natural world. As a result, I gravitate to the organic forms I encounter in nature: seed pods, weathered stones, dune fields, and ocean surfaces. My goal is to abstract the essence of these elements into my works.*
>
> *During the quiet months of the pandemic, I completely reimagined my approach to designing objects. After a decade of making plaster slipcasting moulds from hand-built clay forms, I revisited my engineering background in 3D graphics to forge a new way of working.*
>
> *I begin my design process by creating forms virtually, using Rhinoceros and Grasshopper algorithmic modelling software. I use repetition, interpolation, and the mathematics of natural elements to develop minimal forms with hidden complexity. These tools have allowed me to combine my psyche's creative and analytic dimensions, celebrating the intersection of art and science. At the core, these forms always reference the geometries of nature.*
>
> *I create my mother moulds with a low-cost 3D printer – the bridge between the virtual model and the physical plaster slipcasting mould. Significant refinement is needed to eliminate artefacts from the 3D print, followed by meticulous hand-sanding of the bisque-fired slipcast objects, and again after the final vitrification firing.*
>
> *To design my* Wave Form *series, I use the interaction of different sine wave frequencies to create sharp ridges separated by delicately curved surfaces, as the waves reinforce or oppose each other. These forms echo natural phenomena such as water waves, sand dune ridges, or musical tones. When illuminated from an angle, the ridges and valleys project subtle shadows across these sculptural forms.*
>
> *Inspired by the Japanese art of paper folding, my* Origami *series is a family of objects with repetitive creases that define the form. Within* Grasshopper*, I shape individual creases with elliptic and parabolic curves, usually bent further with sine/cosine waves. These corrugated surfaces are joined together in software to create various multi-sided forms. The resulting slipcast translucent porcelain objects echo the texture and feel of paper with a bit of 'tooth', and the creases are analogous to paper folds.*

RIGHT: *Waveform Object* #1254 (edition 2/10), Mark Goudy, 2021. Unglazed porcelain, soluble metals (cobalt). 24 x 21 cm (diam x h). *Photo: Mark Goudy*

BOTTOM RIGHT: *Waveform Object* #1291 (edition 1/10), Mark Goudy, 2021. Unglazed porcelain, soluble metals (cobalt). 24 x 21 cm (diam x h). *Photo: Mark Goudy.*

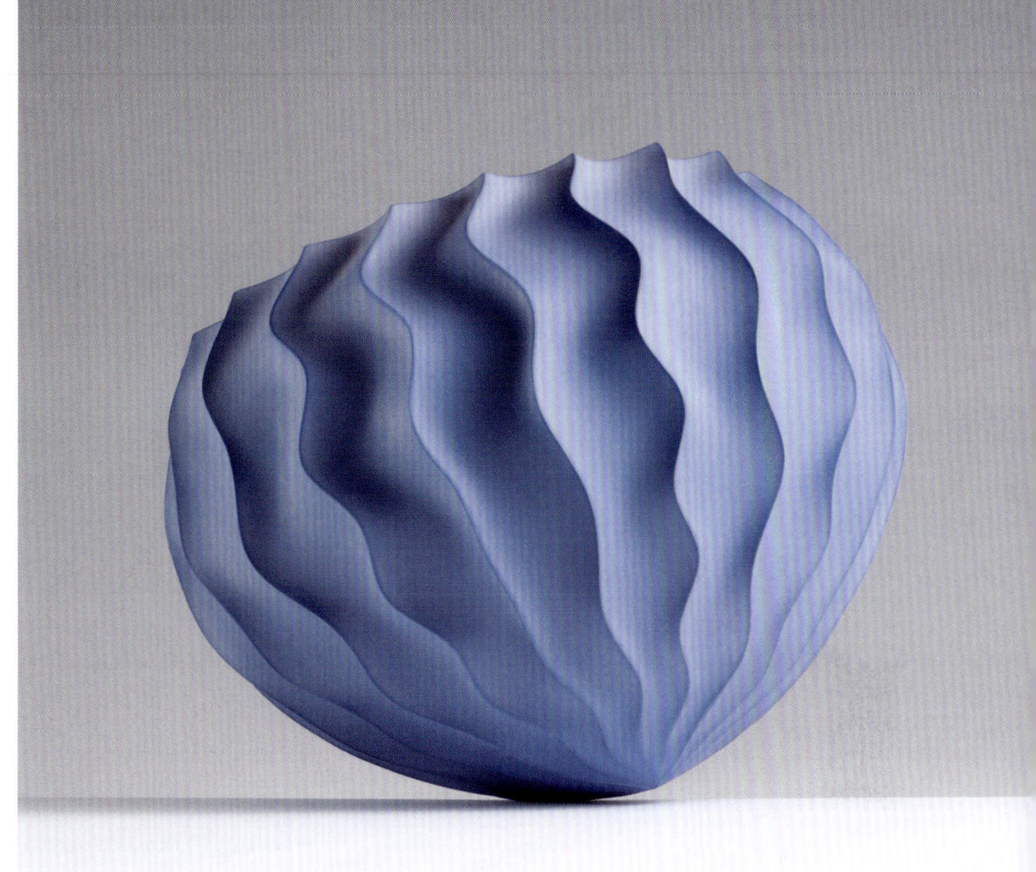

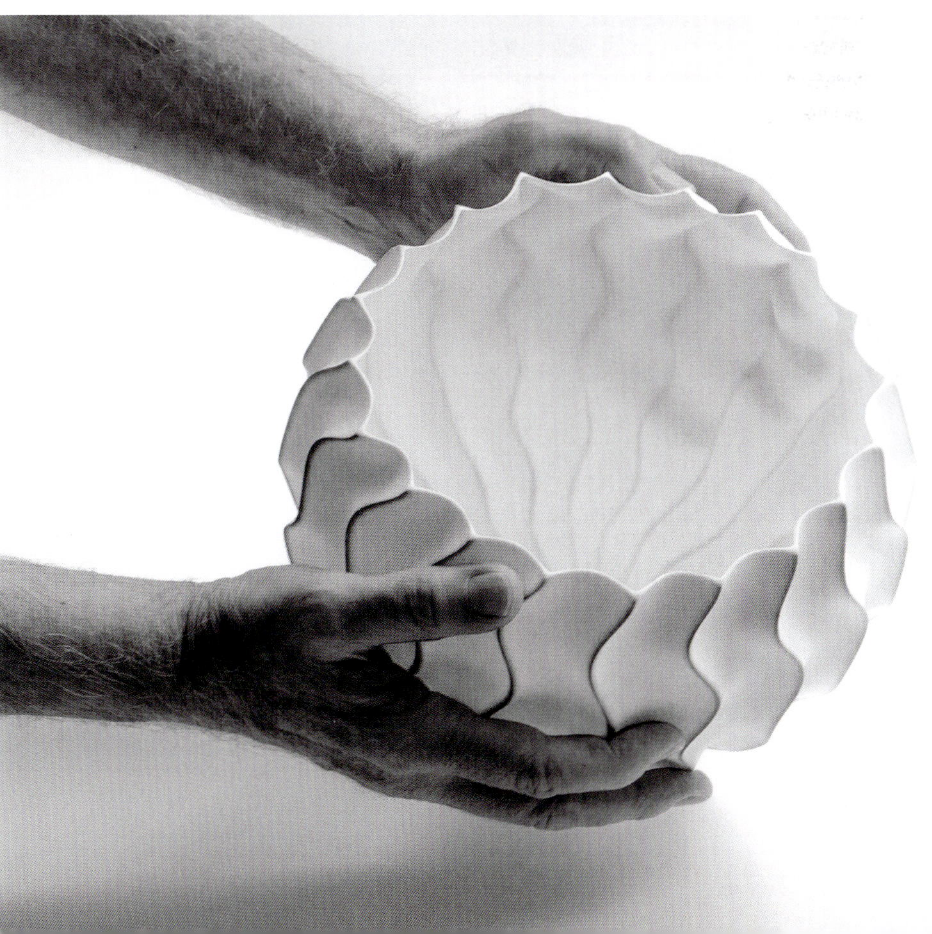

Instead of glazing, I have always used soluble metals to colour my work. I love how these water-soluble colourants inhabit the bisque-fired clay, moving and interacting with each other, concentrating on the ridges during the drying process. These metallic watercolours naturally produce gradations of tone that follow the form. I use refillable felt-tip pens to apply the metals directly on the ridges to enhance the edge effect. My current colour palette involves just three metals: gold, cobalt, and chromium. I combine these metals to yield various shades of red, blue, and green.

With this body of work, I return my focus to form over pattern and am excited to see where it will lead next.

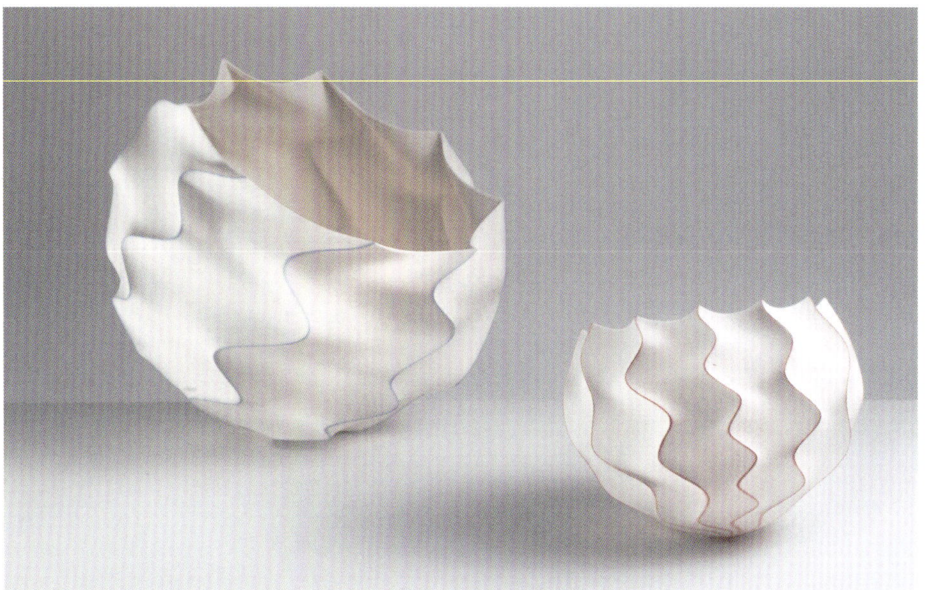

ABOVE: (left) *Waveform Object* #1344 (edition 2/10), Mark Goudy, 2022. Unglazed porcelain, soluble metals (cobalt). 23 x 18 cm (diam x h). (right) *Waveform Object* #1363, Mark Goudy, 2022. Unglazed porcelain, soluble metals (gold). 15 x 13 cm (diam x h). *Photo: Mark Goudy.*

OPPOSITE: *Origami Objects* #1403, #1400, #1407, Mark Goudy, 2024. Unglazed porcelain, soluble metals (gold, cobalt, chromium). (each piece) 14 x 12 cm (diam x h). *Photo: Mark Goudy.*

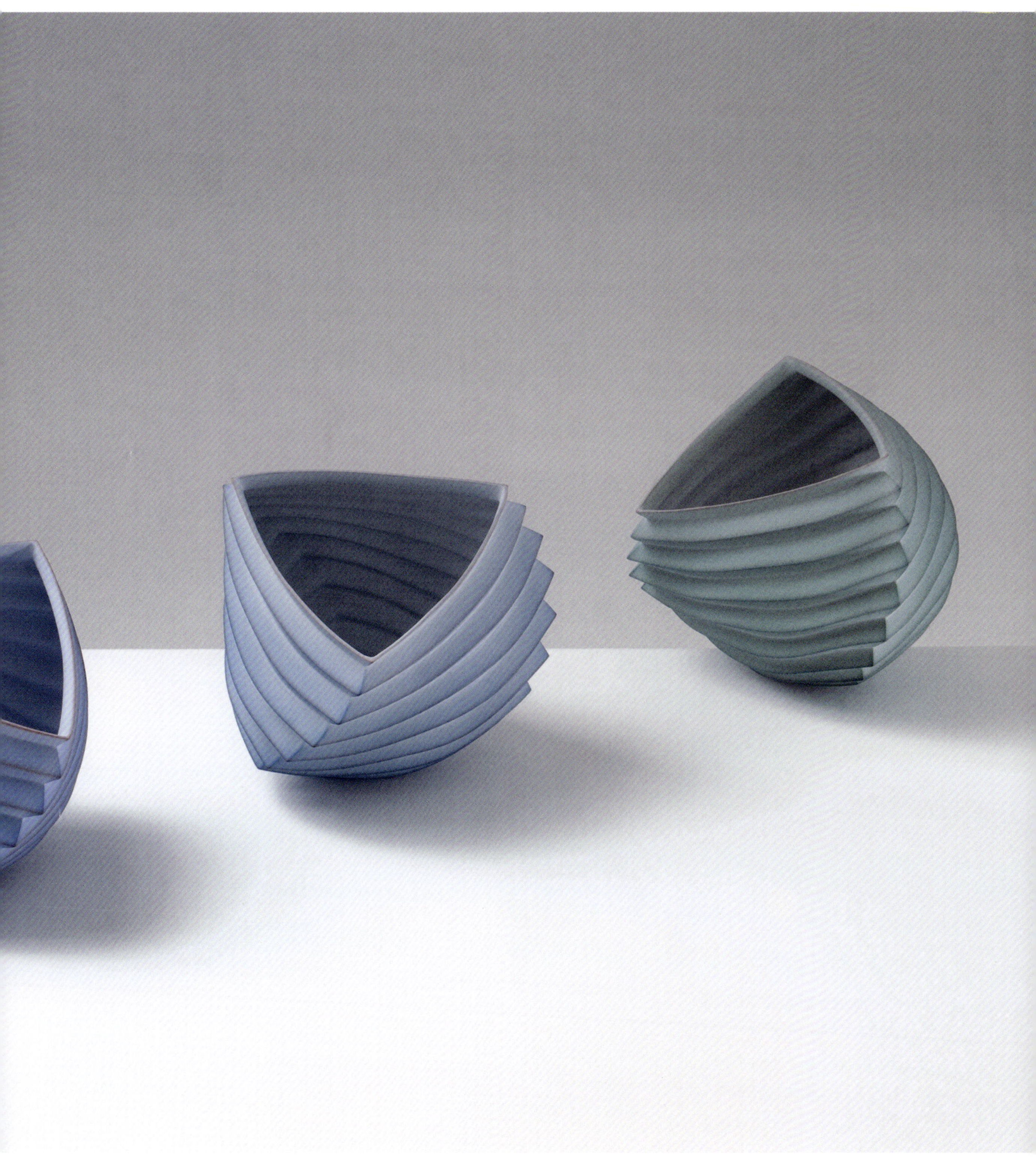

John Shirley (South Africa)

John Shirley was born in 1948 and has been making pots since 1970. In 2000, while studying for his B-tech in Ceramic Design, he started exploring bone china. It was impossible to buy a prepared bone china slip in South Africa, so he began experimenting with locally available materials and, after several months of experimentation, he developed a body that was extremely white and translucent. This experimentation is ongoing because the availability of local raw materials is often unreliable.

Having always been fascinated by the interplay of light with translucent bodies, he wanted to create an ethereal look when it came to adding colour to his ceramics; soluble salts produced the look he wanted.

In his current work with soluble salts, he uses a combination of wax resist and varying solutions of cobalt chloride, potassium dichromate, and ferric chloride, sometimes with small additions of ceramic pigment. The work is bisque-fired to 1000°C. The finished work is unglazed and often subjected to multiple firings of 1000°C, ending with a final firing at 1250°C with a 20-minute soak.

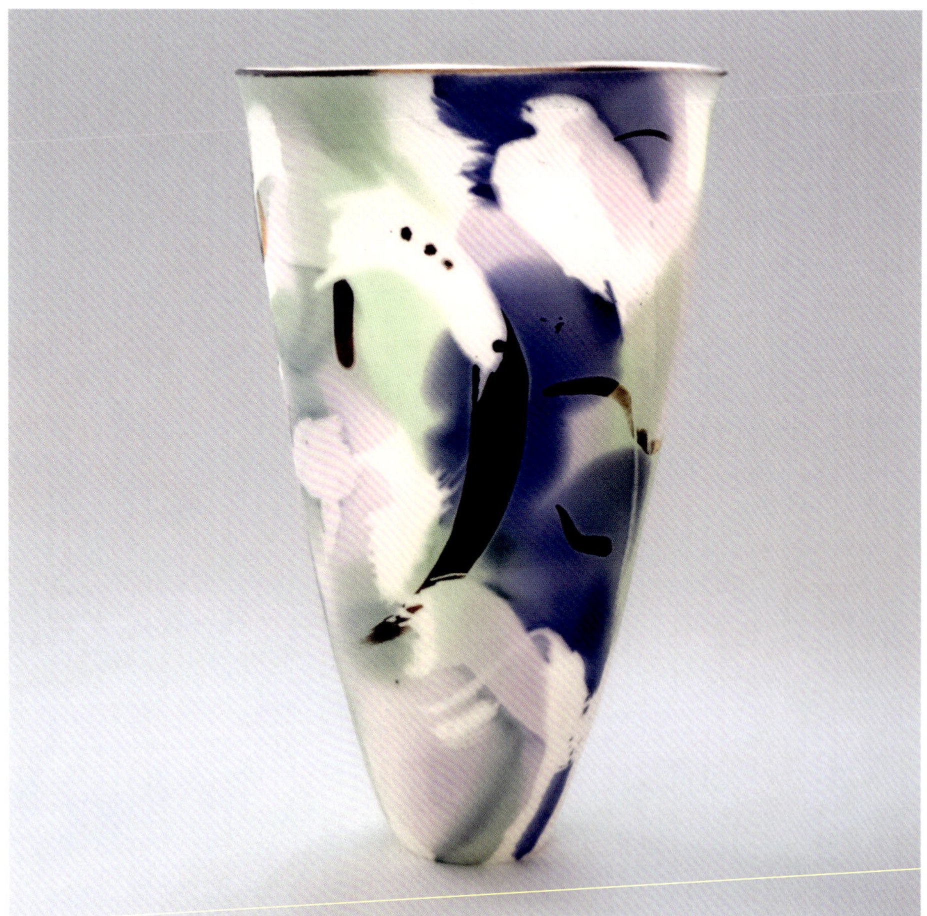

LEFT: *Narrow Vessel*, John Shirley, 2023. Bone china with wax resist and soluble salts. 10 x 15 cm. Fired at 1250°C. *Photo: Lauren Opia.*

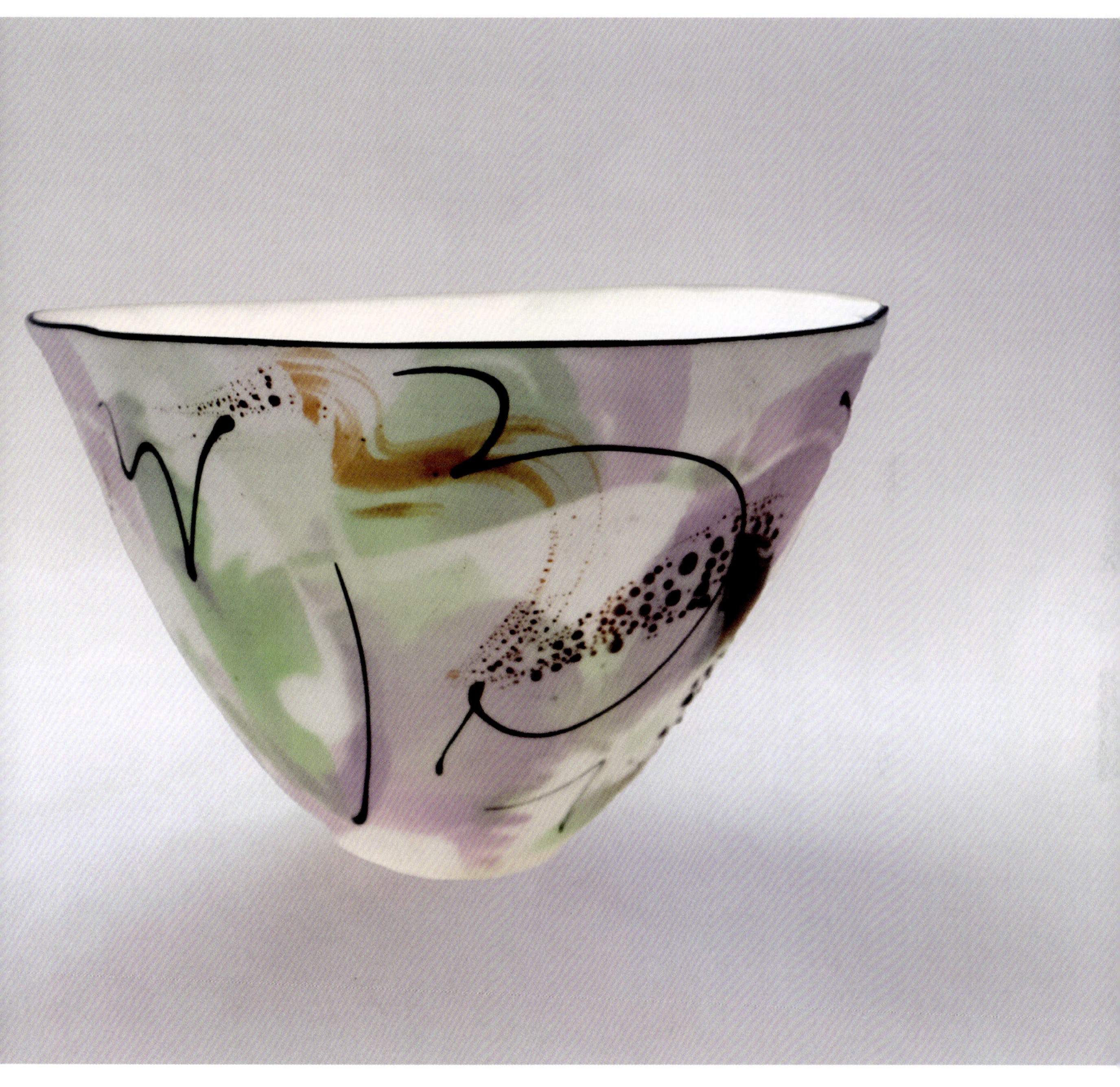

ABOVE: *Bowl*, John Shirley, 2021. Bone china with wax resist and soluble salts. 15 x 10 cm. Fired at 1250°C. *Photo: Lauren Opia.*

Testing water-soluble colourants

My main objective in using soluble salts was to find effective colour concentrations, which would function aesthetically without affecting the translucency.

I first selected colours that I had observed in nature and wanted to portray, and limited my palette to nine colours and two influencers (chemicals that affect how the colours develop). The first test solutions were selected randomly to give me a starting point. They were painted onto bisqueware tiles that had been fired to 1000°C and then to 1250°C in an oxidising atmosphere, without glaze. Where little or no colour appeared, I used a stronger solution, and where the colour was too strong, I tried a weaker solution, until I was satisfied with the results.

Testing colour concentrations

When testing soluble colourants, the first step is to find the ideal colour concentration. Factors involved in producing colours for my work were:

- the actual concentration of colourant in a solution
- the speed at which the colour is brushed onto the ware

These tests were only carried out on bone china, although I did make comparisons with porcelain.

Table 5

SOLUBLE SALTS	COLOUR (g)	WATER (g)
Cobalt chloride	20	100
Cobalt nitrate	20	100
Copper chloride	20	100
Copper sulphate	20	100
Gold chloride	01	100
Iron chloride	25	100
Manganese chloride	50	100
Chromium nitrate	20	100
Uranyl nitrate	25	100
Vanadium sulphate	20	100
INFLUENCERS (TO BE USED OVER THE ABOVE COLOURANTS)		
Tin chloride	50	100
Phosphoric acid (concentrated)	used neat over other soluble salts	

Method

I use 100 g clear jars with plastic screw-top lids.

1 Weigh the container and add 100 g of water.
2 Weigh the percentage of colourant you require, e.g., 20 per cent cobalt chloride = 120 g, i.e., 20 per cent concentration.
3 Label the container, e.g., 'Cobalt chloride 20/100'.

Make test tiles

If I am using slip, I pour discs on to a plaster slab; if I am using paperclay, I spread it onto a textured slab in the form of squares or rectangles.

Alternatively, you can slipcast a 30 cm square and divide it into 10 cm square tiles.

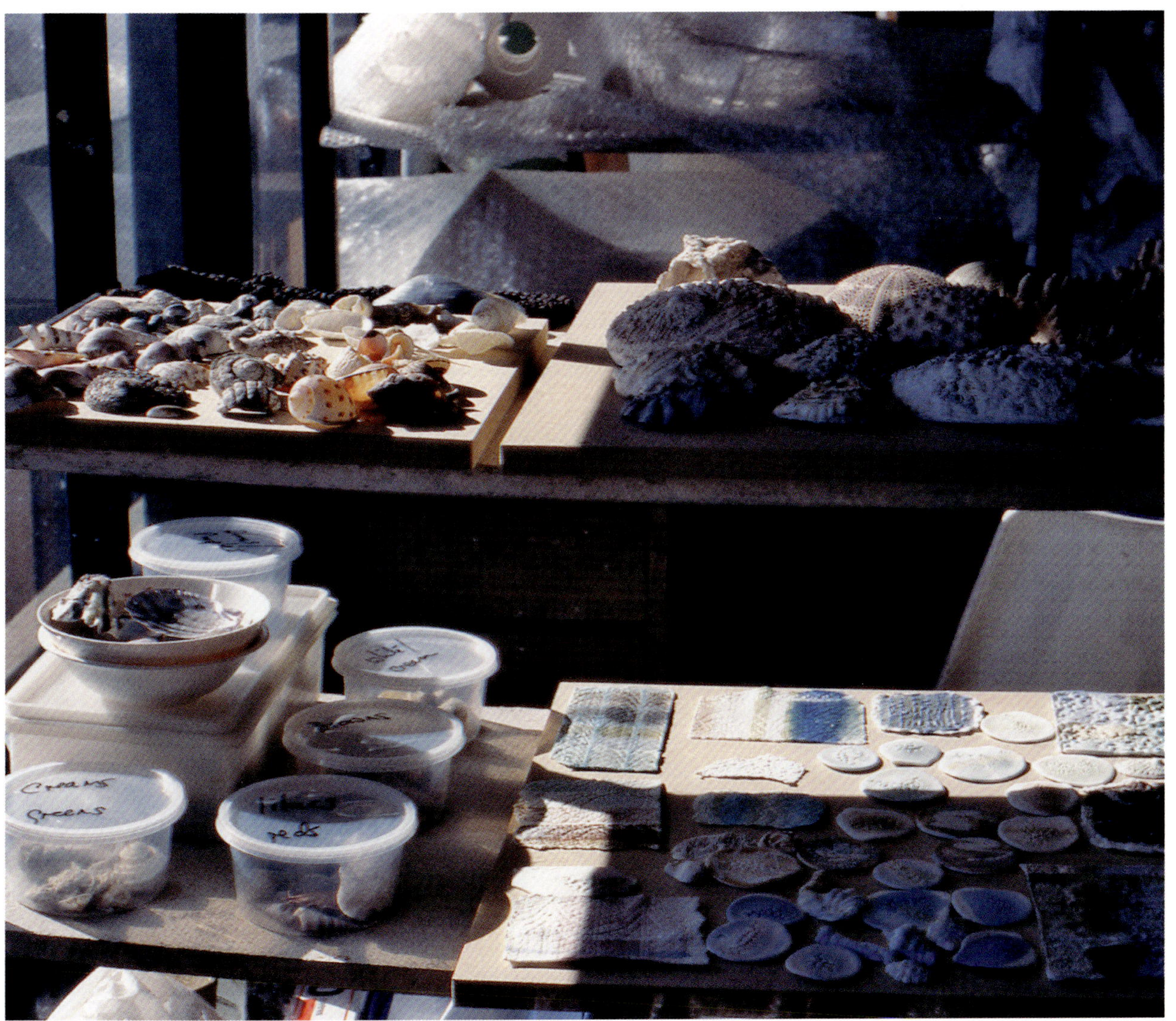

BELOW: Colour tests, Angela Mellor's studio.

Each tile is painted with four stripes of one specific colourant, with equal spaces of unpainted tile in between. The first stripe has one brushstroke; the second, two brushstrokes; the third, three brushstrokes; and the fourth has four brushstrokes. This gives four gradients of colour. These stripes are painted on bisqueware, and act as a control for comparison with subsequent tests.

I find that the concentration of colours can also depend on the speed of application. Light colours are achieved with light, quick brushstrokes, while darker colours are achieved when the colours are applied slowly. The aim is to find the number of concentrations you need to cover a range of hues from the lightest to the darkest. This can lead to experiments with painting techniques, such as applying more than one layer of colourant, and applying colourants at different speeds.

I also painted a stripe of clear translucent glaze onto the test tiles – the result was that all the colours were intensified, and it provided an interesting contrast between a matt and shiny surface. My own preference is still for the subtle matt watercolour effect.

Comparison of colours on porcelain and bone china

Table 6 shows the results of the tests I carried out on porcelain in a reduction atmosphere and bone china in an oxidising atmosphere.

Table 6

COLOURANTS	PORCELAIN	BONE CHINA
Cobalt chloride	Blue	Lilac
Cobalt nitrate	Blue	Lilac
Copper chloride	Pink-grey	Turquoise
Gold chloride	Pink	Pink
Iron chloride	Yellow-brown	Yellow-brown
Manganese chloride	Brown	Brown
Chromium nitrate	Brown-green	Green
Uranyl nitrate	Brown-black	Brown
Vanadium sulphate	Grey	Grey

Influencers

It is possible to add influencers that affect how the colour develops. These are:

- phosphoric acid: creates a bubbling effect when painted over other soluble colours, which I exploited on my sea bowls, creating foam similar to that caused by waves on the seashore
- tin chloride: creates a semi-opaque satin finish over other colours

Applying soluble salts

Soluble colourants may be applied by dipping, spraying, or painting onto greenware or bisqueware. I tend to brush or spray when applying to bisque. I find that too much colour is absorbed when work is dipped into soluble salts, and the colour is rather dense and opaque, instead of producing the translucent effect I require for my work. Having said that, dipping can result in good, even coverage.

Spraying is a good method for applying soluble salts if you require just a delicate hint of colour or a subtle gradation of different colours. Painting is, for me, the most satisfactory way of applying colour. I can increase the depth of colour and experiment with layers of different colours to create new colour combinations, as one would in a watercolour painting.

BELOW: *Coral Bowl*, Angela Mellor, 1997. Bone china with paperclay inlay, painted with soluble colourants. 9.5 cm (w). *Photo: Uffe Schulze.*

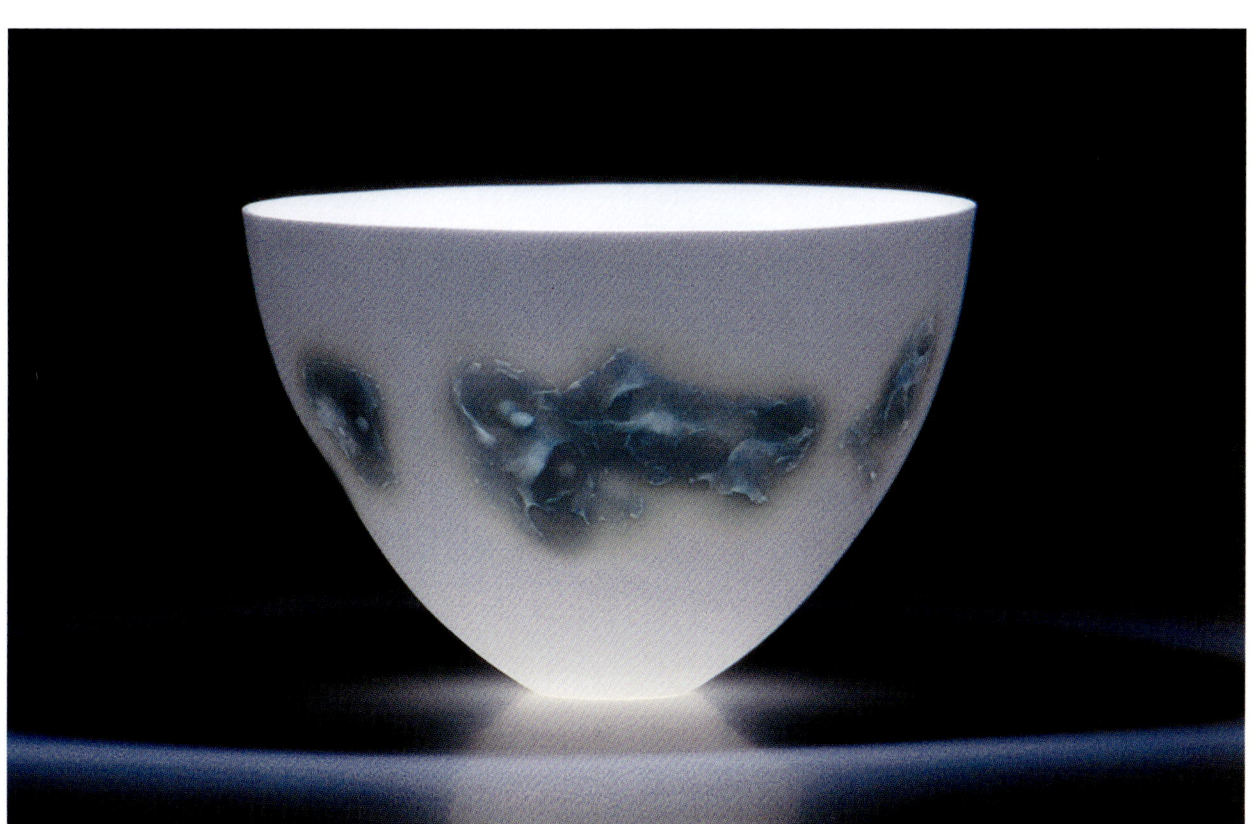

Safety precautions

With any method of application, caution should be taken when applying and handling soluble salts:

- always wear a mask and rubber gloves
- use an extractor, preferably a spray booth
- handle work with rubber gloves until the pieces have been fired (even when placing them in the kiln), as the salts can be absorbed into the skin

Using resist with soluble salts

Tools

- A range of brushes to produce different effects (e.g. a separate brush for each colour)
- Sponges for painting large surfaces
- Paper towels for wiping brushes
- Long rubber gloves that cover your arms
- Fitted goggles
- N95 mask/respirator
- Resist: water-based acrylic varnish (satin finish)
- Food colouring mixed with varnish, to show brush marks on clay surface
- Wet and dry sandpaper, grade 400–600

Method

Remove traces of dust from work with a damp sponge before commencing.

1 Apply the resist to the areas you want to remain white. Leave to dry for ten minutes.
2 Apply a layer of colour and allow to dry. Drying time will depend on the thickness of the body; bone china dries quicker than porcelain.
3 Repeat steps 1 and 2, making sure that the resist has time to dry between coats, otherwise the colours will bleed into each other.
4 Wipe the piece with a soft cloth or a paper towel to remove any colourant that may have settled on the resist. This can be left on to create a textural effect if desired, or it can be removed after firing with wet and dry sandpaper.
5 Fire the piece.
6 If not using glaze, polish with wet and dry sandpaper.
7 If glazing, you can use a neutral feldspathic glaze.
8 I prefer not to glaze, and fire some of the pieces upside down on a setter. These supports can be used only once, as they shrink with the pieces. A layer of alumina is applied to the supports to prevent fusion.

Arne Åse says of the effects of using relief together with soluble salts:

> *Working with relief patterns and translucency makes it possible to produce rich and varied visual effects with quite simple means. When the lights are low but strong enough to produce shadows, only the relief pattern is noticed. When the piece is highlighted, both transparency and the relief will show. [If a dish or bowl is] decorated both on the inside and the outside when the light accentuates the transparency, the pattern on the backside of the bowl shows up diffusely, while the pattern on the front appears quite clearly. These two effects mix to create exciting visual effects.*[7]

To get a really smooth surface, I use wet and dry sandpaper grade 400–600, after bisque and high firing. If a very smooth, shiny surface is required, the surface may be polished in the same way as semi-precious stones. Åse uses tin oxide mixed with water and a polishing machine (a small electric drill with a felt disc as a polishing pad), using water as a lubricant.

Investigating the surface treatments described above has enabled me to explore translucency and colour.

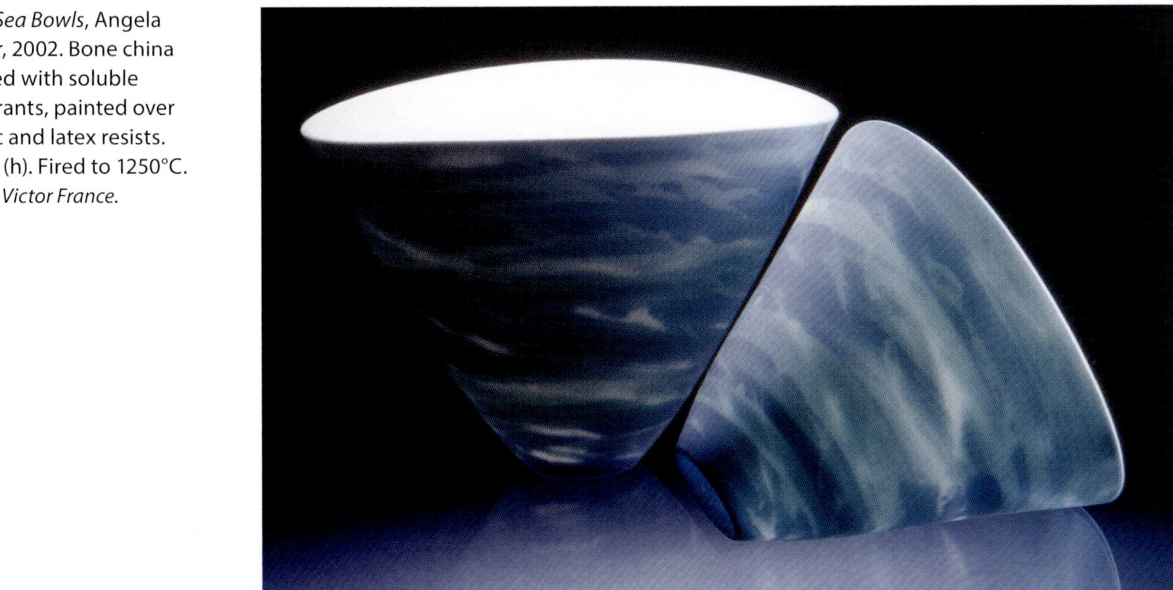

RIGHT: *Sea Bowls*, Angela Mellor, 2002. Bone china painted with soluble colourants, painted over acrylic and latex resists. 10 cm (h). Fired to 1250°C. *Photo: Victor France.*

[7] *Ibid.* p.147.

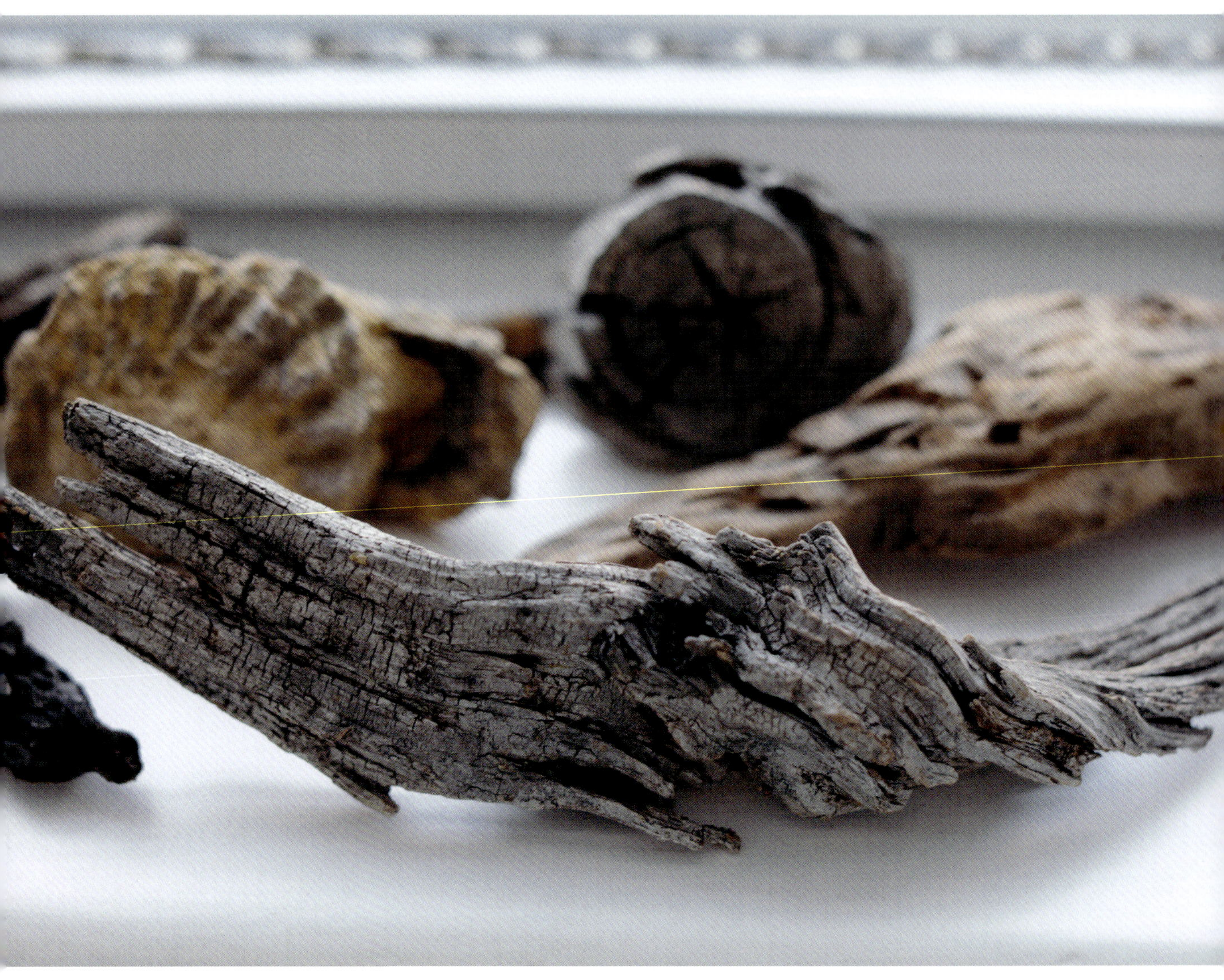

ABOVE: Driftwood, a textural source of inspiration. *Photo: Layton Thompson for* Ceramic Review.

4 Paperclay

During my MA research, I was inspired by the textures of natural objects such as driftwood, stones, coral, and shells, and I wanted to recreate these in my work. In my early work, I impressed textured objects into slabs of porcelain and coloured them with oxides. After moving to slipcasting in bone china, I needed to find another way of applying texture that would not distort the form. Thin sheets of textured paperclay seemed a likely answer.

I was first introduced to paperclay while on a porcelain workshop with Prue Venables at the 8th National Ceramics Conference, which was held in Canberra in 1996. I experimented with folding thin sheets of paperclay into coral-like forms. Later that year, while working in Jakarta, I developed the idea further, creating texture by combing into the thin paperclay sheets while they were still wet. After a few minutes, while the clay was still pliable, I folded the sheets and placed them into bowl-shaped plaster moulds. More sophisticated experiments followed when I was in studying in Tasmania. I started to include tissue paper into the bone china slip so that I could incorporate texture into my work. After many trials, I found a recipe that worked and decided to make some textured plaster slabs from natural objects I found on the beach. I made textured paperclay tiles, 2–3 mm thick, which I used for colour tests.

My MA research enabled me to work on a satisfactory recipe for bone china paperclay, which is very translucent, thereby aiding my exploration of light and organic forms. Thinness of paperclay and firing temperatures have been tested for translucency and were documented in a book called *Working with Paperclay and Other Additives*, by the late Anne Lightwood.

When I first started to write this book, I was the only person using paperclay for translucency. Since I started giving workshops and writing articles for magazines, more people have begun to use this method. It is an easy technique to learn once everything is prepared.

Paperclay is a mixture of clay, paper pulp, and water. The proportion of clay in the mix is greater than the proportion of paper so that objects made in this material can be fired intact in a kiln without disintegrating. The pulp can be made from recycled office paper and newspaper. I use toilet paper because its fine texture and whiteness means it will not stain the clay. Paperclay is as strong as greenware, and can be worked wet over dry and altered in most stages of the making process.

Making paperclay

To make paper pulp

Materials

- 1 toilet roll (paper only)
- 1 bucket of water, two-thirds full with tepid water
- A few drops of bleach
- Paint mixer drill attachment
- Plastic bucket with lid
- 20-litre plastic bucket for mixing
- Household sieve
- Kitchen whisk for mixing

Method

1 Submerge a roll of toilet paper (with the cardboard roll removed) into the bucket of tepid water.
2 Mix well by hand and with a kitchen whisk.
3 Add a few drops of bleach to slow down any bacterial growth.
4 Do not leave pulp soaking too long, or it will develop moulds and smell.
5 If making a small quantity, you can use the bowl of an electric or food mixer to mix thoroughly. For larger quantities, mix with an electric drill bit on the end of a drill, or use a glaze blunger.
6 Use a sieve to drain the water, leaving a medium wet pulp. The pulp should not be too runny, as you will be adding it to liquid slip. More water may be squeezed out by gently pressing the pulp in the sieve if necessary. Equally, the pulp should not be too dry as it will not mix into the slip evenly. The pulp can be stored in airtight plastic buckets for later use. It can be kept in a cool place for a couple of months, or frozen indefinitely.

Adding paper pulp to slip to make paperclay

I use a ratio of one part paper pulp to two parts casting slip, by volume. This ratio has given me the best results for strength while still allowing the paper pulp to burn away and give me the translucent results I like.

Method

1 Fill a bucket with two parts of casting slip.
2 Add one part wet paper pulp slowly by hand, stirring it into the casting slip until it resembles thick oatmeal.
3 Mix well with an electric mixer for 20–30 minutes.
4 Keep in an airtight container. I store it in a 5-litre slip container overnight. The next day, I usually decant a little water from the top before stirring in case the slip is too wet. This can always be added back if the mixture is too thick.

Creating textured plaster slab moulds

I usually create texture by impressing natural objects into a slab of clay, putting a wooden cottle around it, and sealing the joints with clay before pouring plaster in. When the plaster slab is dry, you can pour paperclay onto it. It is good to have several slabs of different textures. Because I work on a small scale, and only use small pieces of texture for decorative effect, my slabs are usually about 30 cm square or, if larger, 30 x 45 cm.

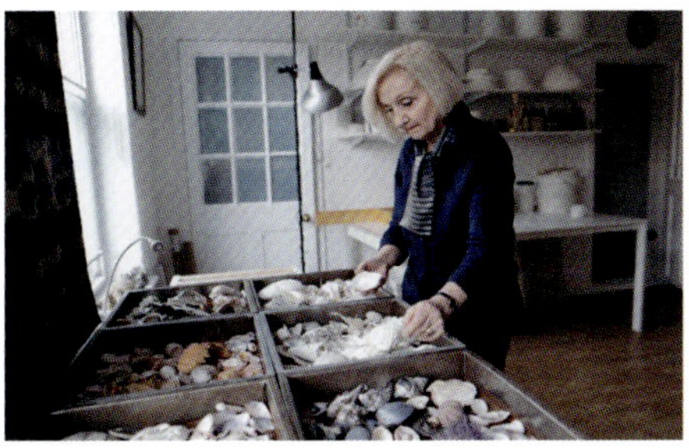

ABOVE: I begin by selecting objects from the natural environment, most of which I collected on my travels around the coast of Australia.

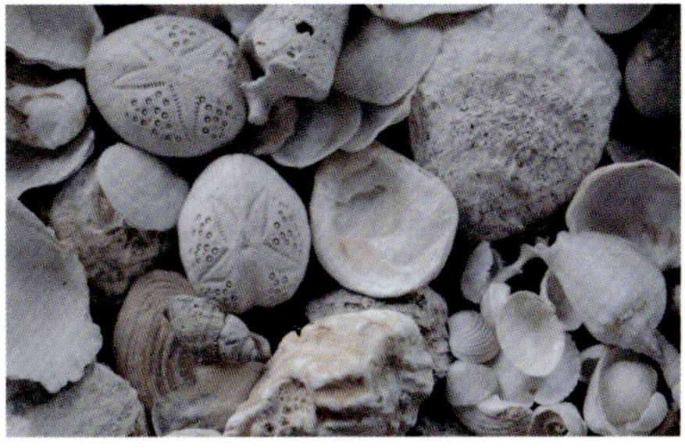

ABOVE: I am inspired by shells, coral, seaweed, and other objects found on the coast.

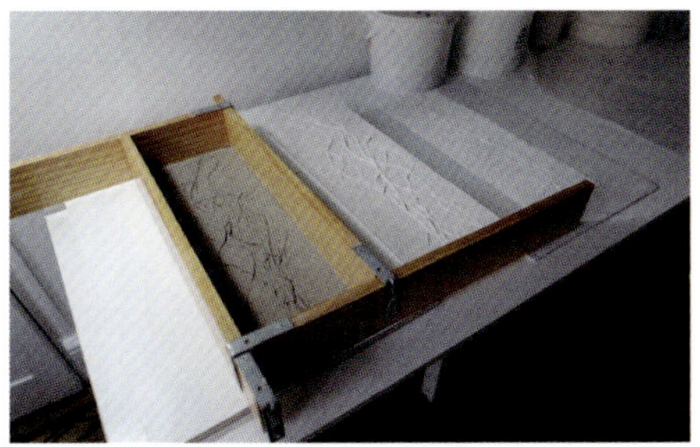

ABOVE: Once selected, I press or roll the objects into clay to create the desired textures. I then cottle up the sides of the clay model with wooden boards, sealing the edges with small coils of clay to prevent the plaster leaking out. I then make a plaster cast, using a fine casting plaster to pick up minute details of the texture.

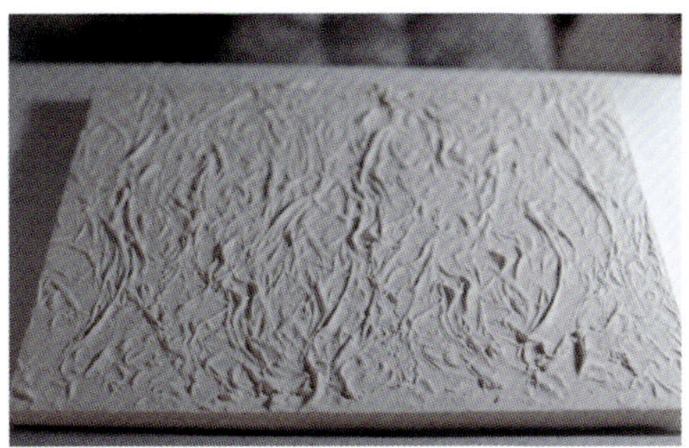

ABOVE: The plaster cast must be allowed to dry thoroughly before applying the paperslip.

All photos pp.73–7: Layton Thompson for Ceramic Review.

Making textured paperclay slabs or sheets

Method

1 Pour fresh paperslip onto a clean, textured plaster slab.
2 Spread the paperslip over the slab with a plastic spatula, scraping off any excess.
3 Wait until the water has evaporated for the desired thickness.
4 Smooth the surface with a rubber kidney, pressing lightly so that the underside will pick up the fine texture.
5 Lift the corner after a few minutes. If it releases easily, you can gently lift off the whole sheet. At this stage, it should be pliable and you can fold or bend it as you wish.

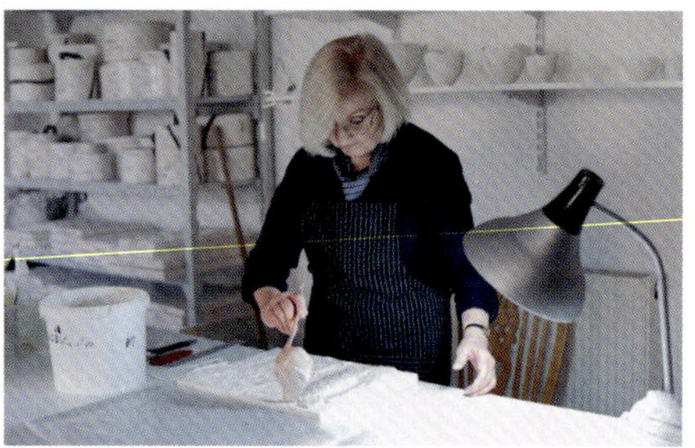

ABOVE: I apply paperslip onto the plaster mould using a ladle. I buy bone china casting slip and make my own paper pulp using toilet roll. I mix one third paper pulp to two-thirds casting slip, mixing the two to make paperslip.

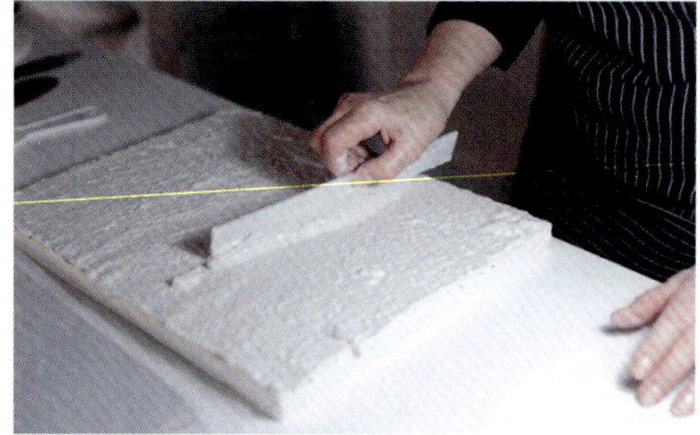

ABOVE: I spread the paperslip across the mould with a metal or plastic rule to create an even coverage.

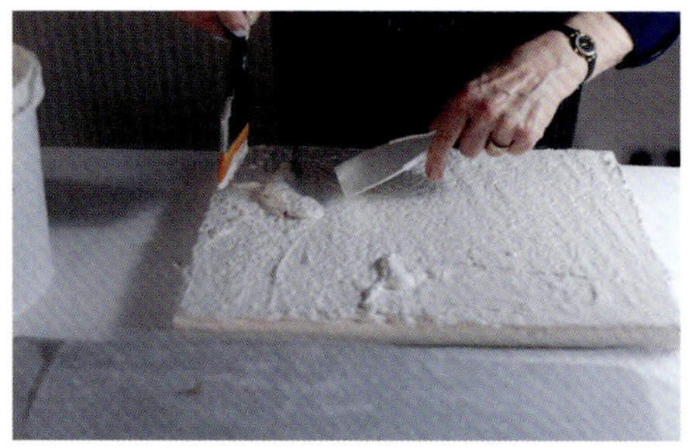

ABOVE: The surplus paperslip is then removed with a wide soft brush and a plastic spatula. The brush works the paperclay into the texture and the spatula is a useful tool to lift off the excess.

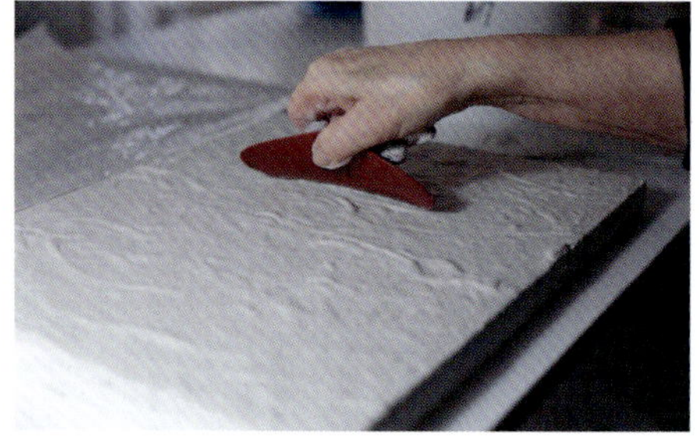

ABOVE: I smooth over the paperslip with a soft rubber kidney to ensure that it has picked up the delicate textures from the mould.

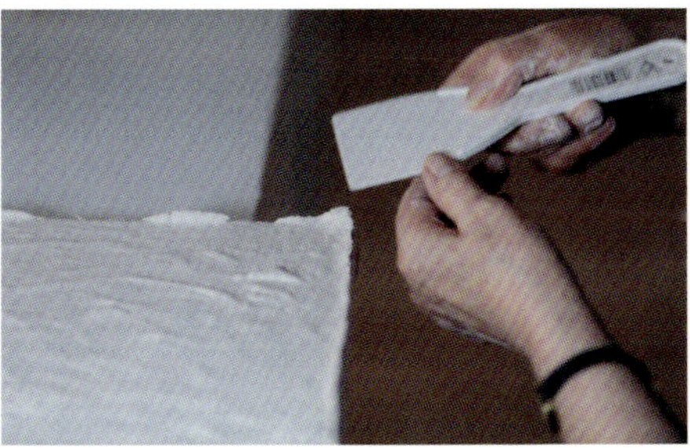

LEFT: Once the paperclay has semi-dried, I release the sheet from the edges of the mould, easing it off gently all the way around. The outside edges dry faster than the centre of the sheet, so it has to be eased off carefully to prevent it from tearing.

Thin sheets of porcelain and bone china paperslip have memory, just like ordinary plastic porcelain. If bent, they will 'remember' this in the firing and revert to that shape. If you want flat slabs without any warping, try to avoid twisting or bending the soft slab. If you do happen to bend it, make sure that you bend it back in the other direction to neutralise the stress and reverse the consequences of the bend.

Slab memory can be controlled to your advantage. If, for instance, you want to make a wavy edge, bend the soft slab as you desire. As it dries, it will tend to move exactly as intended.

Thin, flat sheets can also be left to stiffen on plaster slabs. They can then be cut with scissors or torn. This is the method I use to create areas of delicate texture in my pieces. Small sections of textured paperclay are torn, and the edges are dampened with water before I apply them to the inside of the mould. Bone china slip is then poured into the mould and cast for a few minutes before pouring out. This results in a plain form with a textured inlay.

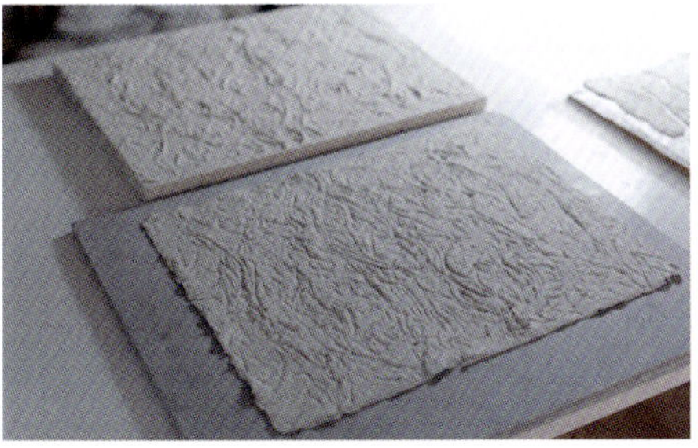

ABOVE: After the edges are released, I lift off the paperclay and place it on a flat plaster bat or builder's board to dry, turning it over occasionally to dry on both sides.

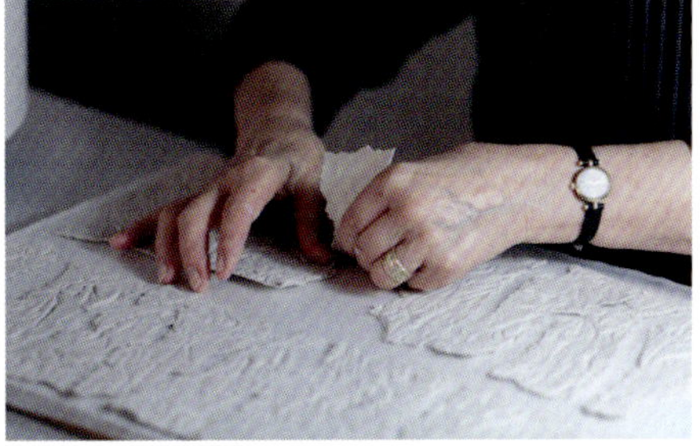

ABOVE: I tear or cut the shapes required to fit the mould. I use moulds I normally use for slipcasting in; in this case, I am using it as a press mould.

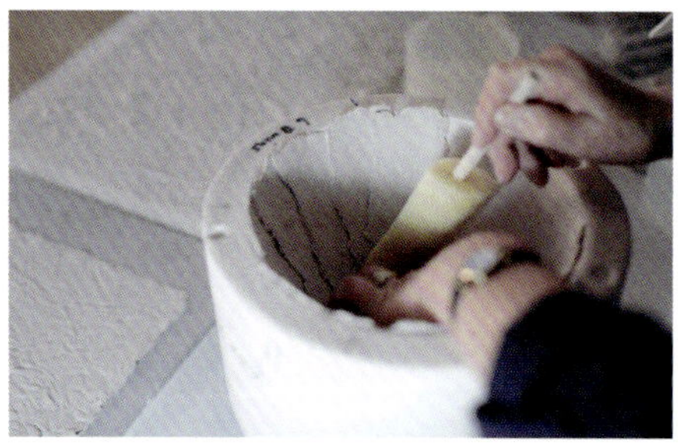

ABOVE: I place each piece into the mould and press it to shape with a sponge. If the mould is deeper, I use a dottle (a sponge on a stick) to reach further into the mould.

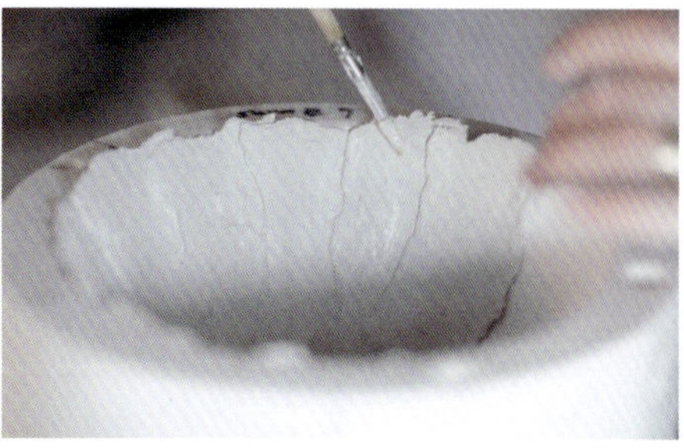

ABOVE: Wet paperslip is used to seal the joins on the inside of the piece to prevent it from coming apart.

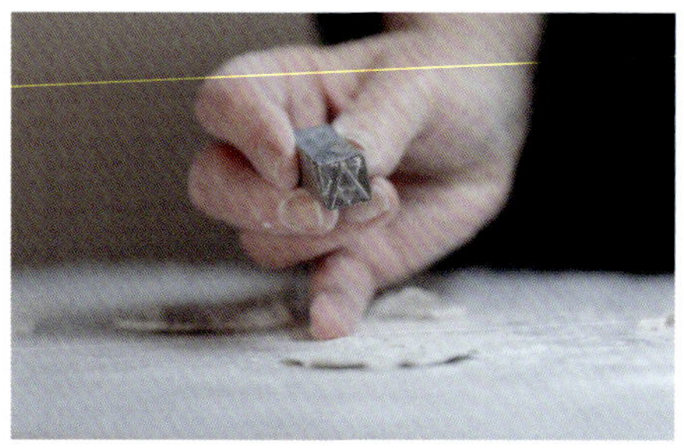

ABOVE: I press my potter's mark into the base of the piece before inserting it into the mould.

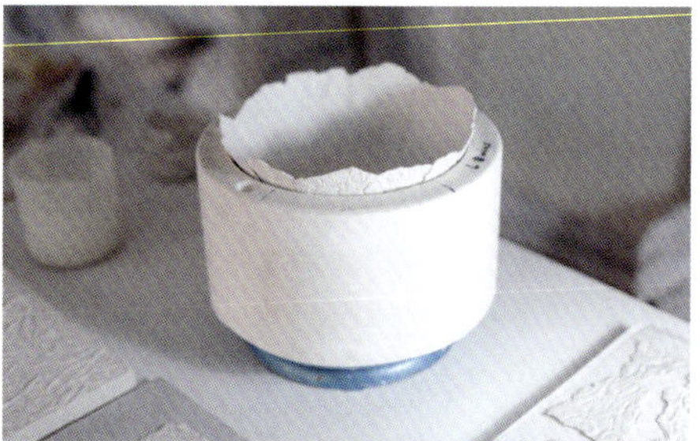

ABOVE: The finished piece is left in the mould to stiffen overnight; this helps to hold its shape until it's dry.

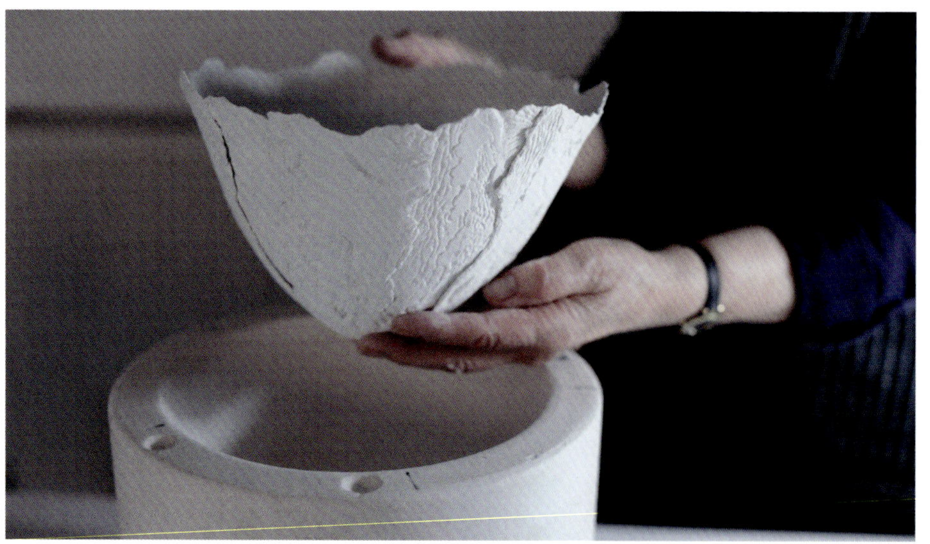

LEFT: I remove the piece from the mould the following day and seal the edges on the outside with paperslip at this stage, which helps to prevent it from coming apart in the firing.

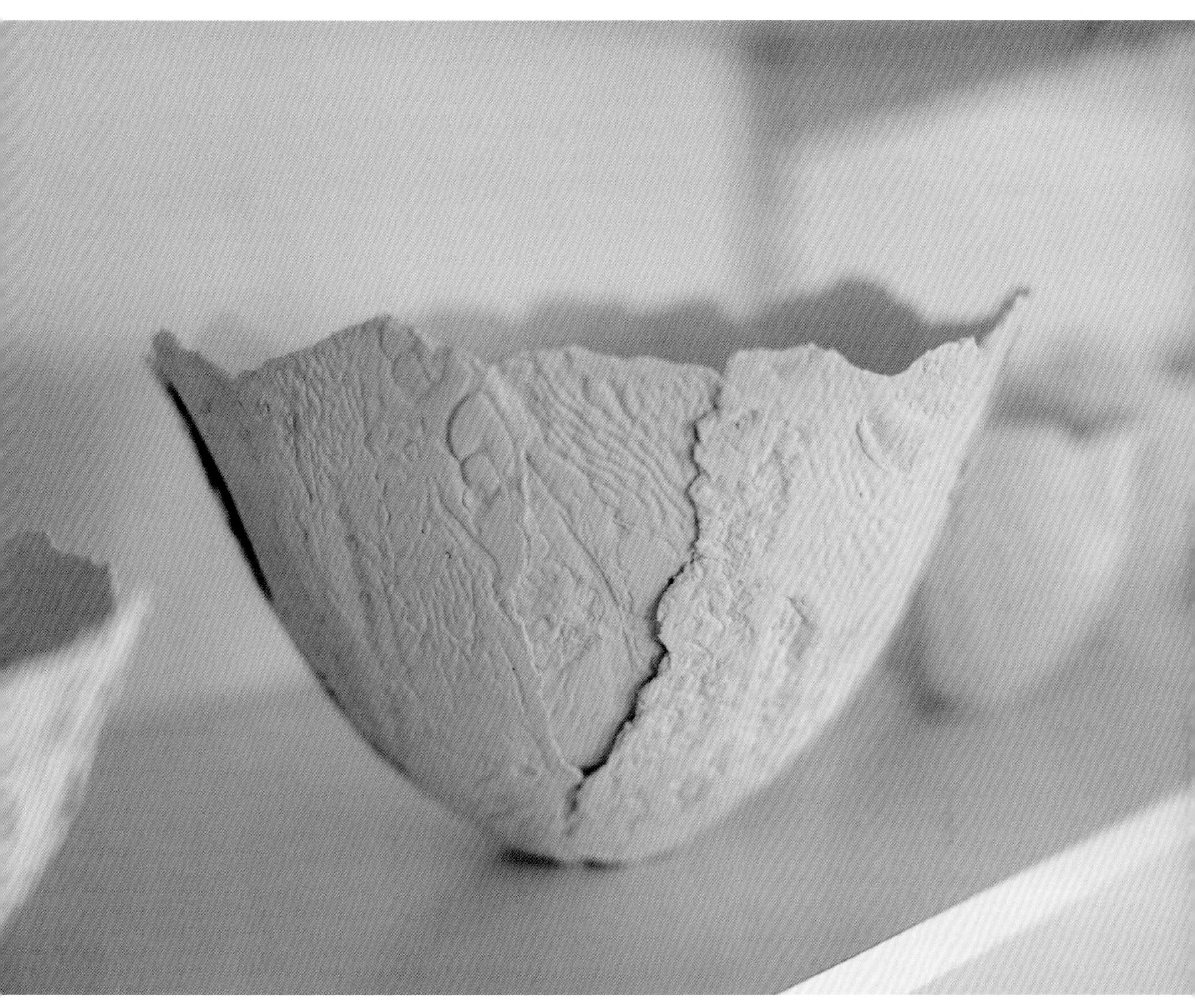

ABOVE: The piece is allowed to dry fully before bisque firing to 990°C or once-firing in an electric kiln to cone 8 with a 20-minute soak for translucency.

Casting porcelain and bone china paperclay

You can add paper or other fibres to casting slip. When the cellulose in paper is added to casting bodies, it affects the chemistry of the slip. In general, cellulose improves the strength of the casting and shortens the waiting time for release and peel off. In addition, work may be repaired, altered, and assembled wet on dry.

Rosette Gault states that certain criteria should be met with basic recipes for casting paperclay recipes. They should:

- pour smoothly
- respond to dispersal easily
- set up quickly in a plaster mould
- release easily from a plaster mould
- be as strong as possible as a greenware
- be able to be fettled or sponged easily
- have the right fired colour[1]

Studio-made paperslip must be of the right consistency (like oatmeal) to cast up well. If it is too thick to start with, the casting may lack detail in some areas. The plaster should also be of the right moisture content to get smooth detail on the surface of the casting. The studio-made paperslip can be poured or smeared into open moulds. Trimming is best done with scissors or a sharp fettling knife, when removed from the mould at the leather-hard stage. If it is too dry, fettling is more difficult. The casting will be quite strong and usually releases sooner than normal casting slip.

Plaster slabs are good for drying out paperclay. They should be clean, smooth, and water-absorbent. If the plaster is too dry, the paperclay dries too quickly and may stick. To avoid this, wipe the dry slab with a damp sponge before use. If you require a thicker slab or sheet, pour a thick layer of paperslip onto very dry slabs. If the plaster slab is too wet or uncured, the paperclay will not dry out.

Paperclay has created many possibilities for my work in fine bone china that would have been impossible with regular bone china casting slip. The main advantage I have found is its flexibility. Being able to move and bend the paperclay without cracking it is an enormous advantage. It is also useful when joining pieces together, whether wet or dry. Complex forms can be made with freedom, assembled at any stage and even added to bisqueware. When joining regular clay porcelain slabs, there is a tendency for the joints to open up in a high firing. This is one of the main reasons I decided to slipcast my work.

[1] Rosette Gault, *Paper Clay*, A & C Black, London, 1998, pp.40-1.

Firing porcelain and bone china paperclay

Porcelain and bone china paperclay require high bisque firing of at least 1000°C. I have found no problem in high firing bone china paperclay to 1250°C. It retains the shape well and gives good translucency. If glazing bone china, I usually do it at the bisque stage as I would porcelain, but the glaze must be very thin. Sometimes warping can happen, as with regular porcelain or bone china. I have found that this is usually associated with the form.

Using paperclay for inserting texture into my work has been my greatest discovery. The ability to cast paper-thin sheets, which accentuates the translucency of the objects, is my prime concern.

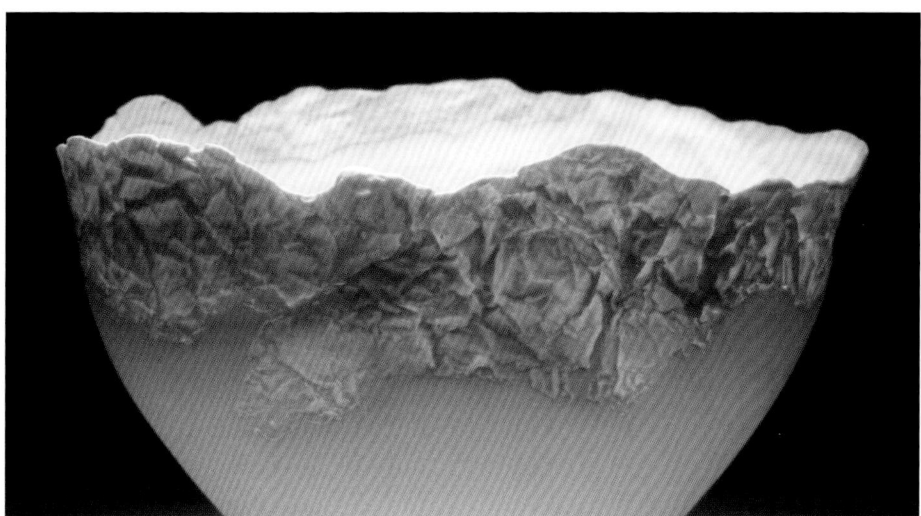

RIGHT: *Glacial Light* bowl (detail), Angela Mellor, 1999. Textured bone china paperclay. 10 x 8 cm (h x w). Oxidised fired to 1250°C. *Photo: Victor France.*

International ceramists using paperclay

Mette Maya Gregersen (Denmark)

Mette Maya Gregersen trained in ceramics in London before taking an MA in Art Psychology at Sheffield University.

Psychological theories form an integral part of her process, helping her to understand the immediate world. She likes to find structural connections between mind and nature. In studying the patterns created by water, her work process becomes an investigation into understanding cycles in nature. When I asked her about her work, Mette Maya said:

> *My work explores the intricate patterns found in nature. Especially inspired by wave movement and the dynamic essence of the ocean's waves, by mimicking the organic pattern and dynamic forms in natural phenomena, I seek to create pieces that reflect a moment of motion.*

Through fluid forms and undulating lines, I aim to embody the rhythm and energy of water in motion and its partial transparency. Each curve and crest is meticulously crafted to reflect the natural ebb and flow, inviting viewers to experience the serene yet powerful force of the sea. I create sculptures not only to showcase the beauty of wave patterns but also to evoke a sense of tranquillity and continuity found in nature's cycles.

Mette Maya's waves are built upon wooden structures layered with porcelain paperclay. The wood burns away during the first firing, leaving an impression, like a memory. She uses a combination of earthenware and stoneware glazes. They are then glazed and fired to 1,280°C, sometimes with multiple firings.

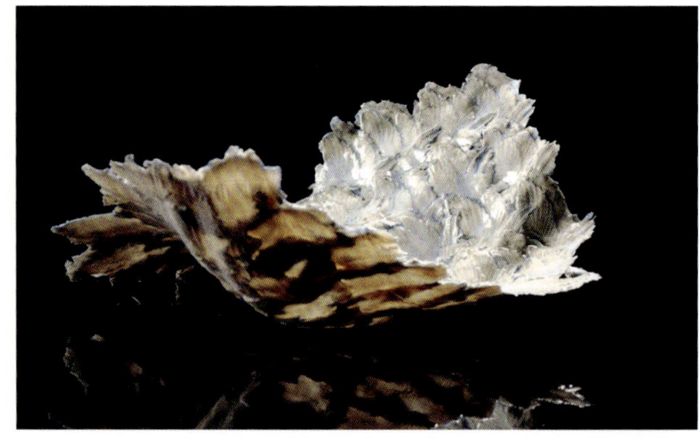

RIGHT: *Wave of Wings*, Mette Maya Gregersen, 2014. Paper porcelain with inlaid pattern and image, glaze. 30 x 50 cm. *Photo: Thomas Juul.*

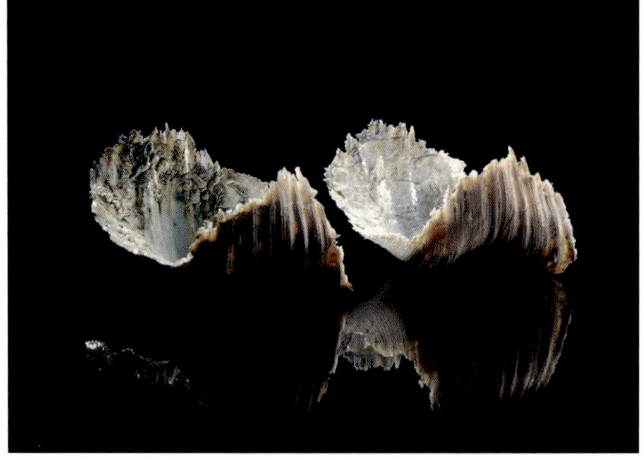

RIGHT: *Day & Night*, Mette Maya Gregersen, 2014. Paper porcelain with inlaid pattern and image, glaze. (each piece) 35 x 48 cm. *Photo: Thomas Juul.*

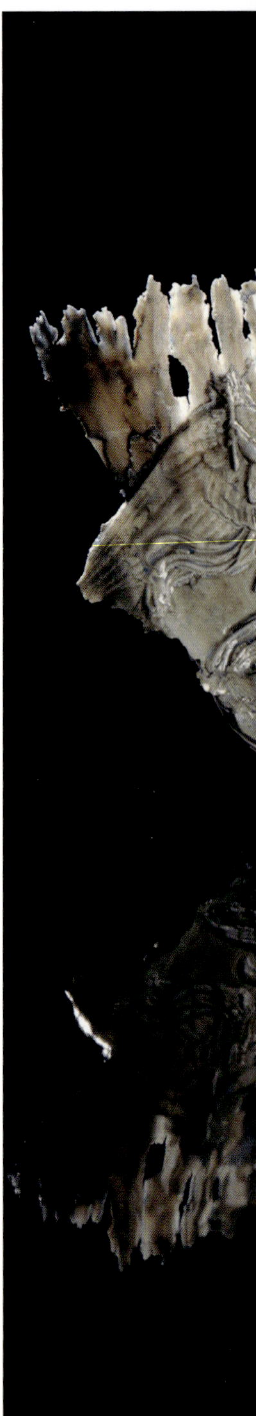

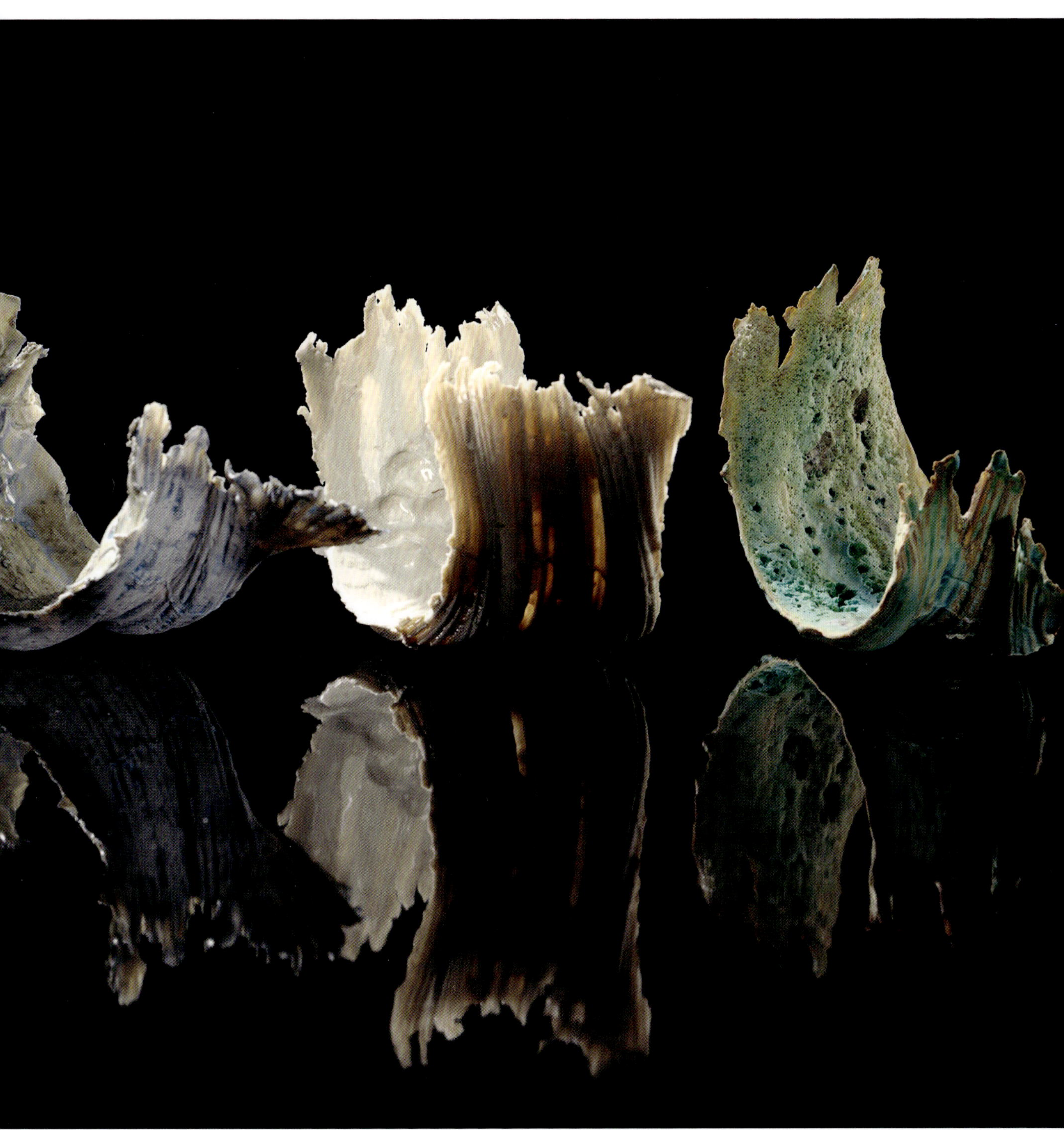

ABOVE: *Three Waves*, 2014. Paper porcelain, glazed. (each piece) 16 x 24 cm. *Photo: Thomas Juul.*

Françoise Joris (Belgium)

Françoise Joris is a porcelain artist whose work is identifiable by its ethereal, fragile appearance, and suggestions of marine and plant living forms. She uses cellulose or flax fibres in porcelain. She aims to achieve aerial artworks without the support of a shell, so she uses a balloon and porcelain slip to build her artwork.

Françoise prepares paper-thin porcelain sheets, cutting and tearing them to create forms. These are assembled according to her creative ideas. They are fired to 1,250°C in an electric kiln. The whiteness and translucency of the porcelain creates a surreal appearance.

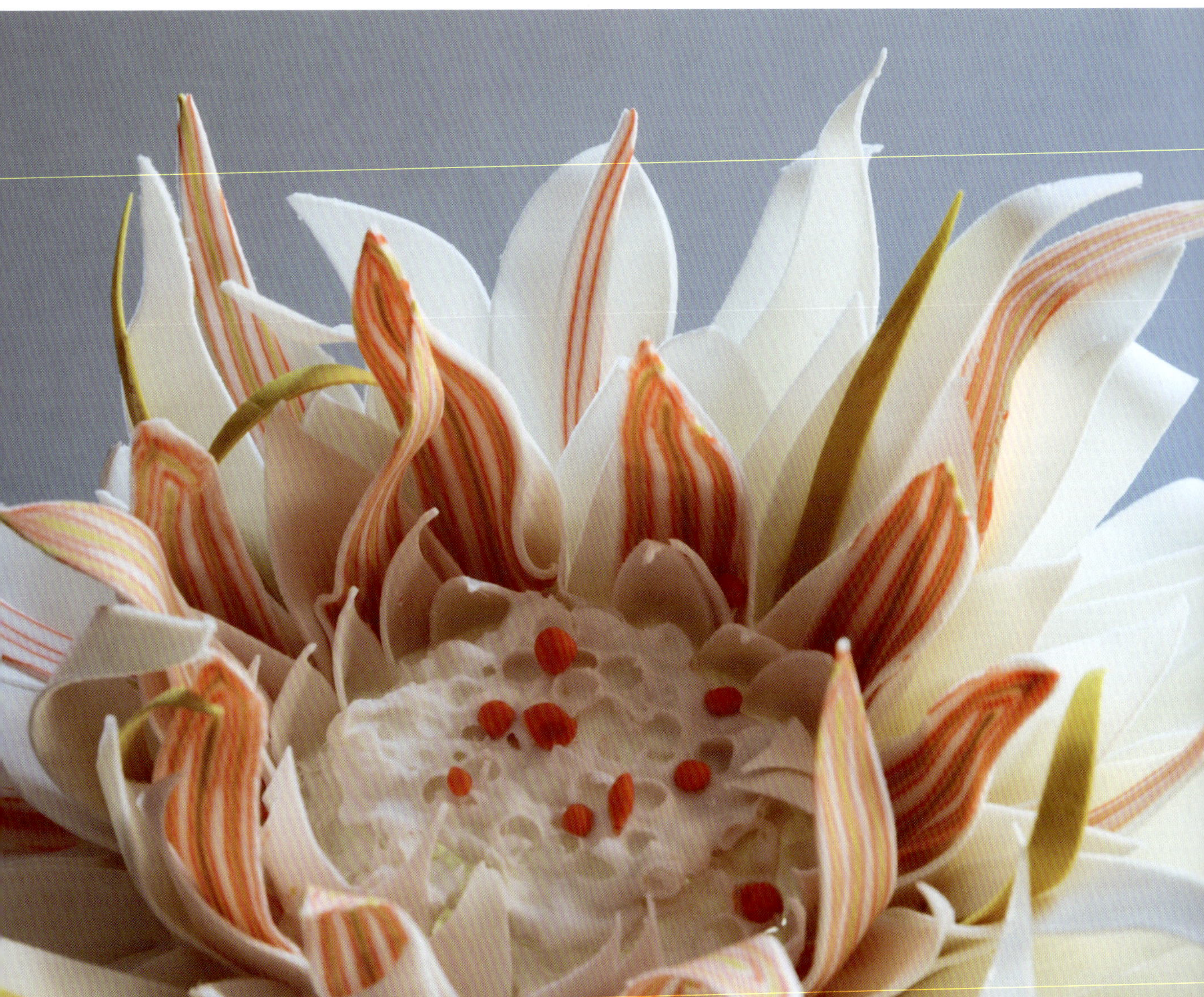

OPPOSITE BOTTOM: *Revival*, Françoise Joris, 2021. Paperclay. 15 x 17 cm. Fired at 1250°C. *Photo: Jean-Claude Pierloot.*

RIGHT: Untitled, Françoise Joris, 2018. Paperclay, fired at 1250°C. 19 x 29 cm. *Photo: Jean-Claude Pierloot.*

BOTTOM RIGHT: Untitled, Françoise Joris, 2020. Paperclay, fired at 1250°C. 21 x 15 cm. *Photo: Jean-Claude Pierloot.*

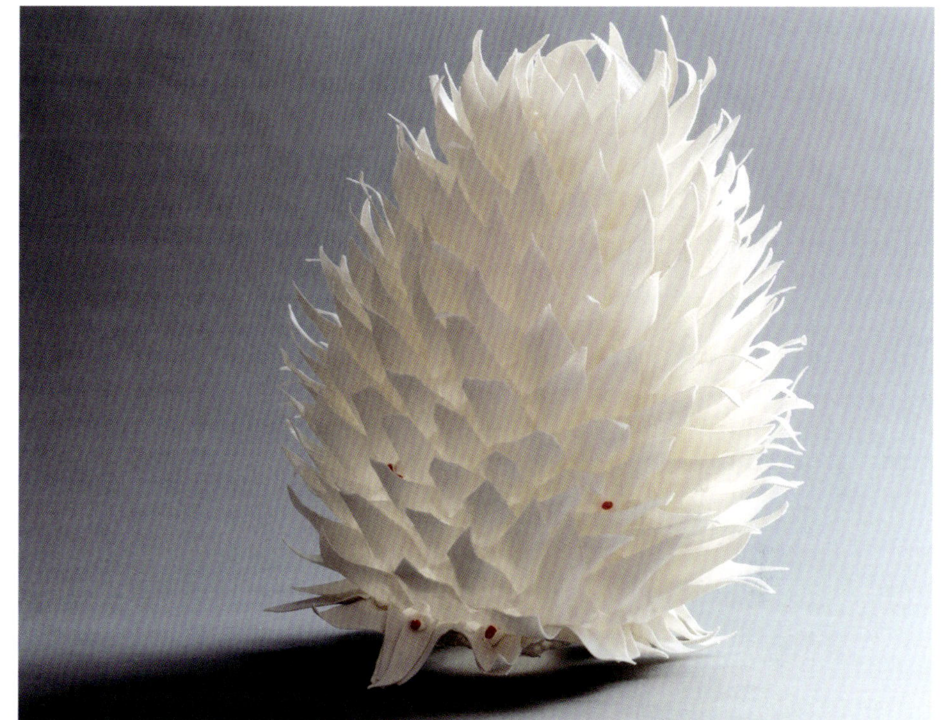

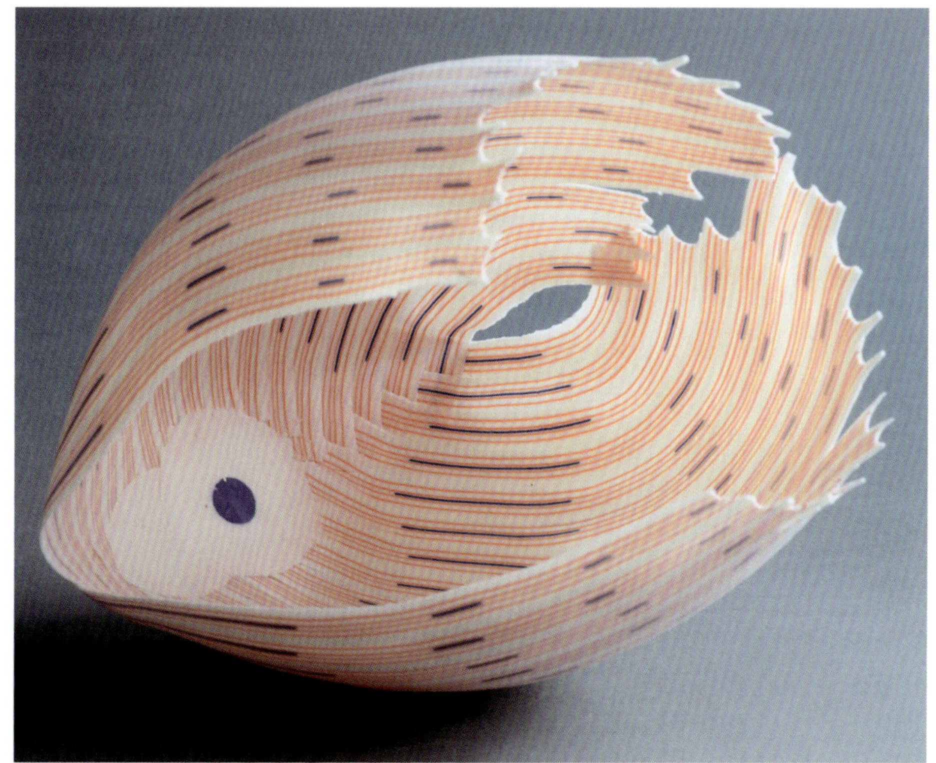

Dorothy Feibleman (UK/Japan)

Dorothy Feibleman started her laminated porcelain expression in 1969 while at Rochester Institute of Technology, New York, and moved to the UK in 1973, after graduating. Her ceramic laminated translucent porcelain and gold jewellery is in the collection of the Victoria and Albert Museum in London.

Her translucent *nerikomi* appeared in the 1st International Ceramics Competition Mino'86 in Gifu, Japan. In the late 1970s and early 1980s, she worked in white. Excited by new colour stains available, she formulated a range of mixable coloured porcelain now sold by Potterycrafts Ltd. In 1993, she won an Inax Design Prize, and took part in the Inax Design Prize Exhibition in Tokoname, Japan. In 1996, she attended the International Academy of Ceramics (IAC) meeting in Saga, Japan, where she showed her work.

BELOW: *ICE*, Dorothy Feibleman, 2004–5. Two laminated white porcelains in several layers. Translucent clay is Dorothy's recipe. (each piece) 40 cm (diam). Fired at 1190°C. *Photo: Ken Yanoviak.*

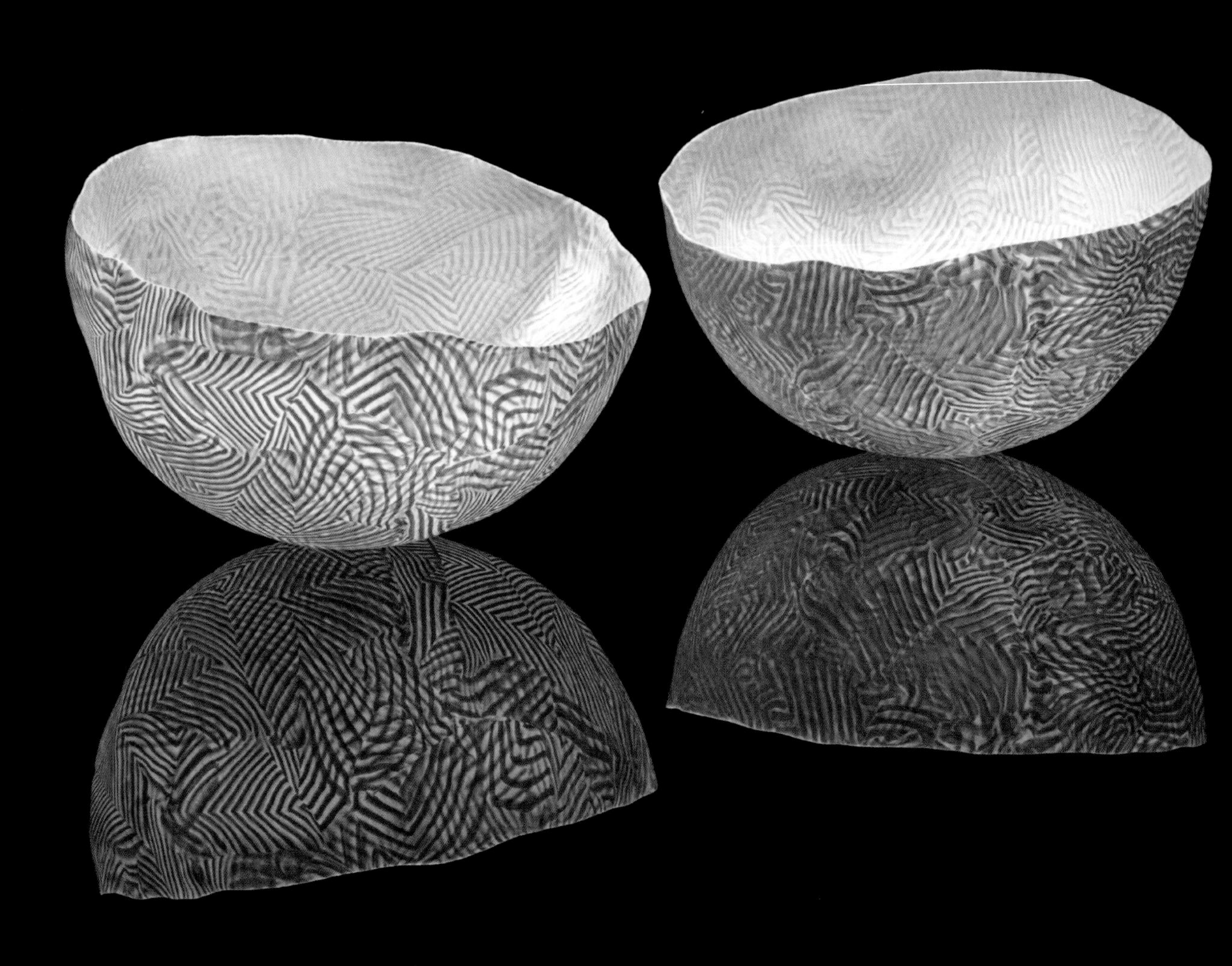

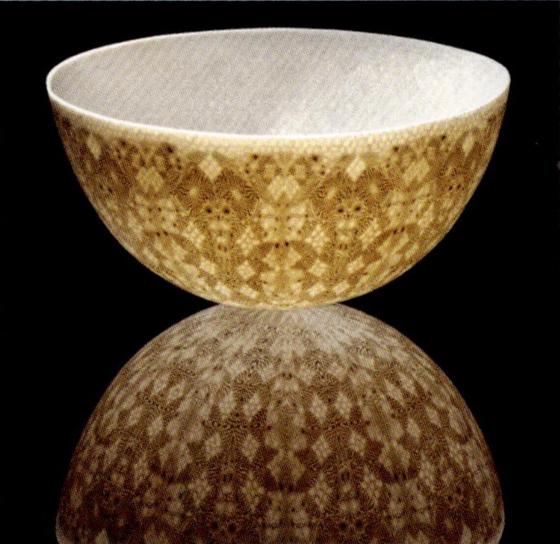

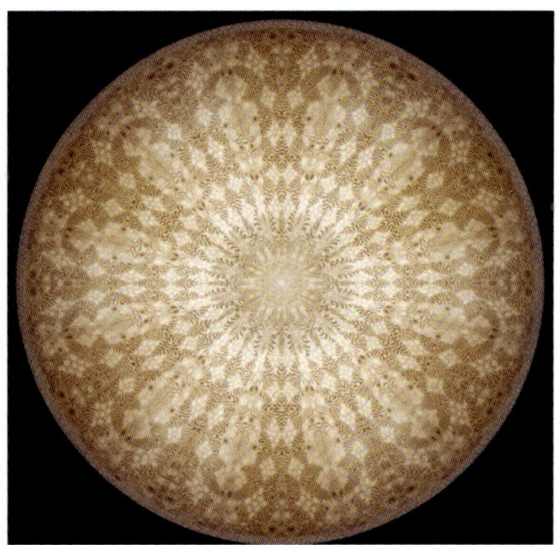

TOP LEFT: *White White and Pink Geometric*, Dorothy Feibleman, 1999. Two white porcelains laminated with different pink laminated images. Translucent clay is Dorothy's recipe. 26 cm (diam). Fired at 1190°C. *Photo: Ken Yanoviak.*

TOP RIGHT: *White White Geometric*, Dorothy Feibleman, 2003. Two white porcelains laminated. One is Dorothy's April 2nd translucent porcelain. Fired at 1190°C. *Photo: Tetsuro Inagaki.*

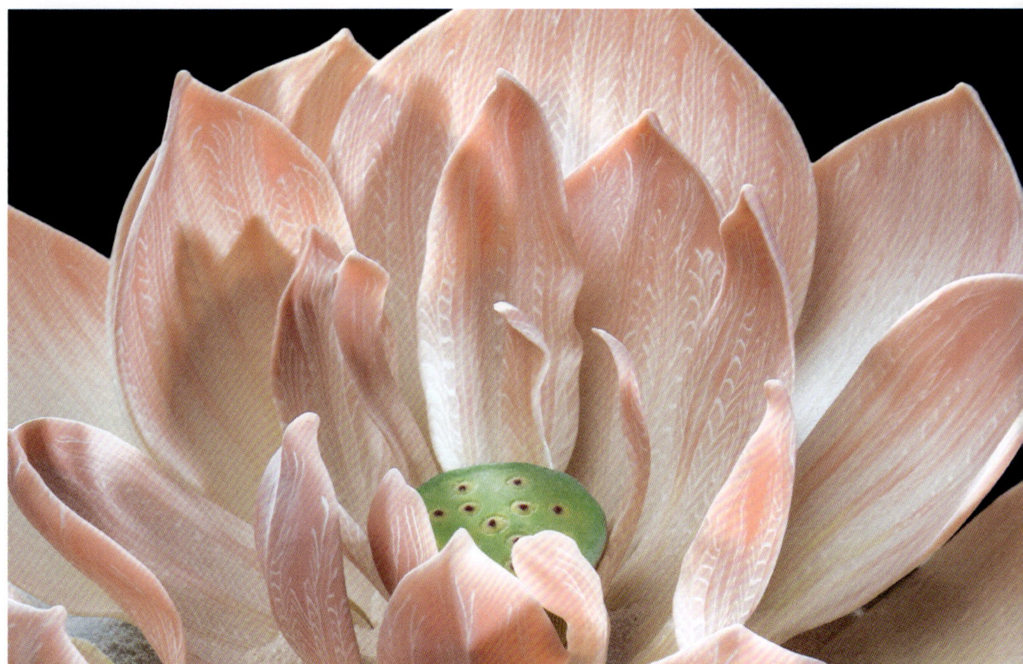

ABOVE: *Lotus in Sand* (detail), Dorothy Feibleman, 2005–6. Two white porcelains laminated. One is Dorothy's April 2nd translucent porcelain. Colour gradations. (full installation) 1.2 m x 60 cm (l x w). Fired at 1190°C. *Photo: Ken Yanoviak.*

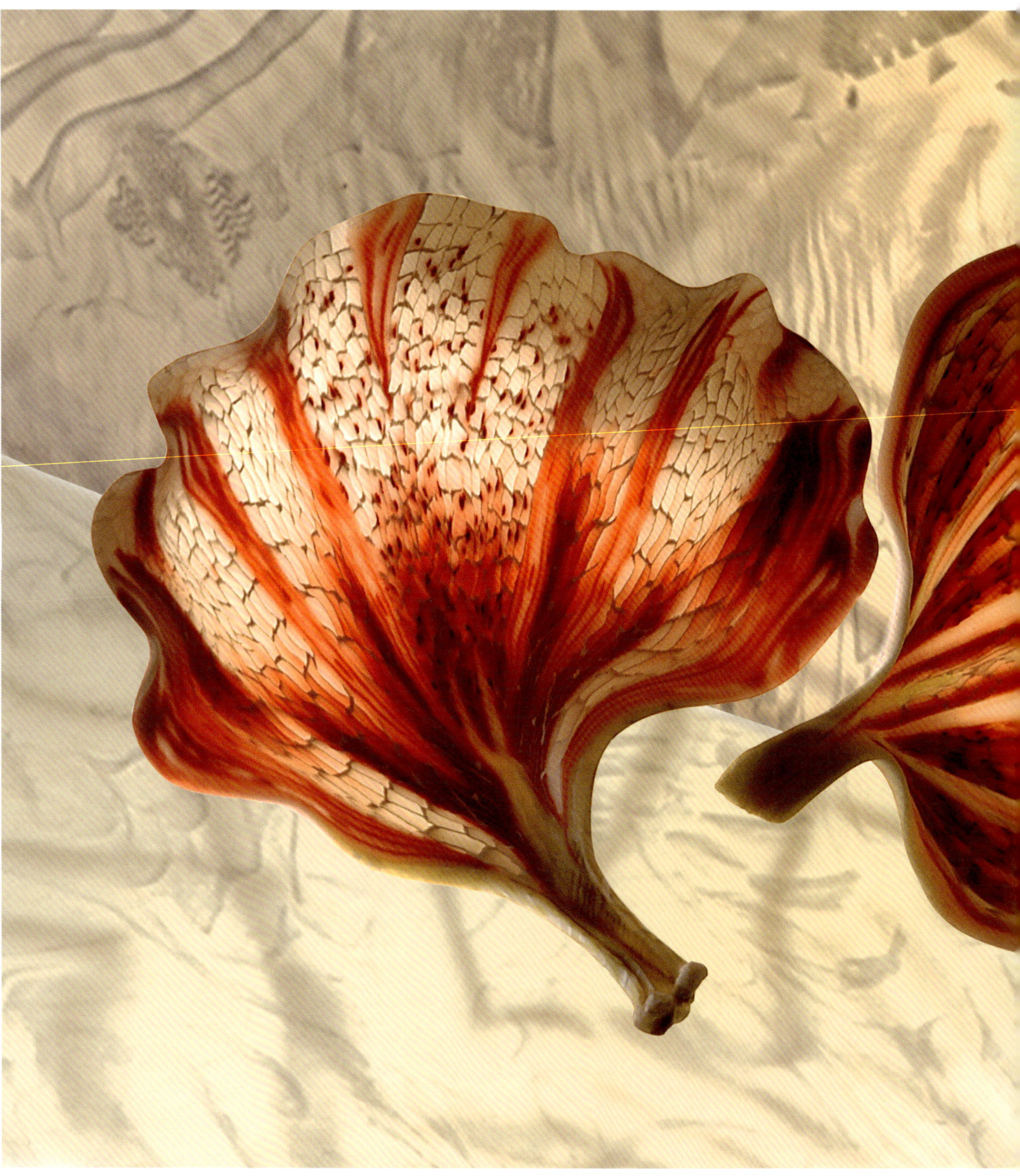

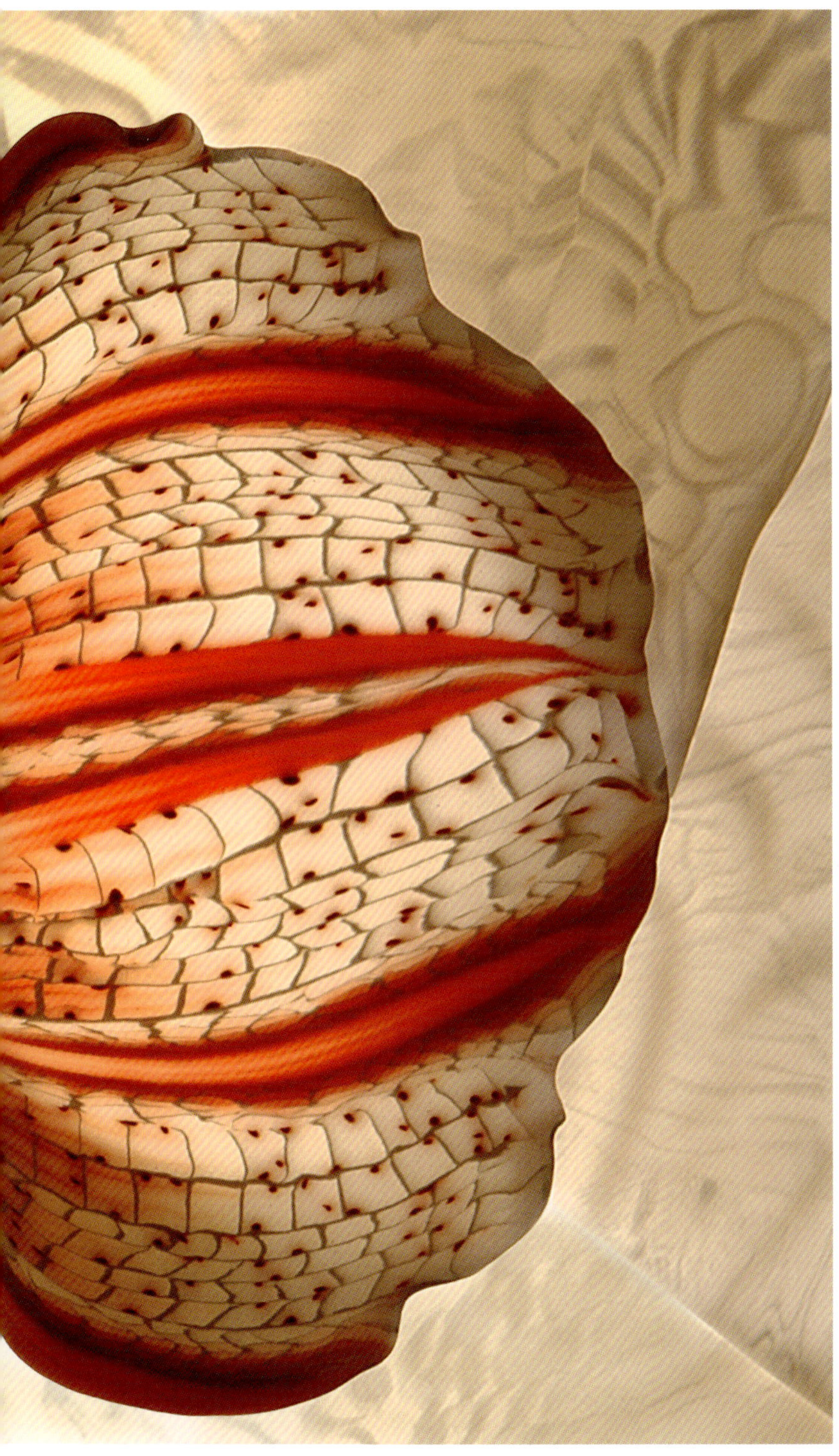

LEFT: *Petals and Ice* (detail), Dorothy Feibleman, 2004–5. Two laminated white porcelains in several layers and two white porcelains with colour gradations. Laminated translucent clay is Dorothy's recipe. (full installation) 1 x 1 m floating in the air on a suspended glass sheet. Fired at 1190°C. *Photo: Yamada.*

Maggie Barnes (UK)

Maggie is a ceramic artist and member of the Society of Designer Craftsmen. She is known for her work in porcelain and takes inspiration from her lifelong interest in marine life, geology, and palaeontology.

Of her working methods and influences, Maggie says:

My childhood was spent by the sea, scouring rock pools for soft-shelled crabs for my father and uncle who were fishermen. I took home my own 'treasures' in damp pockets to be stored and studied. My creative work is the legacy of those years.

My primary school teacher taught me to be selective and choose only the natural objects which I found the most interesting. He taught me to see, select, and record what I found in words and drawings – a habit that has remained with me. My porcelain forms reference qualities found in the natural world; the material is sympathetic to the subject and gives me the results I require.

My early influence in paperclay was when I exhibited Carol Farrow's work in my gallery in Knaresborough and joined forces with Harrogate College to introduce paperclay to students. Sadly, illness deprived the world of her creative talent.

Training as a designer-maker, I was awarded an Arts Council Research and Development grant and I began to consider how to extend my private porcelain practice. I started reading up on paperclay and experienced an unexpected urge to experiment and reconsider future possibilities. New paperclay items developed alongside established porcelain work and were fired. I use Southern Ice porcelain paperclay fired to 1240°C.

When first introduced to paperclay, I was advised to 'keep it simple', and despite the years between, I have found it remains good advice.

My method begins with three basics: quality, 100 per cent cotton, handmade paper, a slab of porcelain, and clear, clean boiled water. Hot water helps to speed up what can be a tedious, time-consuming process, softening both paper and clay.

Once dampened, the paper tears easily into smaller pieces, boiled water speeding the process once the mixture is broken down further using an electric kitchen grinder. The block of porcelain is shaved into small shards using an ordinary metal grater; both ingredients are then transferred into a clean container, covered with freshly boiled water, and gently stirred to combine the ingredients. Continued stirring helps to speed the softening, but it requires patience and should not be hurried because steady stirring is necessary to create the smoothness that is essential for liquid paperclay.

Once the correct texture is reached and any surface water drained, the remaining mixture is wedged, and slabs rolled in readiness for use. These are gently pressed onto previously carved and fired porcelain tiles and slabs, carefully removed when stiffened, slow-dried under gentle pressure, then fired to 1240°C. The paper disappears, leaving translucent porcelain images of the original carvings.

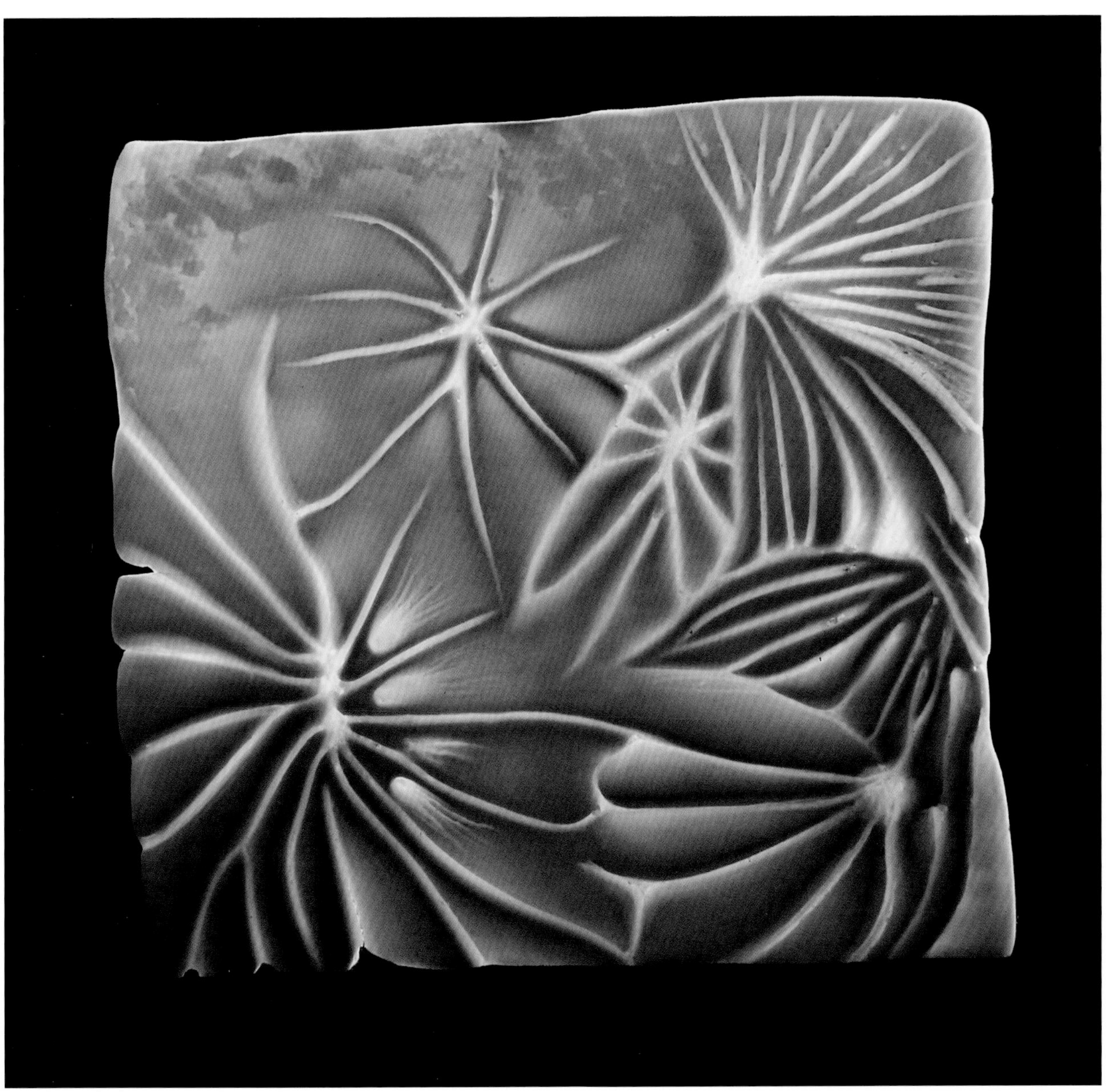

ABOVE: *Bouquet*, Maggie Barnes, 2020. Southern Ice porcelain paperclay tile. 10 x 10 cm. Fired at 1240°C. *Photo: David Chalmers.*

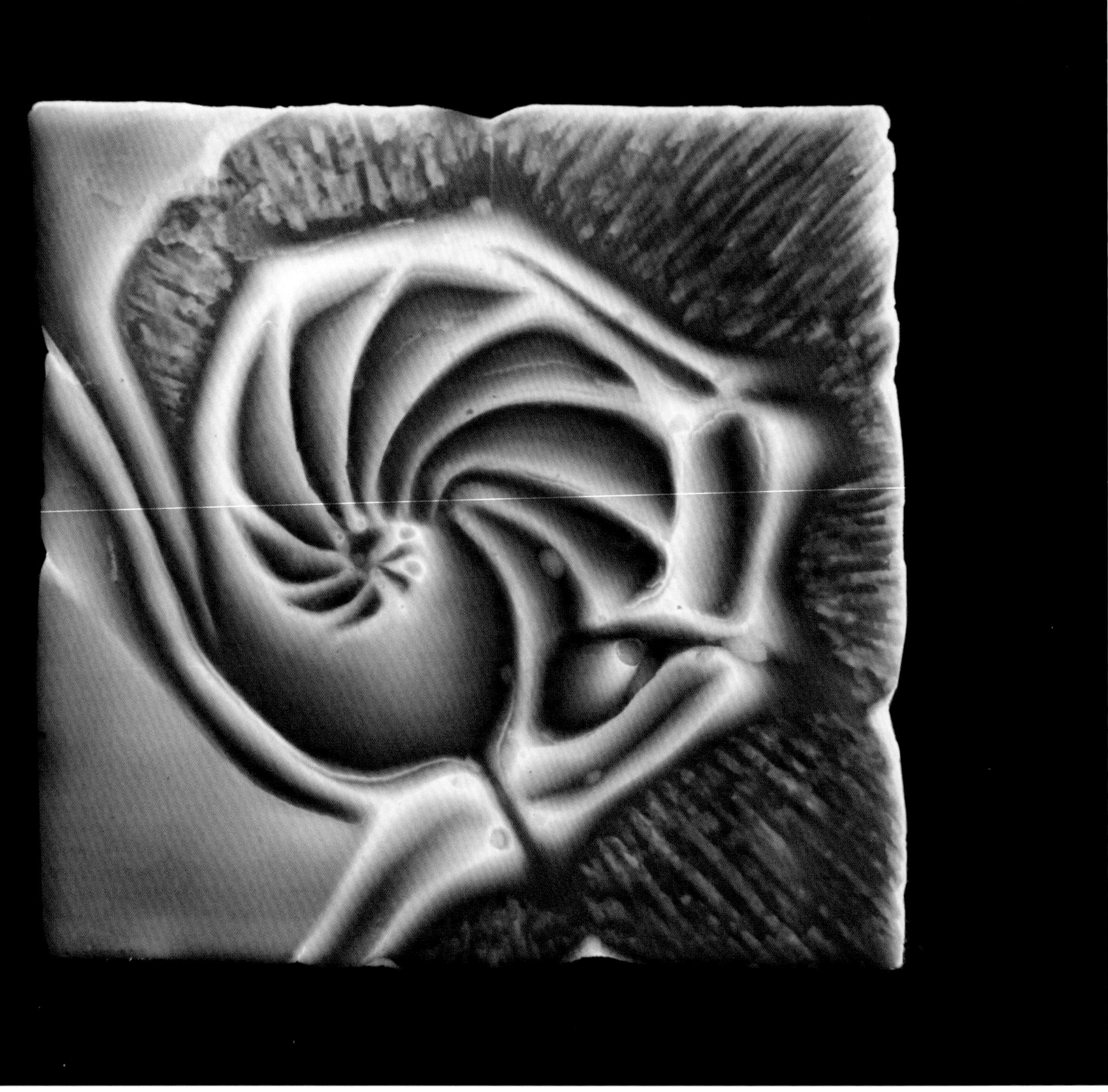

ABOVE: *Nautilus*, Maggie Barnes, 2022. Southern Ice porcelain paperclay tile. 9 x 9 cm. Fired at 1240°C. *Photo: David Chalmers.*

OPPOSITE: *Fracture*, Maggie Barnes, 2022. Southern Ice porcelain paperclay wall piece. 10 x 20 cm. Fired at 1240°C. *Photo: David Chalmers.*

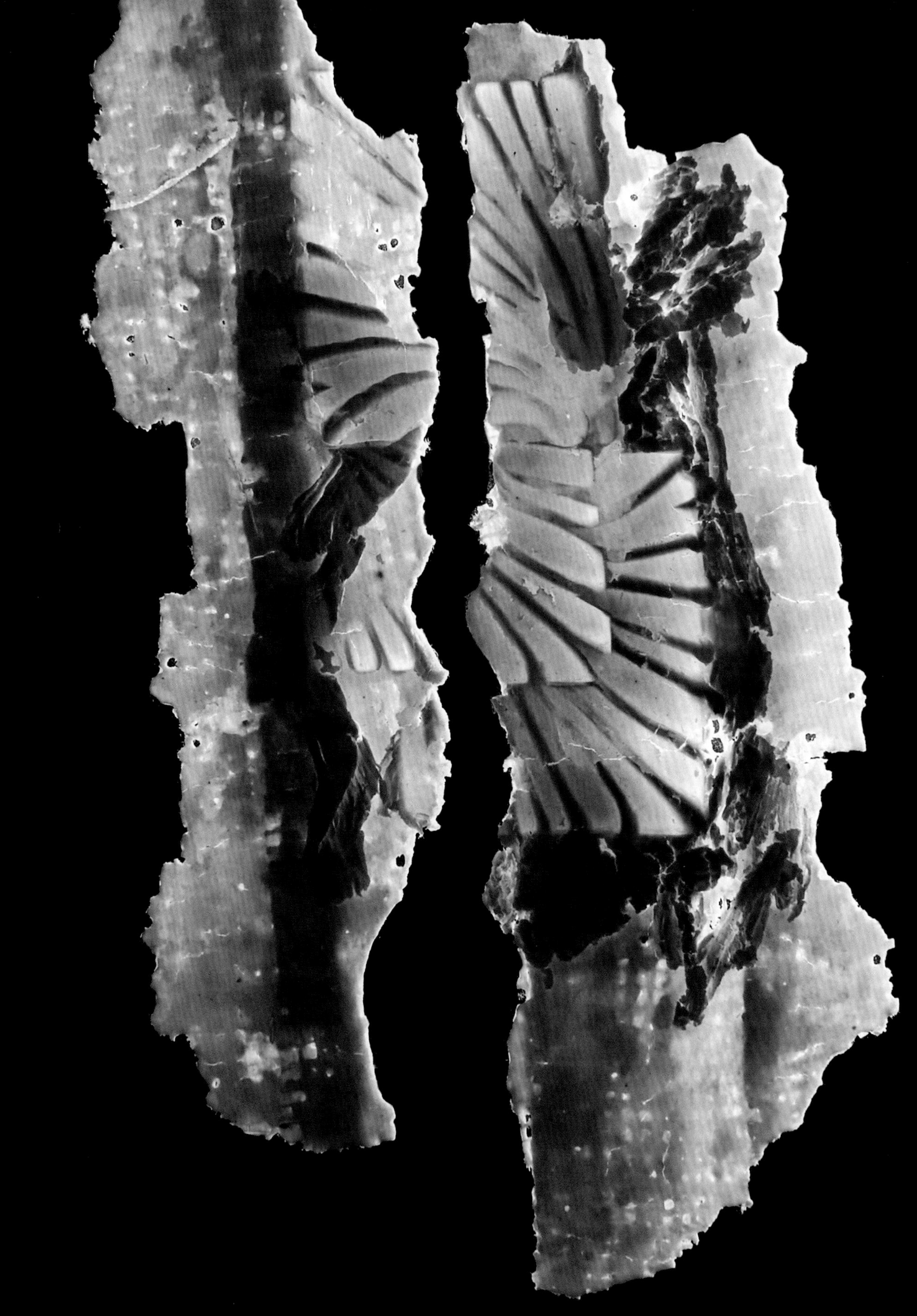

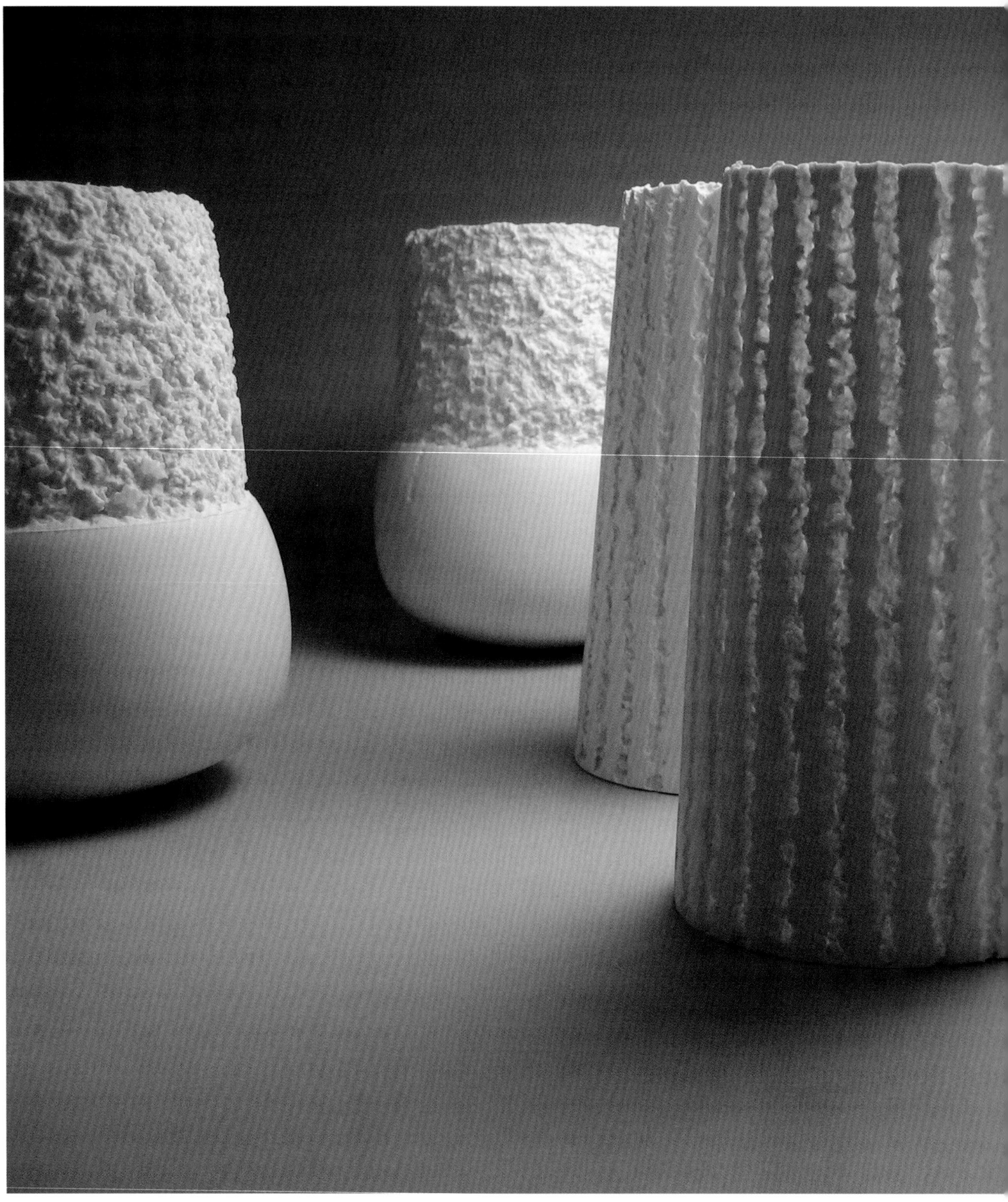

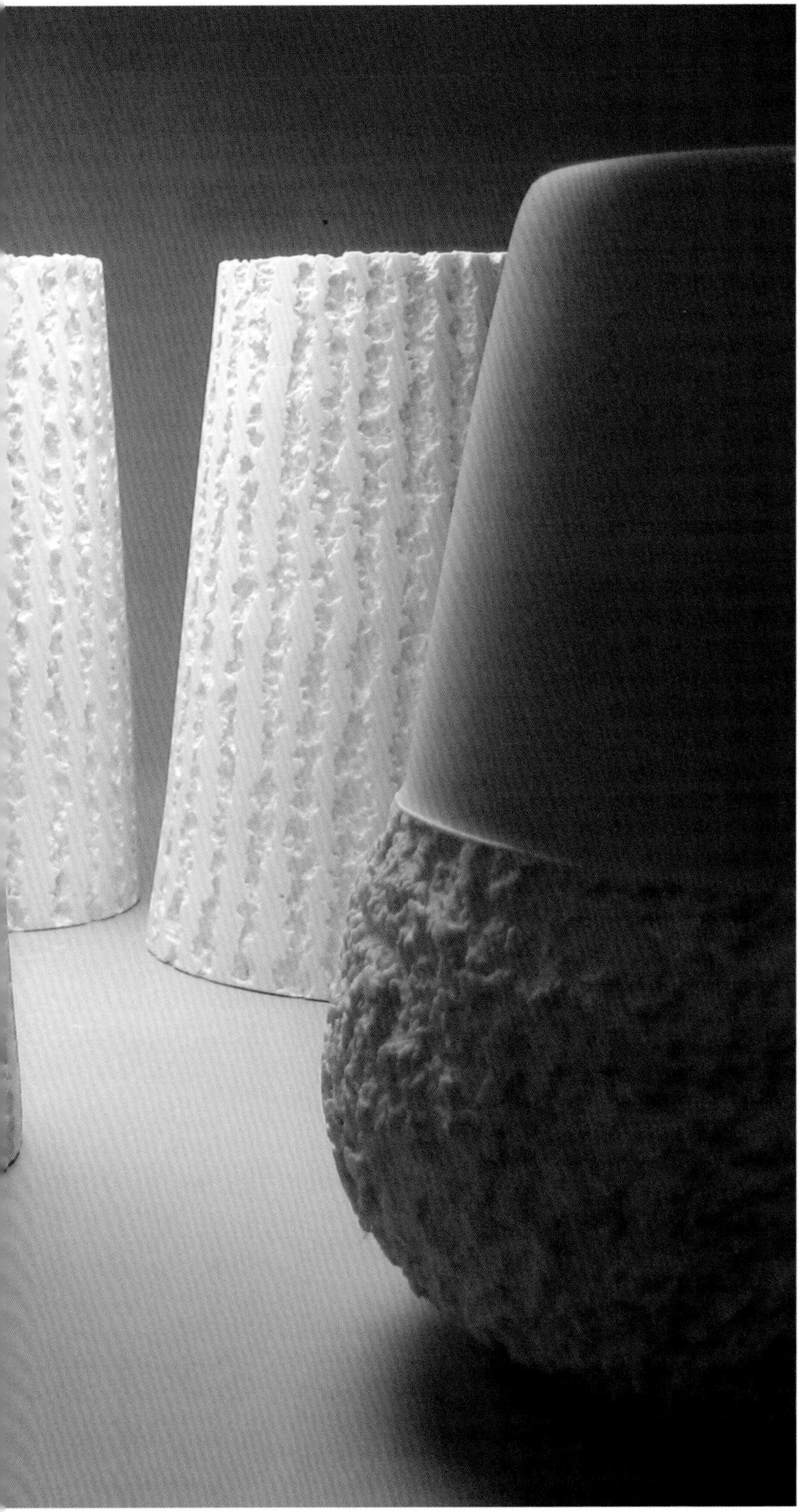

Monika Patuszyńska (Poland)

MA Ceramics, Academy of Fine Arts, Wrocław, Poland Member of the IAC.

Monika describes herself as 'a curator and explorer of abandoned spaces and untried paths'.

In 1999, she graduated from the Faculty of Ceramics and Glass at the Eugeniusz Geppert Academy of Art and Design in Wrocław, Poland. She was the last designer at the Pruszków Table China Factory, and the last president of the International Ceramics Symposium *Porcelain Another Way*. Between 2014 and 2016 she was Artistic Director at the Institute of Design in Kielce.

She has been commissioned by the Maison Guerlain salon in Paris, Chanel in Singapore, and Tiffany in New York. She has won numerous international awards and is represented in public collections worldwide.

LEFT: From the series *PapForms*, Monika Patuszyńska, 2007. Porcelain. 34 cm (h). Fired at 1240°C. *Photo: Monika Patuszyńska.*

BELOW: From the series *PapForms* (detail), Monika Patuszyńska, 2007. Porcelain. 34 cm (h). Fired at 1240°C. *Photo: Monika Patuszyńska.*

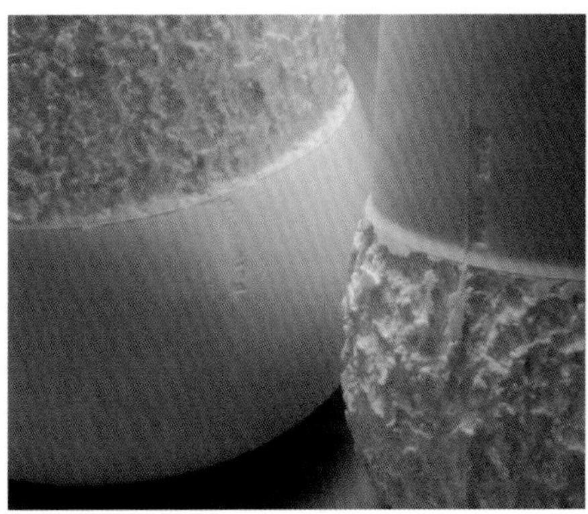

In her early work, shown here, she uses traditional slipcasting and mould-making methods. She allows the seam lines to show, rather than disguising them, and along with her mark, they become part of the design. She has used paper fibres to give translucent structure to these slipcast forms. In her latest work, she exploits traditional methods and seeks out difficulties posed by working with porcelain. She smashes and saws moulds into shards and reassembles them to create new formations for new moulds. She learns through the process and enjoys the discoveries she makes through her art. She actively seeks out seams, edges, and broken textures to create large-scale dynamic, fluid vessel forms.

When I asked her about her work and process, Monika explained:

> *Both* PapForms *and* NonForms *come from my passion for playing with negative spaces and creating structures that burn out and disappear during firing.*
>
> *In the case of* PapForms, *it is paper which I use to line the insides of plaster moulds and in the case of* NonForms, *chocolate Nesquik® balls, which I fill selected shapes with and then pour the casting slip into the empty spaces among them.*

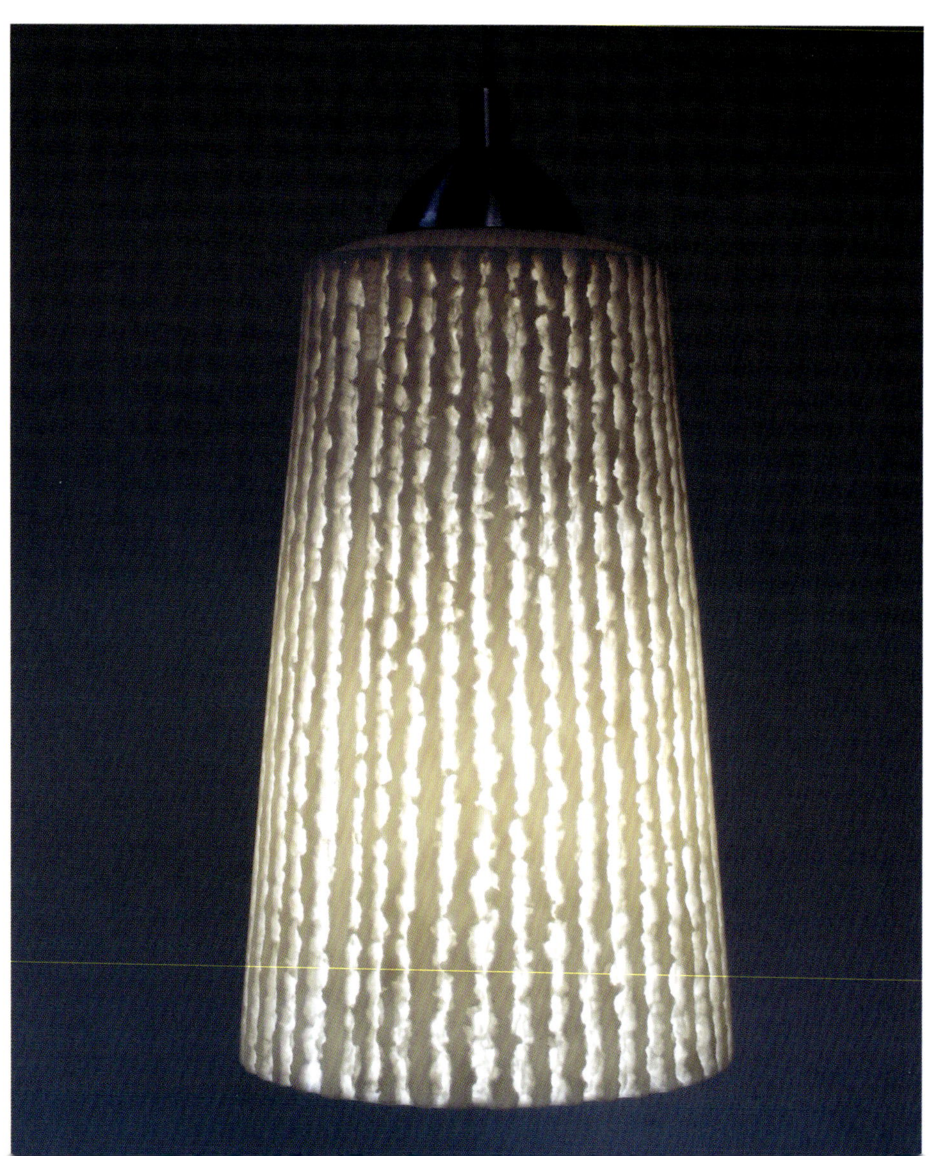

BOTTOM LEFT: From the series *PapForms*, Monika Patuszyńska, 2007. Porcelain. 34 cm (h). Fired at 1240°C. *Photo: Monika Patuszyńska.*

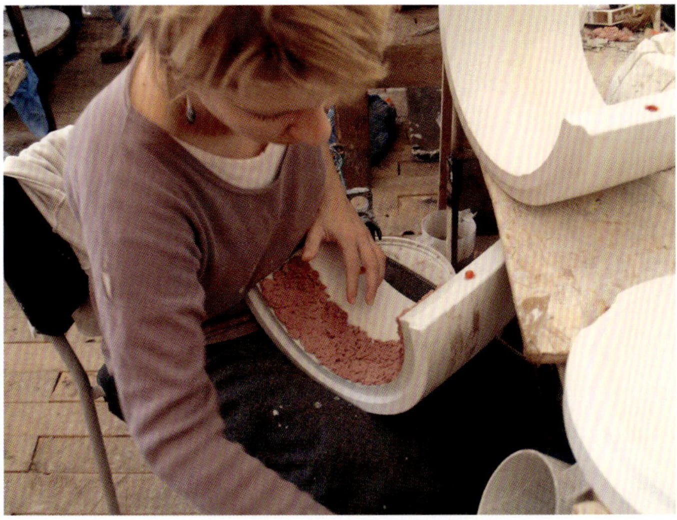

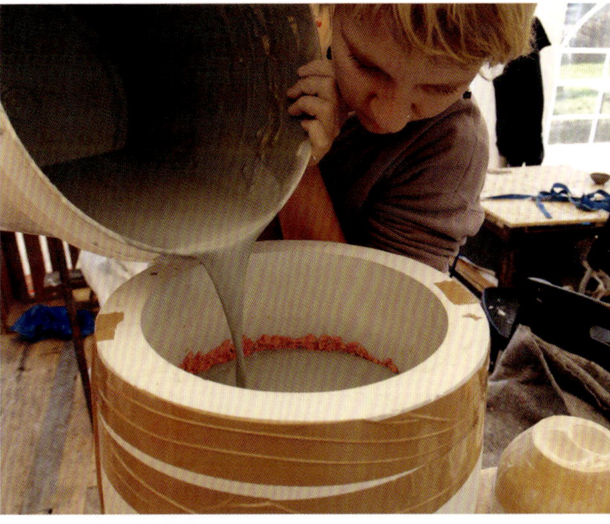

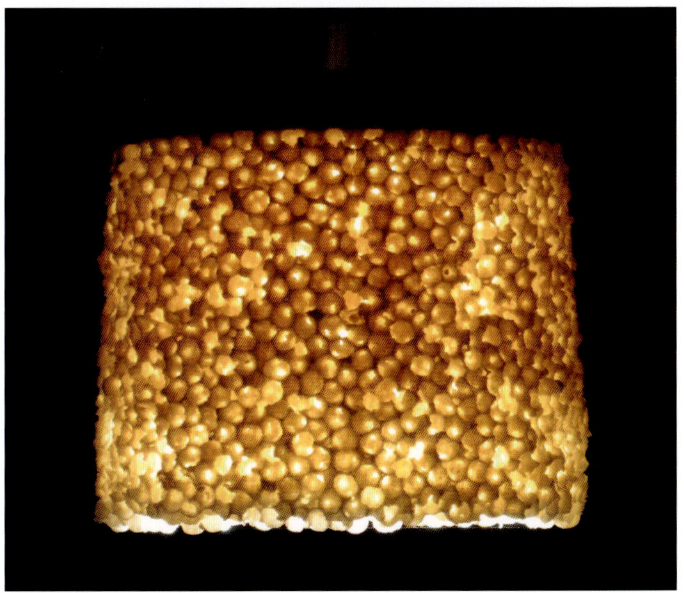

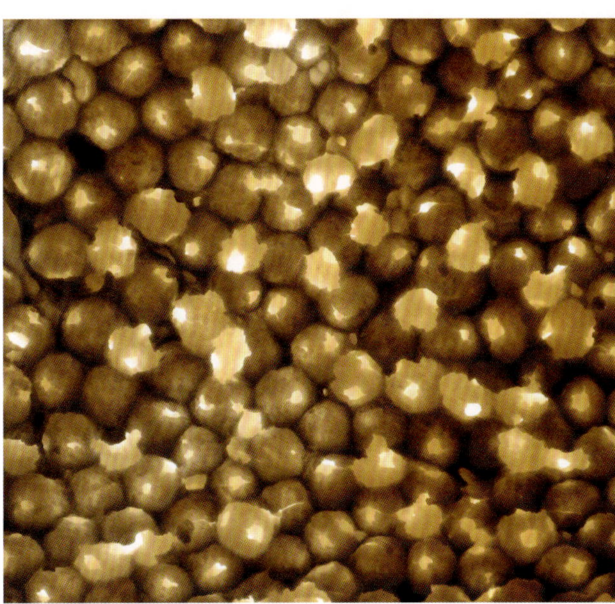

TOP LEFT AND TOP RIGHT: Monika Patuszyńska, making *PapForms*, 2007. *Photo: Aleksandra Dobrowolska.*

BOTTOM LEFT: From the series *NonForms*, Monika Patuszyńska, 2009. Porcelain. 34 cm (h). Fired at 1240°C. *Photo: Monika Patuszyńska.*

BOTTOM RIGHT: From the series *NonForms* (detail), Monika Patuszyńska, 2009. Porcelain. 34 cm (h). Fired at 1240°C. *Photo: Monika Patuszyńska.*

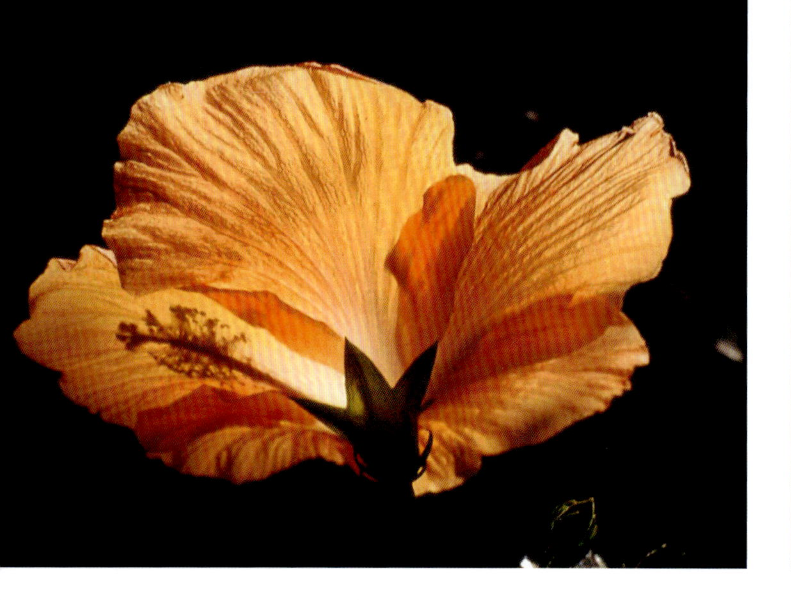

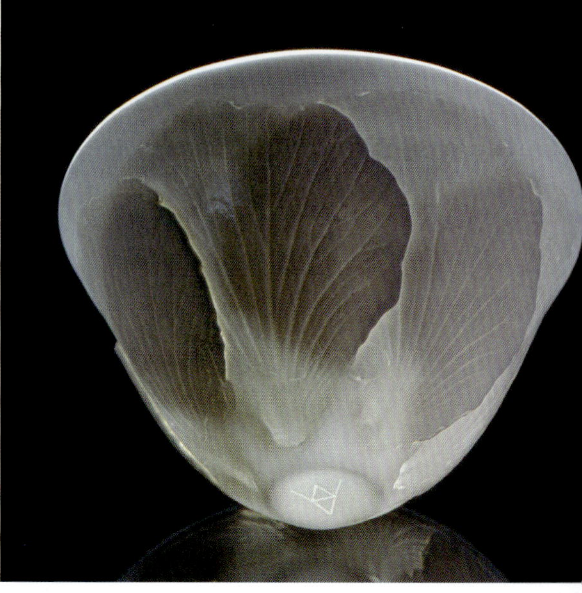

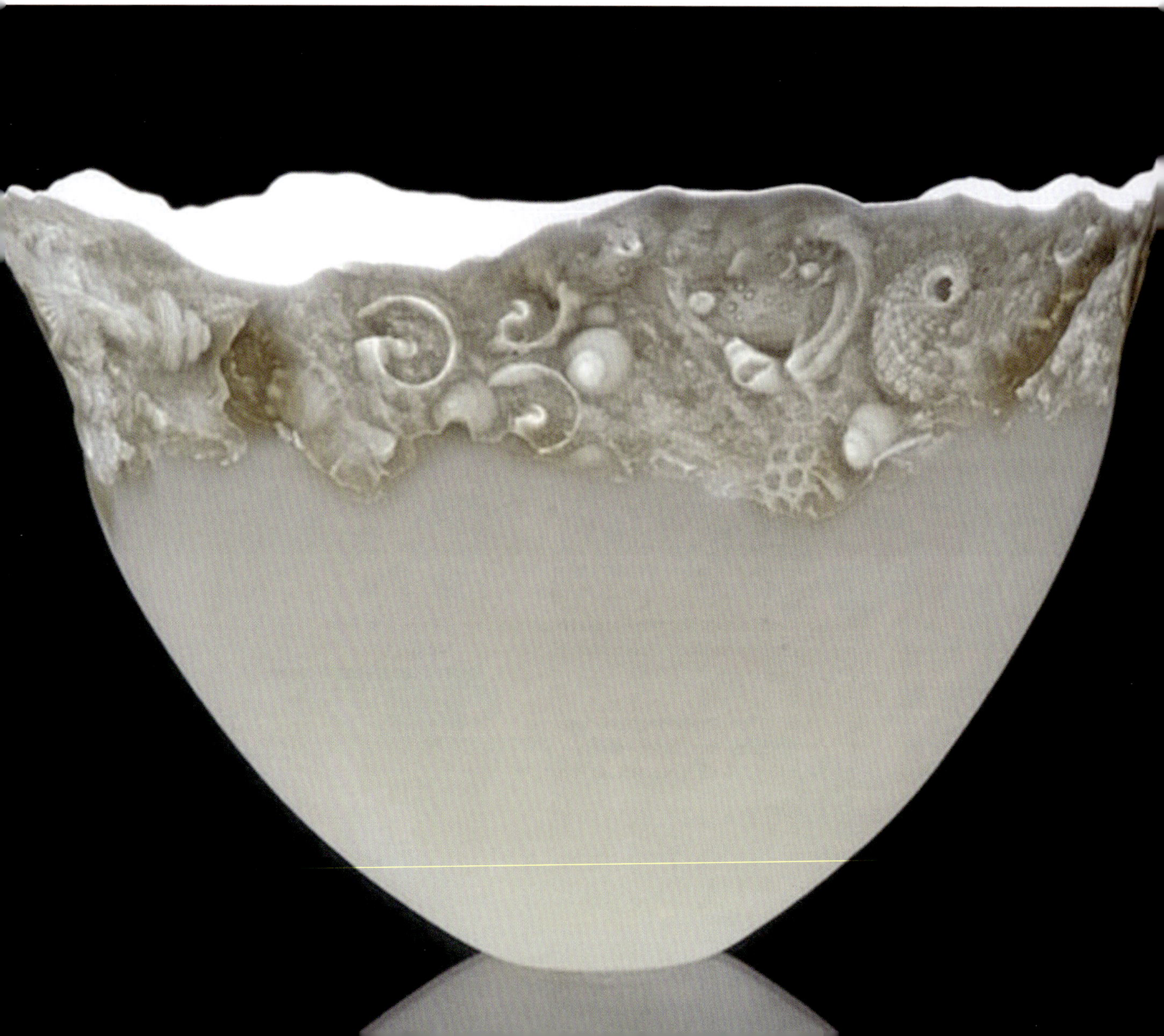

5

Inspiration and design

OPPOSITE TOP LEFT: Hibiscus flower.

OPPOSITE TOP RIGHT: *Petal Bowl*, Angela Mellor, 2002. Bone china with paperclay inlay. 10 x 18 cm (h x w). Oxidised fired to 1250°C. *Photo: Victor France.*

OPPOSITE LEFT: *Cretaceous Bowl*, Angela Mellor, 2003. Bone china with paperclay inlay. 12 x 18 cm (h x w). Oxidised fired to 1250°C. *Photo: Victor France.*

While living in Australia, I became deeply interested in the environment. Landscapes, light, and coastal environments were reflected in my work. Photography enabled me to capture not only line and colour, but also the unique and mysterious quality of light.

One sunny afternoon, Manfred brought me a hibiscus flower from our garden because he'd noticed how the light was passing though the petals and he knew I would love it. I photographed the plant, capturing how the sun caused the petals to glow, and this was the inspiration for my *Petal Bowl*.

The study of rocks, driftwood, coral, and shells led to many ideas for design work. The textures of these organic forms were recreated from plaster moulds using paperclay. Fragments of these were incorporated into my work.

My early training was in art and design. The basic principles I learnt then have remained with me and still influence my work today. Leon Nigrosh, in his book *Claywork*, says:

> *Good design is a combination of vision and common sense. It is akin to the difference between looking and seeing. Looking both ways before crossing the street involves merely a quick glance to ensure survival. Seeing an object involves a conscious effort to focus not only the eyes but the mind as well; to actively discern the object parts; to compare its shape with objects of similar nature; and to decide its individual merits. While the public cannot see an object until it is a* fait accompli, *for the artist, a work must be visible in the mind's eye before it is ever constructed.*
>
> *Although technical facility is an important part in the execution of a ceramic form, no amount of expertise can substitute for the lack of a well-conceived design.*[1]

I take the following considerations into account when I am designing an object:

- *Fitness for purpose*: this should be evident when designing a utilitarian object. When designing a sake jug, I had to consider that it would pour well and that the slight depressions in the side of the form (instead of a handle) would be comfortable to hold. Experiencing the tactile quality of an object is also important. Eating and drinking from or handling an object is more pleasurable when the surface is smooth, rather than rough.
- *Balance:* this has also been a consideration in my work, especially when using found objects with bone china

[1] Leon I. Nigrosh, *Claywork: Form and Idea in Ceramic Design*, Davis Publications, Worcester, Mass, 1986, p.113.

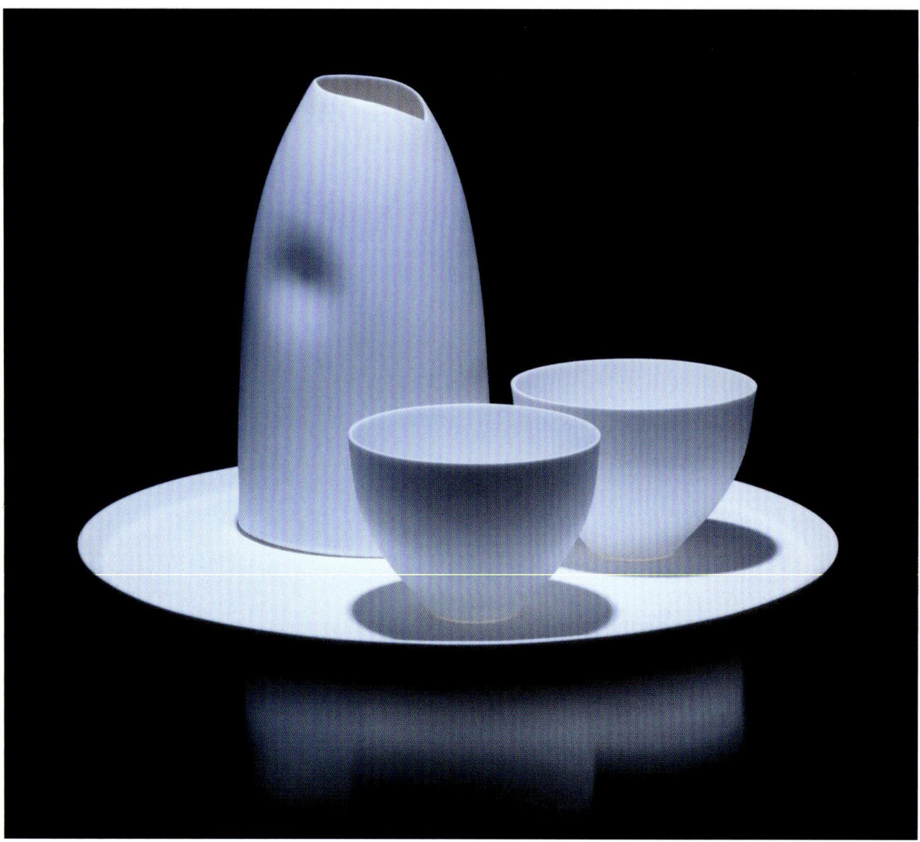

LEFT: *Sake Set*, Angela Mellor, 2015. Jug and two bone china cups, with tray. Bone china. 22 x 15 cm (h x w). *Photo: Stephen Bond.*

Contrast

When designing narrow-based flute forms, I needed a base on which to rest the pieces for stability. My initial idea was to use shells that I had picked up on the beach. As the shell form itself was not very durable, I decided to cast them in glass, and took a short course on glass casting with Helen Stokes at Monash University. I felt that the glass forms would provide a good contrast and support for the bone china vessels, and fulfil my conceptual aims.

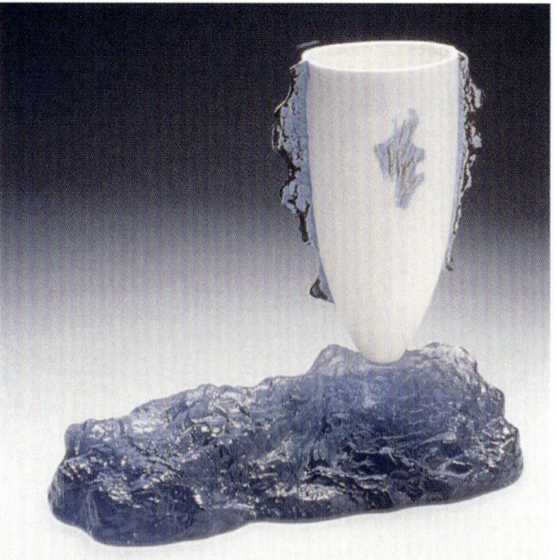

LEFT: *Coral Vessel*, Angela Mellor, 1998. Bone china painted with soluble colourants on cast glass. 15 x 17 x 10 cm. *Photo: Isamu Sawa.*

Afterwards, I did not feel that glass was the right medium to complement the work. The clarity of the glass seemed to attract the light, making the translucent bone china appear more opaque instead of continuing the light through the piece, which was my original intention. My aim throughout had been to explore the translucency of bone china, so I abandoned this idea and looked for an alternative that would help me achieve my aims.

I always take my inspiration from natural forms, and my ideas turned to rock: a strong, dense, opaque material might present a more suitable foil for bone china. A piece of black rock I found on the beach provided a suitable base for three bone china flute forms. The bone china coral forms were painted with soluble colourants over latex resist, the design and colour taken from an underwater photograph of coral. I named this piece *Coral Reef.*

The contrast in colour and texture of the rock was sympathetic to the coral forms and portrayed my conceptual idea of a coral reef.

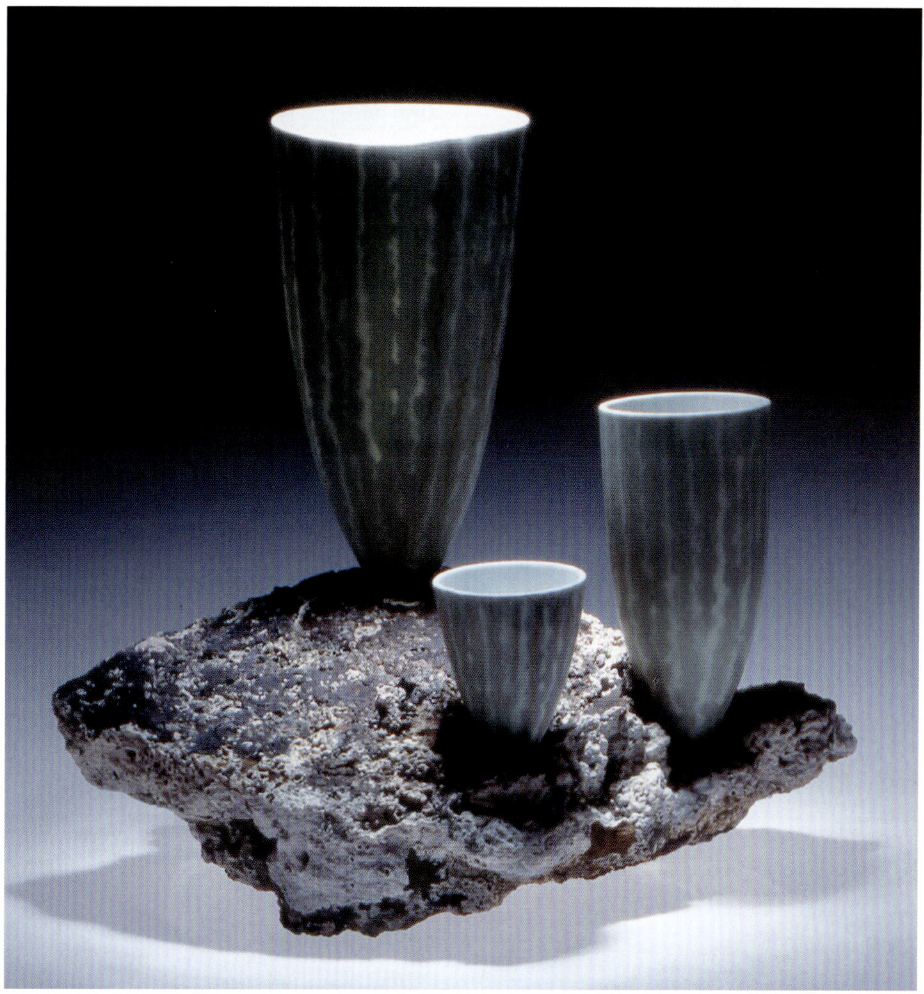

RIGHT: *Coral Reef,* Angela Mellor, 1999. Bone china with soluble colourants on black rock. 30 x 26 x 22 cm. *Photo: Victor France.*

Another idea that followed on from the flute forms I used in *Coral Reef* came from the stalactites and stalagmites in the tranquil Lake Cave in Western Australia. These impressive sculptural structures made a great impression on me, and became the basis of a new series of work called *Passage of Time.*

Starting with the flute form, I decided to elongate it, making it taller and narrower. This meant redesigning the shape and making new plaster models and moulds on the lathe. I liked the idea of grouping the forms together, and so designed models of different heights.

These forms were to be left white to complement some white quartz stones that my husband had found in a quarry in Sydney. I still felt that something was needed to relieve the somewhat stark white bone china form while enhancing its translucency. Struck by the beautiful glistening texture of the quartz as it caught the light, I felt that I should like to emulate a fragment of this texture in the bone china to allow the light to pass through.

I studied quartz rocks for their form and suitability to balance with the bone china forms. I brought my idea to life by inserting a small piece of textured paperclay into the mould before casting. Sculptural forms, which provide rough areas against smooth and capture light, offer both a visual and tactile experience. This group of works brings back the feeling of wonderment and tranquillity I experienced in Lake Cave.

Exaggerating the form with strong lines and curves gives it life. For example, it is important when designing the form to consider whether a narrow base will elevate it. This was my intention with the stalactite forms that inspired *Passage of Time*: I wanted the tall narrow pieces to continue their upward movement.

Passage of Time received an Honourable Mention in the Ceramics as Expression category at the 1st World Ceramic Biennale 2001 Korea International Competition.

These two stalactite forms were originally part of the larger *Passage of Time* installation seen on the page opposite.

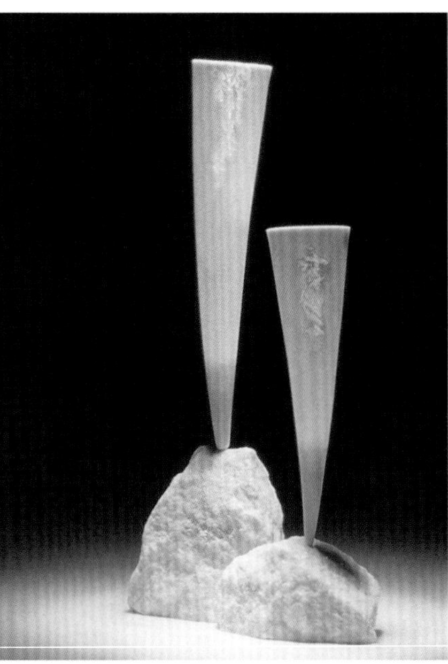

LEFT: *Passage of Time,* Angela Mellor, 2000. Bone china with paperclay inlay, on quartz rock, inspired by stalactites and stalagmites. 28.5 x 8 cm (h x w); 20.5 x 9 cm (h x w). *Photo: Victor France.*

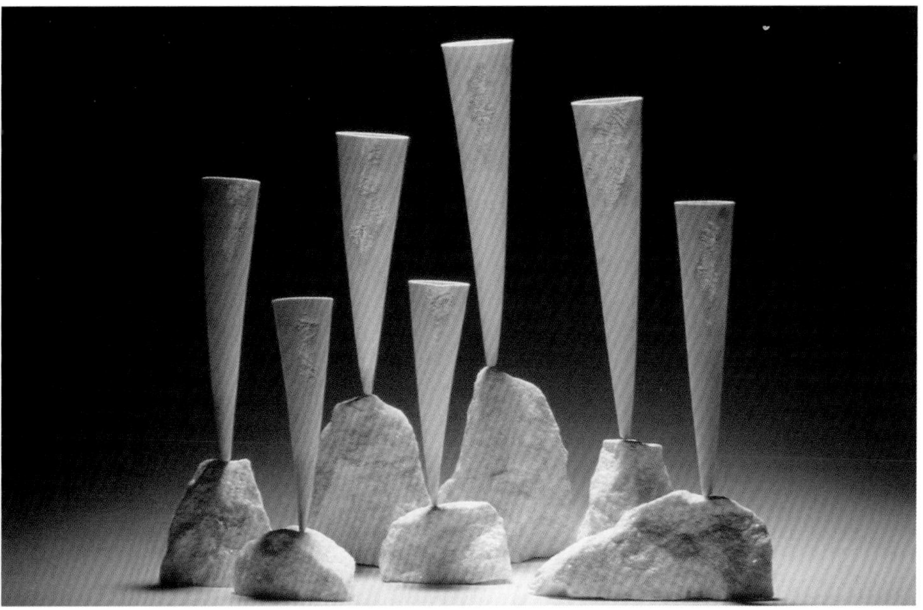

RIGHT: *Passage of Time*, Angela Mellor, 2000. Bone china with paperclay inlay on quartz rock, inspired by stalactites and stalagmites. (installation) 60 x 38 cm (h x w). *Photo: Victor France.*

While I was visiting South Korea for the exhibition, a group of us spent a day in North Korea, which for me was the most inspirational part of the trip. We explored the unspoilt natural beauty of Mount Kumgang, also known as Diamond Mountain. We walked four kilometres up the steep mountain in the pouring rain, seeing the tempestuous waterfalls, gigantic rocks, and a wonderful display of autumn leaves along the way. It was well worth the climb, as it inspired my *Waterfall Vessel* and the *Waterfall Vases*, which were subsequently selected for the 6th International Ceramics Competition Mino'02, Japan, where they gained an Honourable Mention.

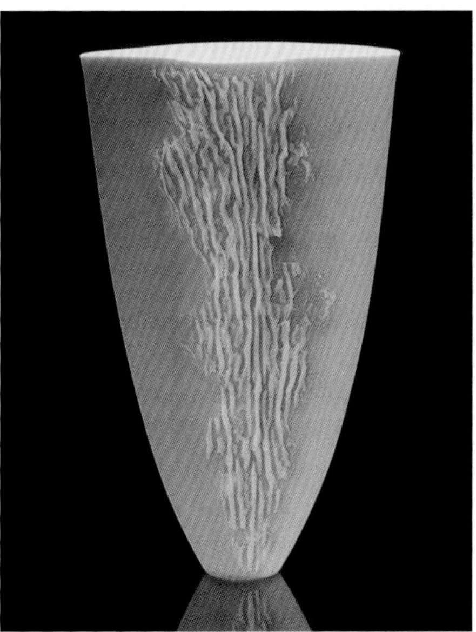

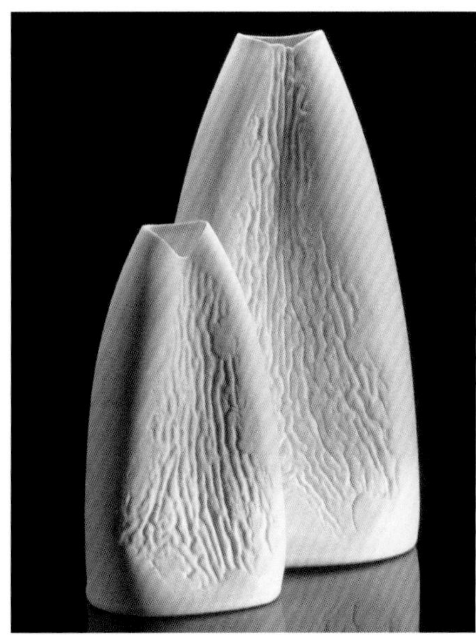

RIGHT: *Waterfall*, Angela Mellor, 2002. Bone china with paperclay inlay. 18 x 10.5 cm (h x w) *Photo: Victor France.*

FAR RIGHT: *Waterfall* vases, Angela Mellor, 2002. Bone china with paperclay inlay. *Photo: Victor France.*

Contrast in materials is interesting in itself, but I would also like to draw attention to the contrast of the bone china slip and bone china paperclay. Slipcast bone china can take on a very smooth surface, in contrast to textured paperclay. I took advantage of this material contrast in the *Glacial Light* series, where I used texture to enhance the surface and allow the light to pass though at the thinner textured areas. Here, contrast has a dual purpose: creating light against dark and smooth against rough.

BELOW: *Glacial Light* bowls, Angela Mellor, 2002. Bone china paperclay inlay. 10 x 12 cm (h x w); 12 x 18 cm (h x w); 8 x 10 cm (h x w). Oxidised fired to 1250°C. *Photo: Victor France.*

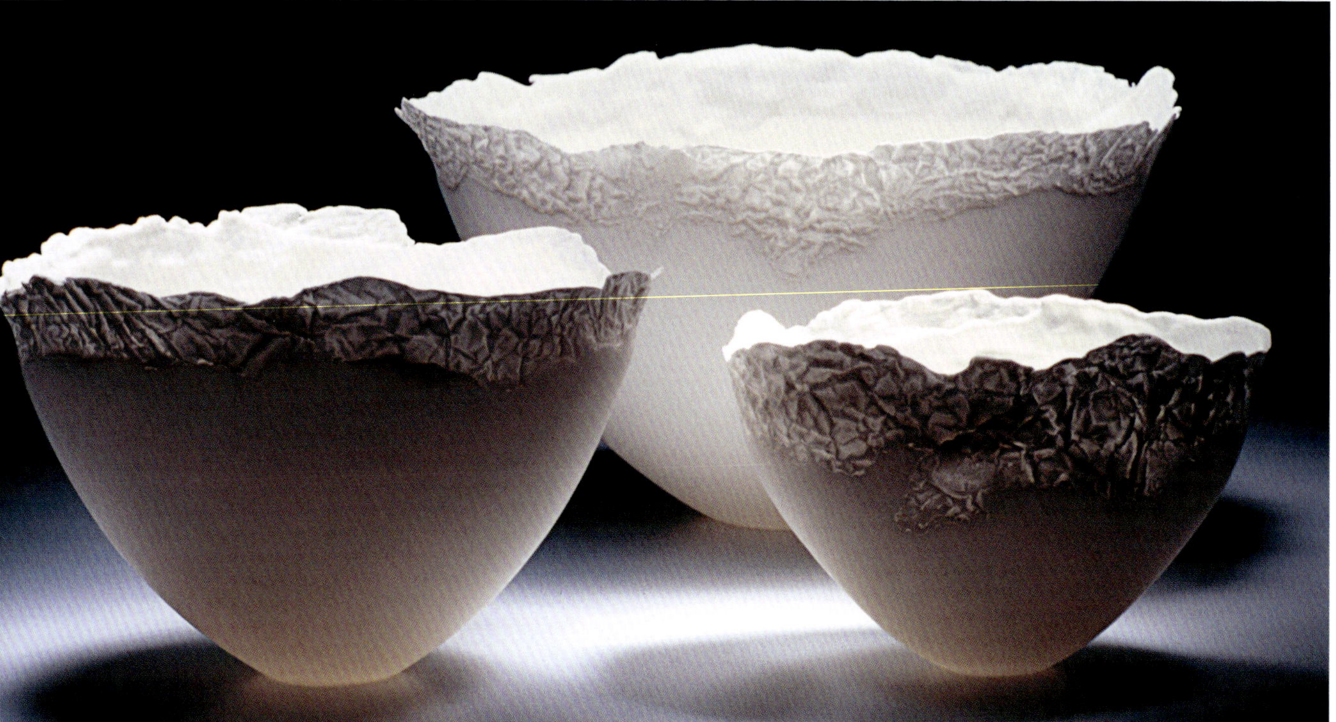

The *Glacial Light* series was inspired by a film I watched about Antarctica during my time at Hobart University. The dazzling light of Antarctica emphasises nature's contrasts, and I recall thinking how bone china would be a perfect medium to portray this.

The art critic David Bromfield described my *Glacial Light* series as:

> *A work glowing with the presence of rocks and shells in shards of ice tumbled, slowly through rock for millennia. Like other works these pieces catch the light in their thin shells so that it leaks infinitely slowly from an irregular fragile lip. The crumbling circlet of fragile indentations just below the top edge of each bowl emphasizes this crisp, solid luminosity, so that it seems to have been trapped there forever. The unique presence of the* Glacial Light *series is not finessed from some flower or shell. It is established directly between the peerless white china, pure light and the passage of time.*[2]

[2] David Bromfield, 'Timeless Light', *Ceramics: Art and Perception*, 44, 2001, pp.6-10.

RIGHT: *Glacial Light* vessels, Angela Mellor, 2001. Bone china with paperclay inlay. 13 x 8.5 cm (h x w); 19 x 11 cm (h x w); 10.5 x 7 cm (h x w). Oxidised fired to 1250°C. *Photo: Victor France.*

NEXT PAGE LEFT: *Glacial Light* vessel, Angela Mellor, 2001. Bone china with paperclay inlay. 18 x 11 x 7 cm (h x w x d). Oxidised fired to 1250°C. *Photo: Victor France.*

NEXT PAGE RIGHT: *Arctic Rose*, Angela Mellor, 2001. Bone china with paperclay inlay. 11 x 19 x 7 cm (h x w x d). Oxidised fired to 1250°C. *Photo: Victor France.*

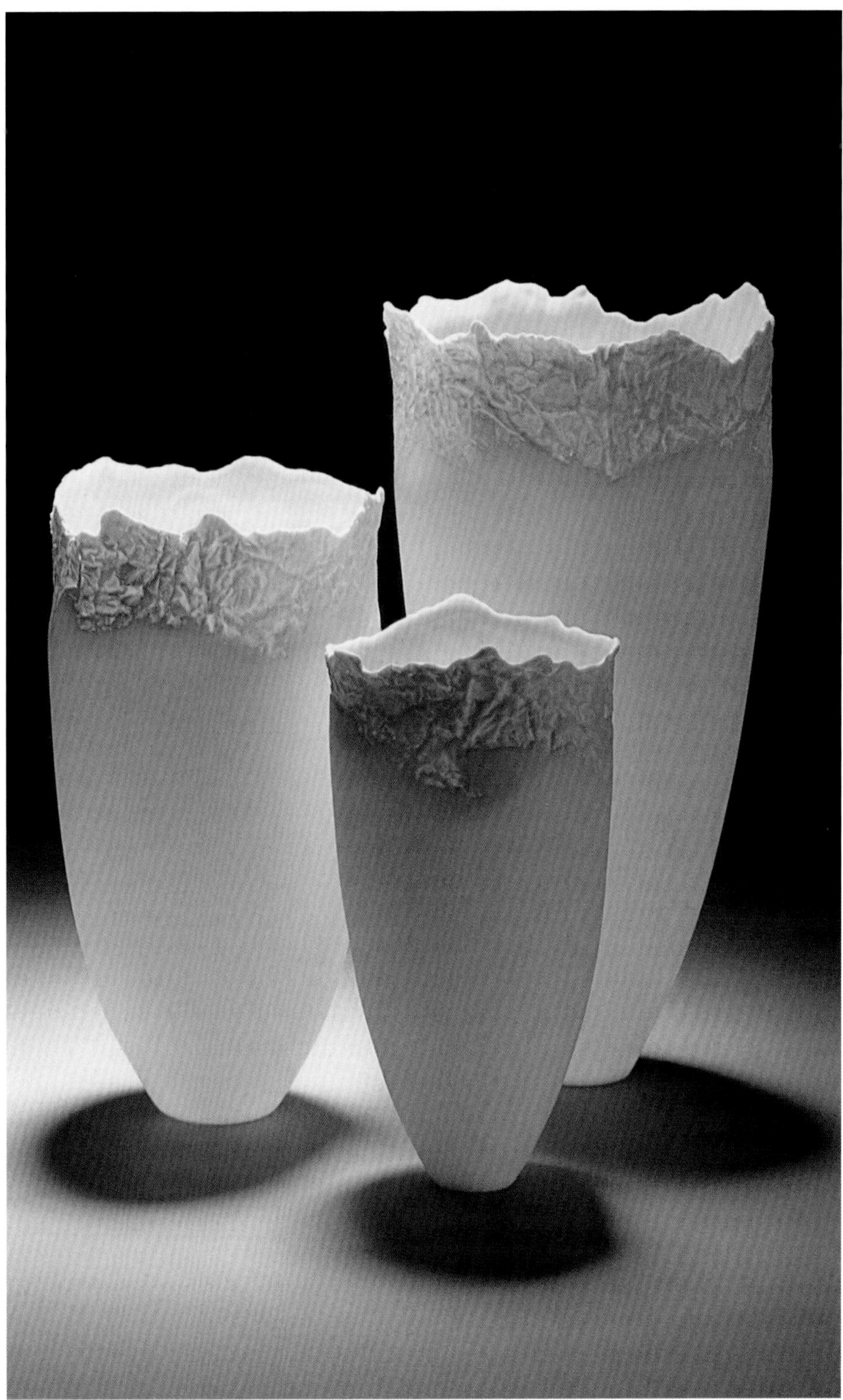

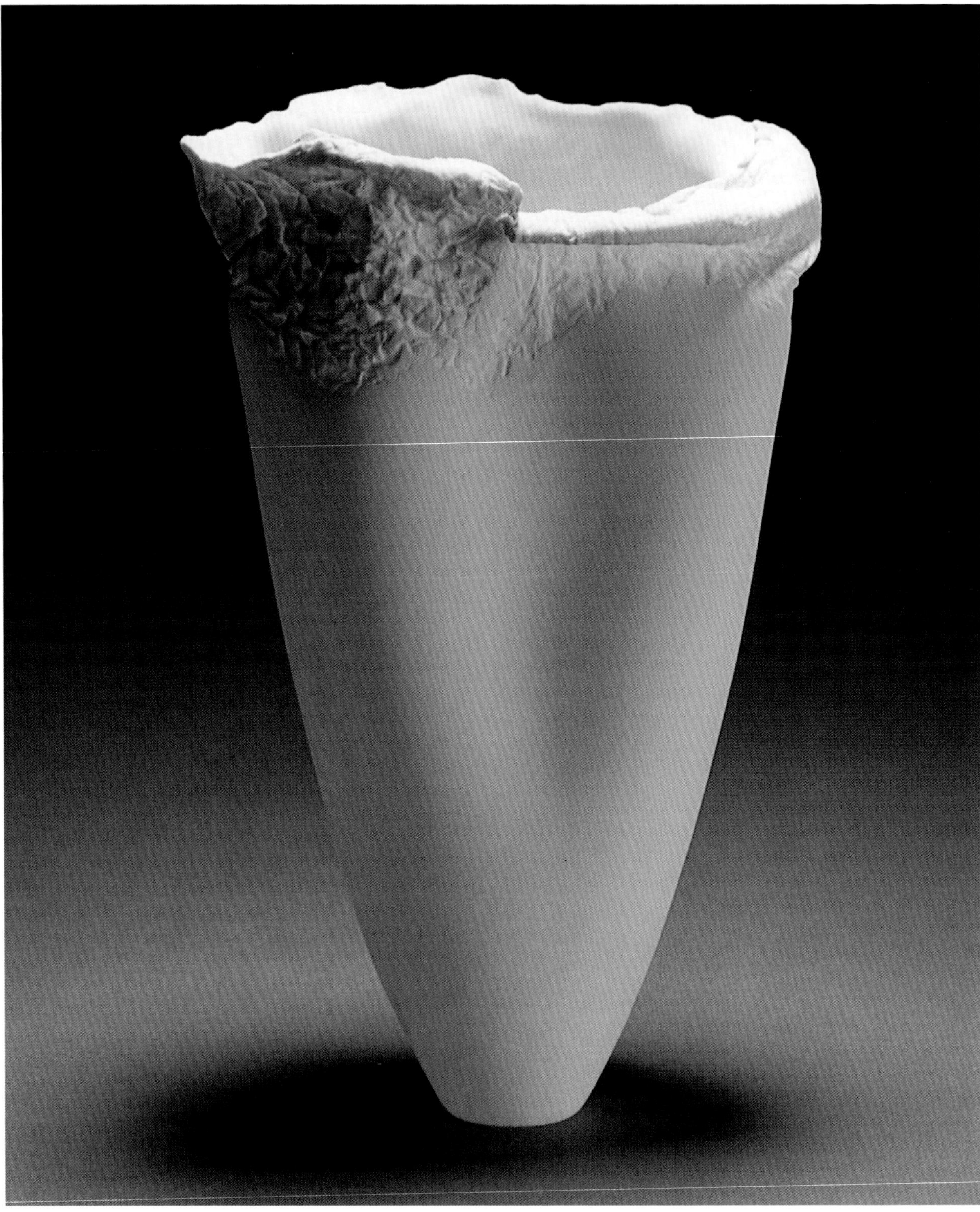

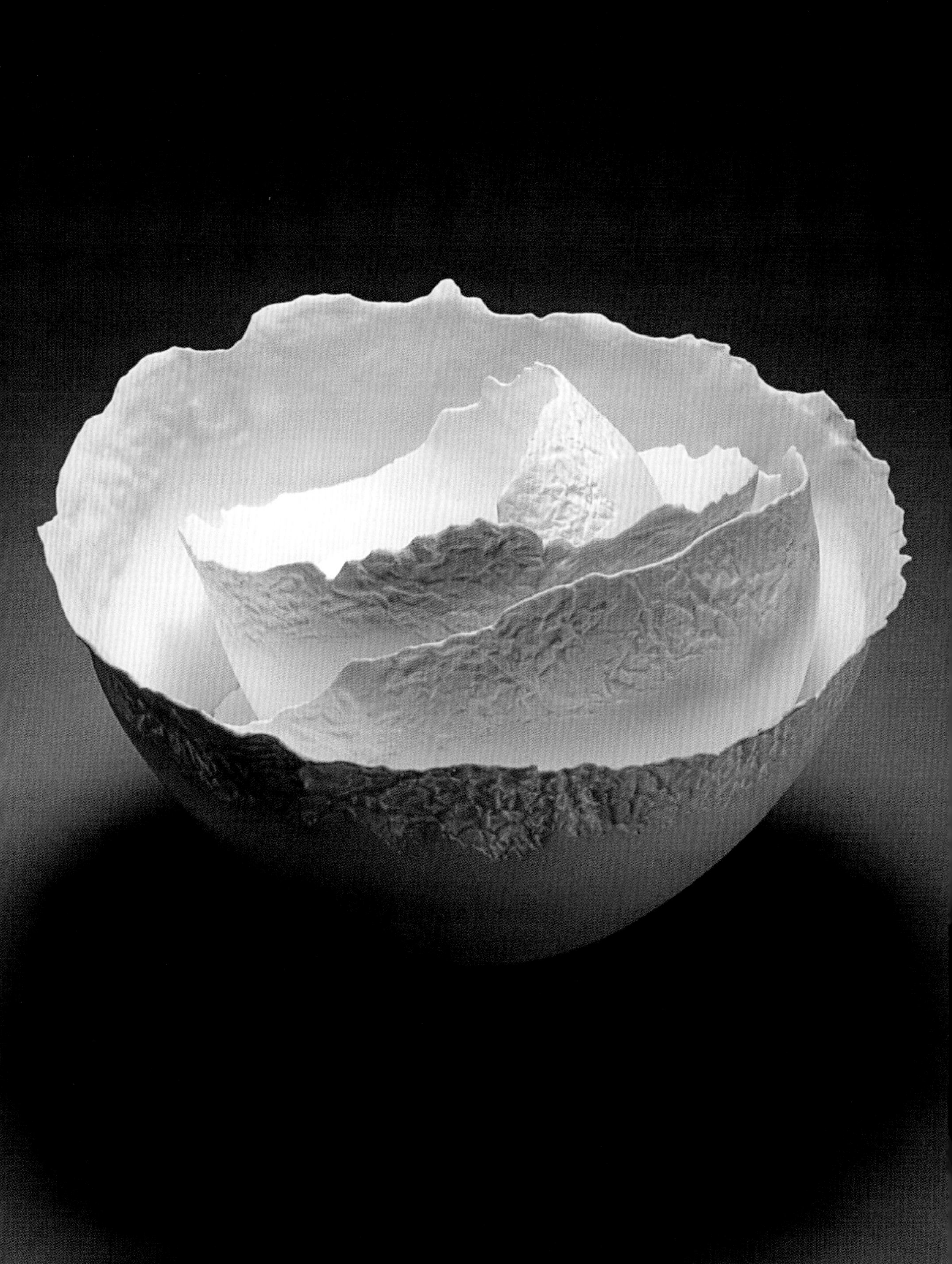

Light

How pure is the sky. How bright is the sun.
Christoph Gluck, *Orpheus and Eurydice*

These were my thoughts as I stepped off the plane in Western Australia for the first time. I had never experienced such clarity of light before. Little did I know then what a major influence Australia's light was to have on my ceramics. The clean Antarctic light of Hobart has been a source of inspiration to painters, potters, photographers, and writers alike, and during my studies there, I also fell under its spell. I wanted to show this clean beauty through my work in bone china.

BELOW: *Sky Bowl*, Angela Mellor, 2016. Bone china with soluble colourants. *Photo: Stephen Bond.*

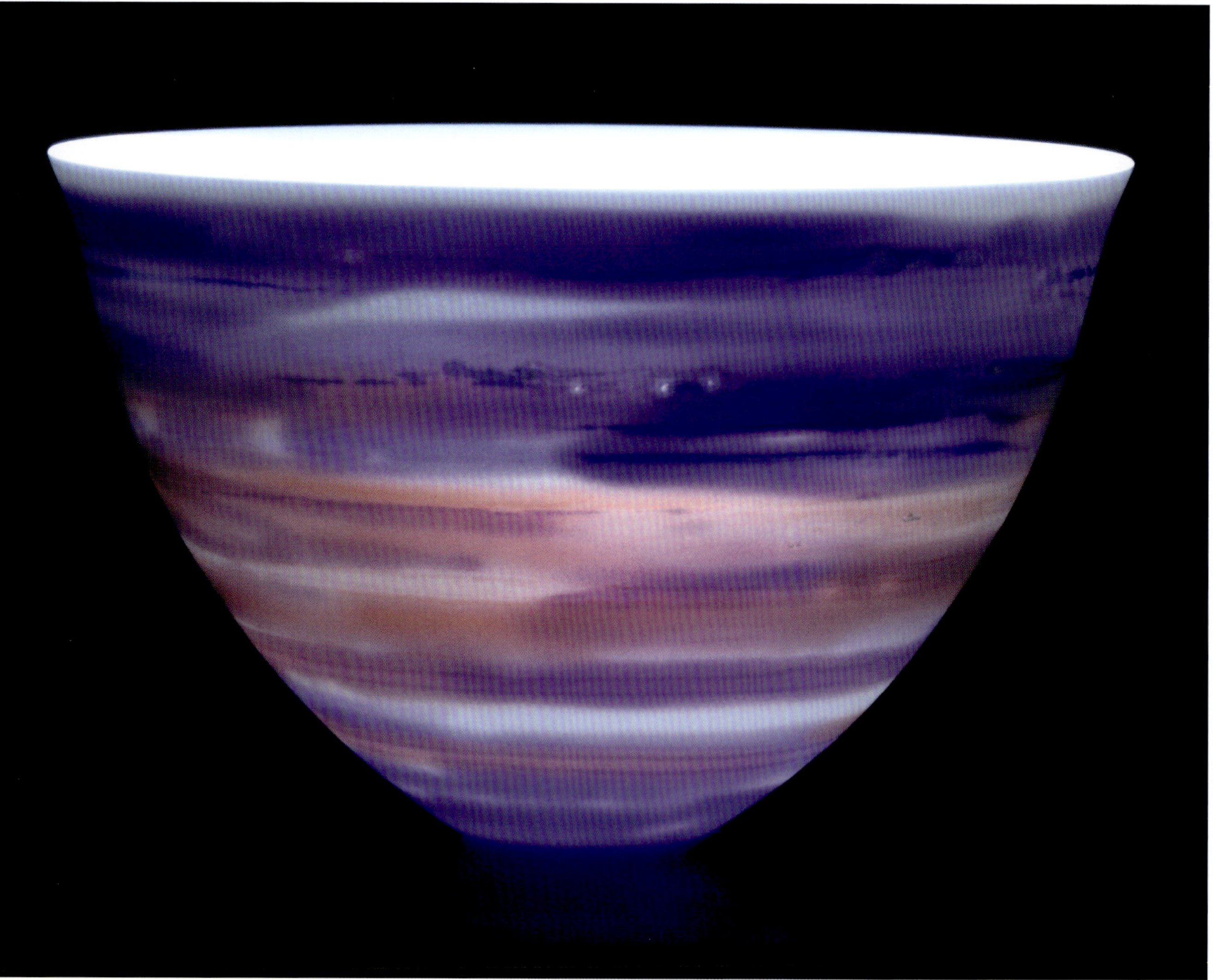

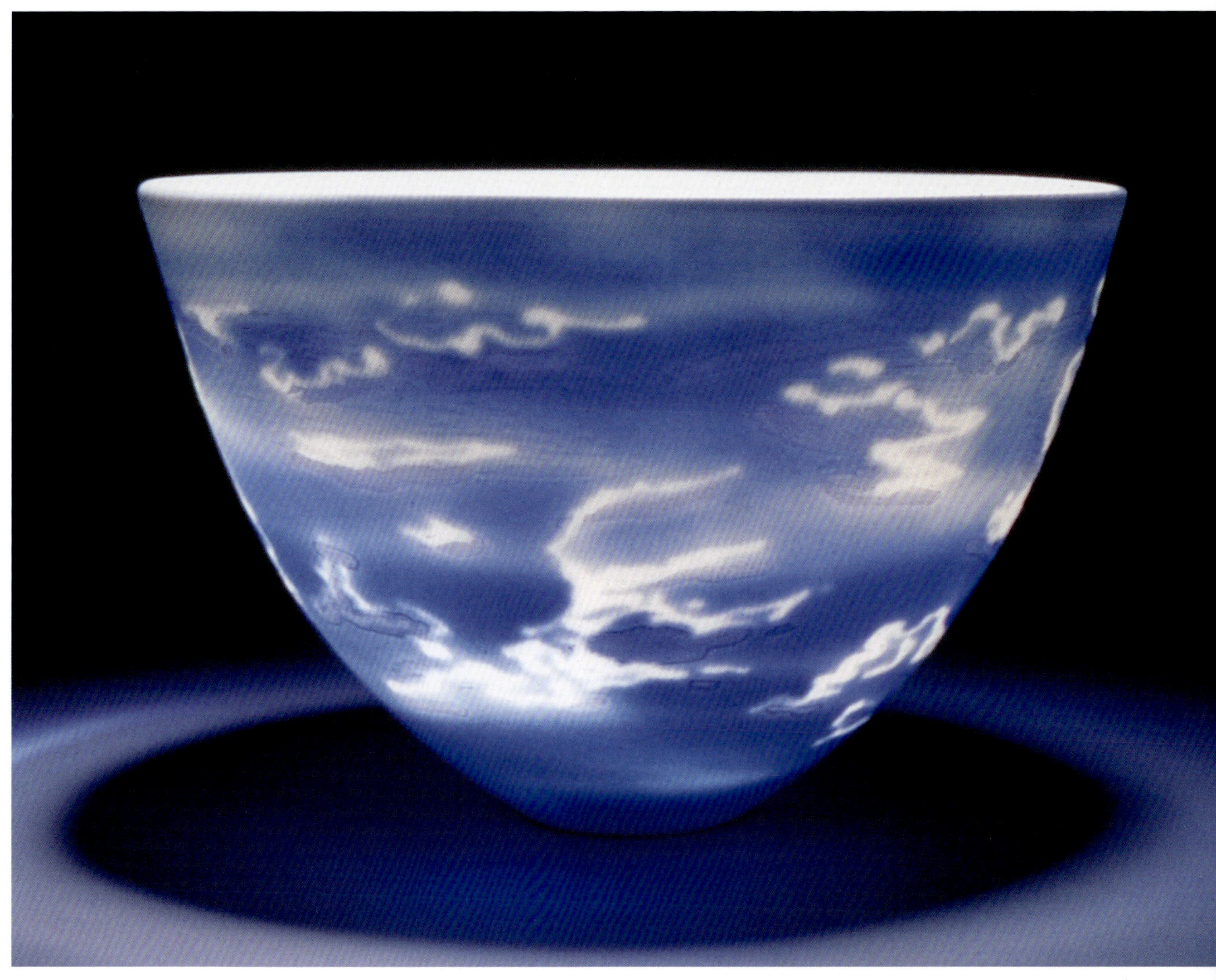

ABOVE: *Sky Bowl*, Angela Mellor, 2001. Bone china with soluble colourants and acrylic resist. *Photo: Victor France.*

When I studied at Monash University, the sky was a constant source of inspiration. My studio, situated on the sixth floor, was the window to the most beautiful, ever-changing skies. The radiant colours and cloud formations became the decoration for a series entitled *Sky Bowls*.

Equally, the effect of light on water has always fascinated me. Australia's coast is breathtaking. The clear, crystal-blue sea revealed a whole new palette to me. Painting with water-soluble colourants enabled me to capture some of these colours in my *Sea Bowls* (see p.57 and 69).

The effect of light on water also drew me to the idea of juxtaposing glass with bone china. I liked the idea of a bone china object floating on the waves.

Colour and texture

Initially, colour played a secondary role in my work, my preference being black and white. During my research, I became fascinated with the watercolour effects obtained by soluble colourants, which, if applied thinly, do not noticeably affect the translucency of bone china. In fact, when lit from above, the coloured pieces become more luminous and ephemeral.

Studying the colour of natural objects such as shells played an important part in my experiments with soluble colourants. The textures of natural objects have all helped to add interesting elements to my work throughout the years.

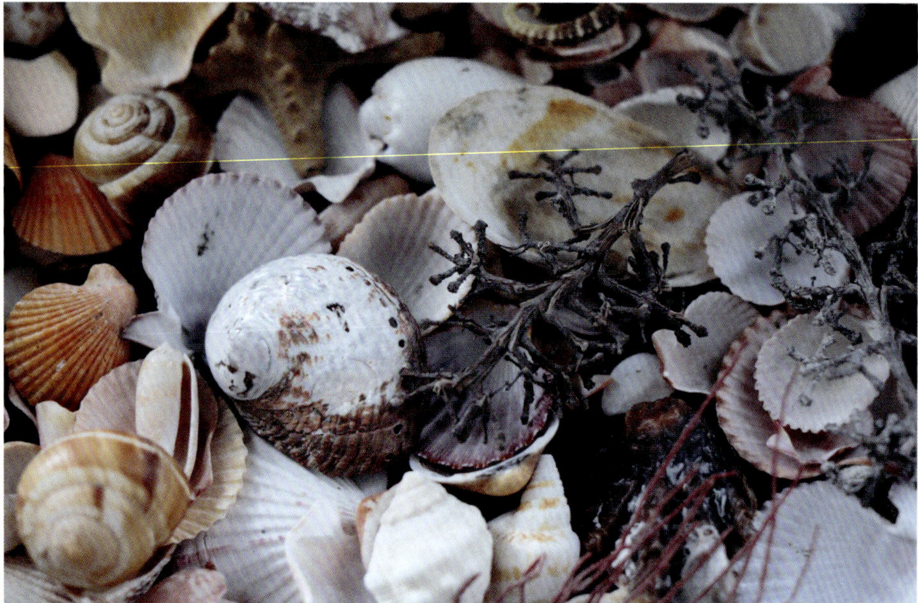

LEFT: Inspirational source: the colour and texture of shells. *Photo: Layton Thompson for* Ceramic Review.

Form

The inspiration that directs my train of thought has developed slowly over the years. Ideas grow and develop with life's changing situations, but my design considerations have been with me for many years.

I have predominantly used the basic forms of the cone and the sphere. I use conical forms for bowls, jugs, and flutes, and spherical forms for cups and bowls.

Alongside the individual vessels and sculptural works produced during my investigations into translucency and organic forms, I was inspired to produce a range of functional ware. I was using a small cup mould as the basis for all my decorating and colour tests, and an idea came to me when they were all set out in my studio that these cups would be suitable for entry into the 5th International Ceramics Competition Mino'98 in Japan. I realised I needed a container for them and collaborated with a student at the University of Tasmania, who made the box.

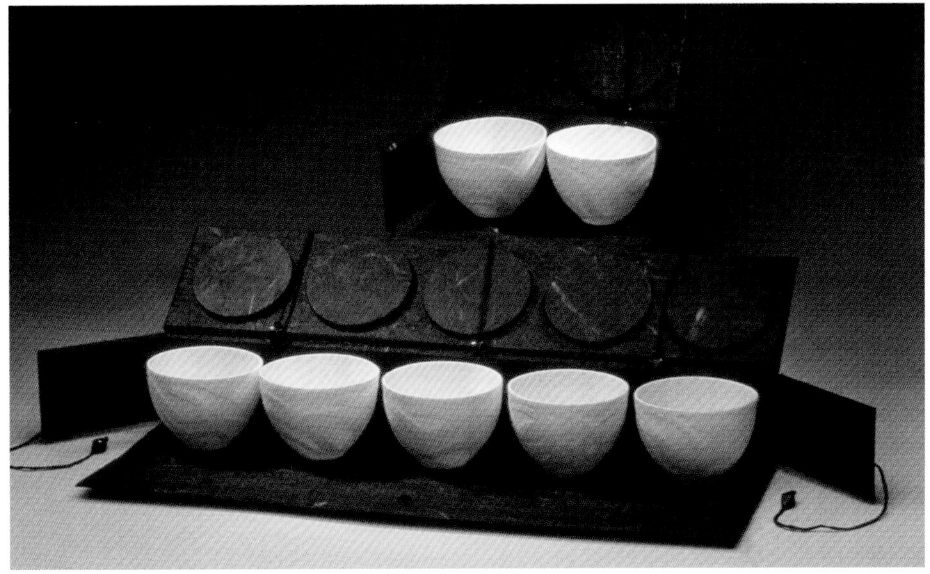

RIGHT: *Sake Sets*, Angela Mellor, 1997. Set of two and set of five bone china cups in black *momagami* box. Etched decoration. 6 x 38 cm (h x w). *Photo: Uffe Schultze.*

I am very proud that my *Sake Cups in Black Momagami Box* received an Honourable Mention at the festival.

This led to the development of a *Sake Set*, comprising a jug and five cups on a black porcelain tray, which received an Honourable Mention in the 1st World Ceramic Biennale 2001 Korea International Competition.

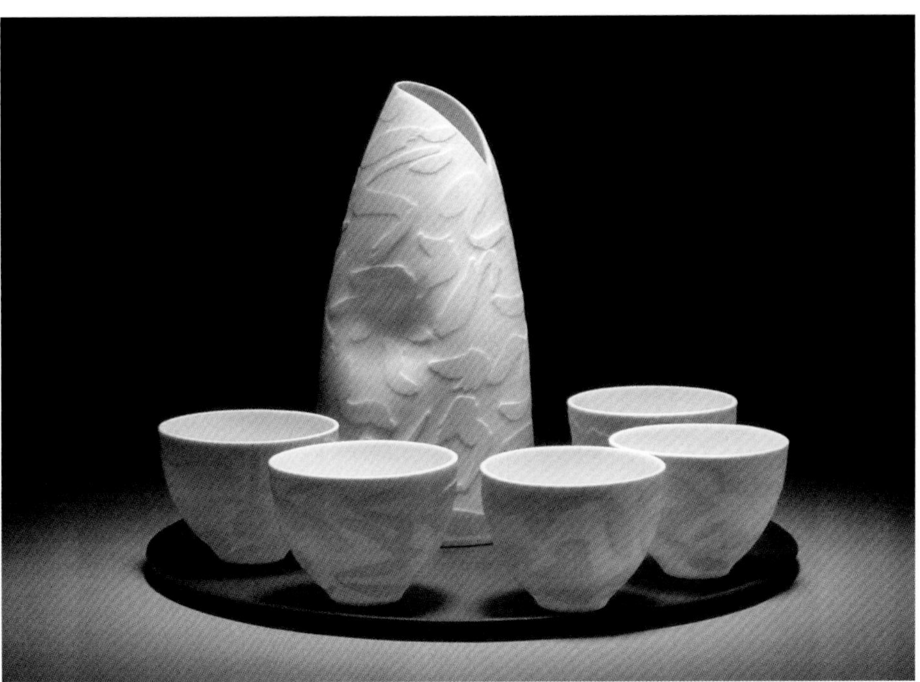

RIGHT: *Sake Set*, Angela Mellor, 2000. Jug and five bone china cups on black porcelain tray. (jug) 23 cm (h); (cup) 7 cm (h); (tray) 25 cm (w). *Photo: Victor France.*

In turn, I expanded on the *Sake Set* and produced my *Japanache* tableware. This comprised a set of six table settings, which were displayed at Planet Furniture in Sydney.

The slipcast bone china paperclay plates were worked so that they would ripple and bend the light to mimic wave patterns in sand. These landscape rhythms inform the shape of each piece. The textured surface is reminiscent of circles left by raindrops on the ocean, their stillness embodying the Japanese concept of *Ma* (the space and silence between things). The same texture was incorporated into the bowls as a motif.

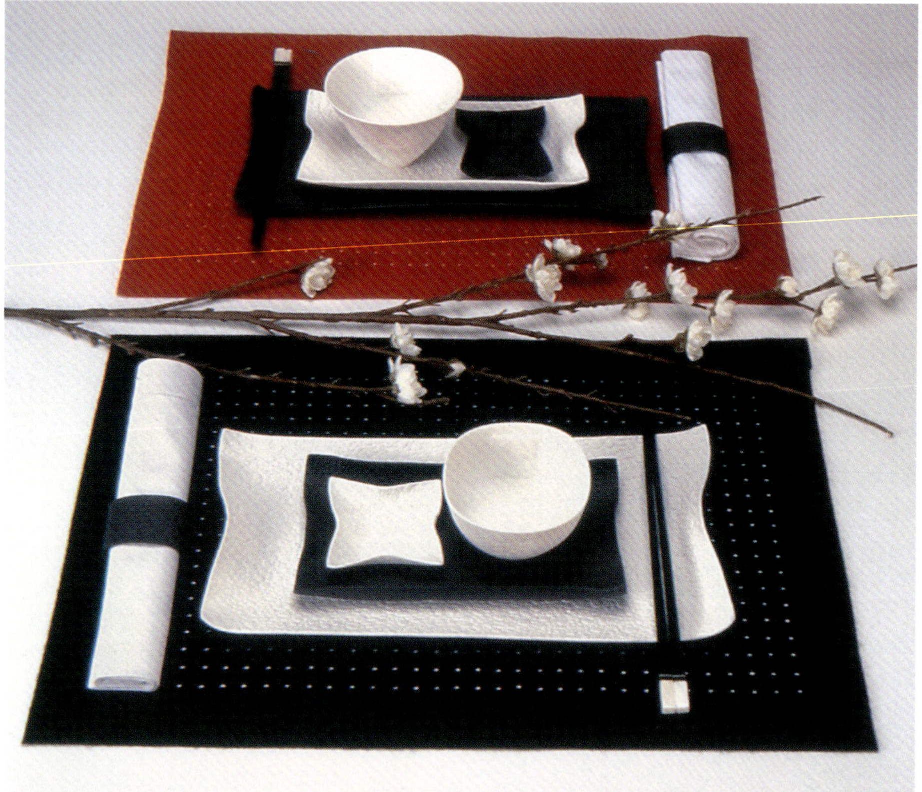

LEFT: *Japanache* table setting, Angela Mellor, 2003. Bone china. *Photo: Victor France.*

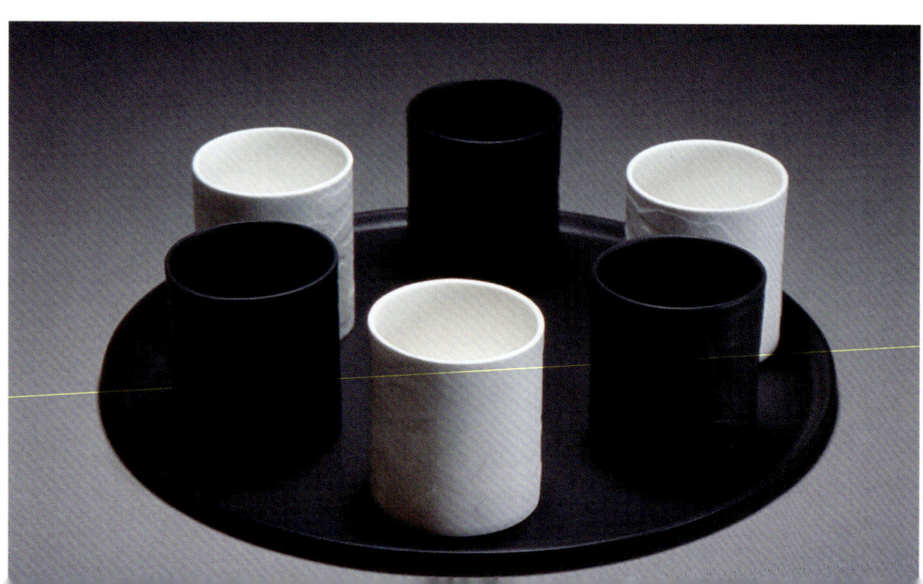

BOTTOM LEFT: *Six Cups with a Tray*, Angela Mellor, 2010. Bone china. *Photo: Victor France.*

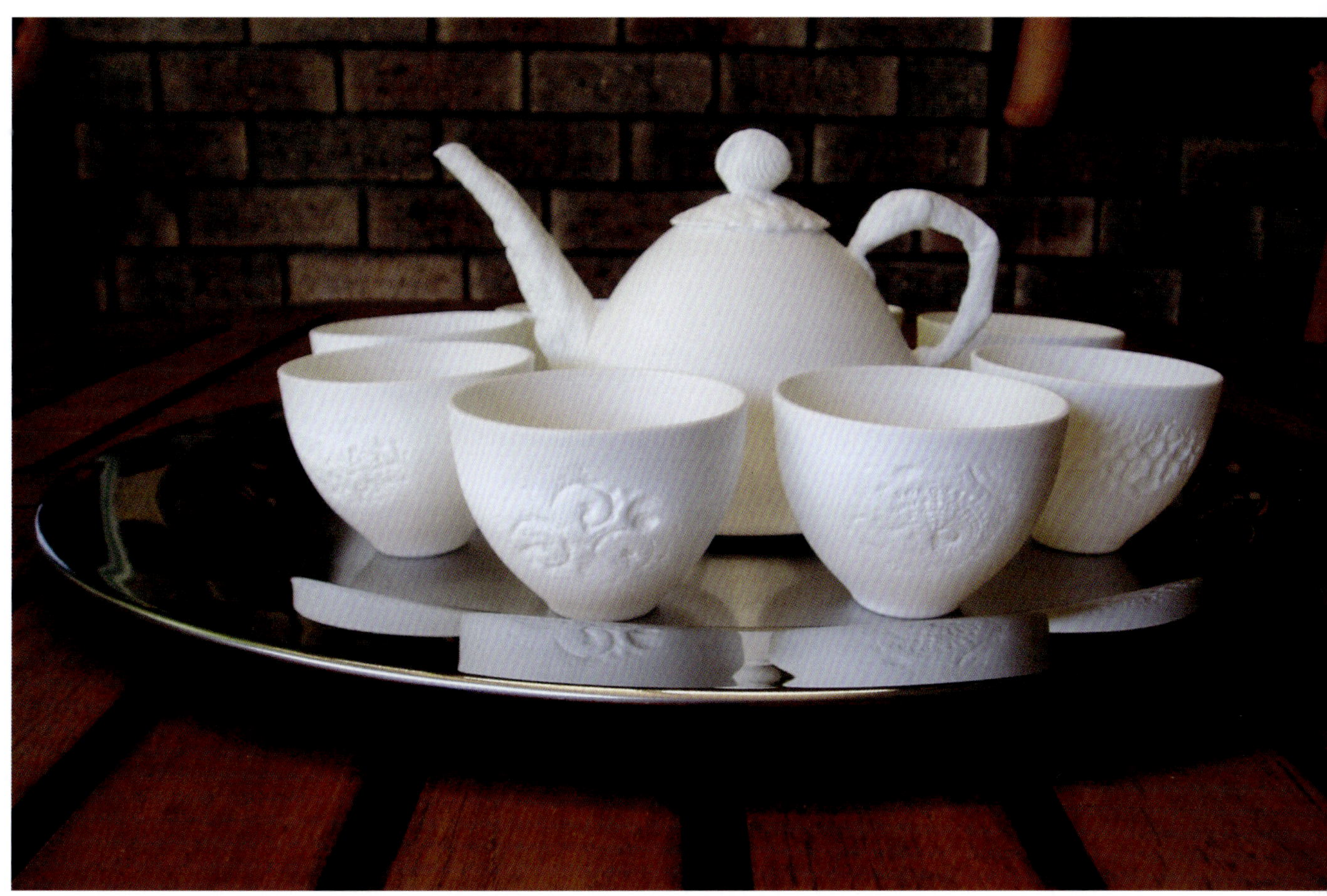

ABOVE: *Tea Set*, Angela Mellor, 2011. Porcelain tea pot, eight cups.

Six cups with a tray were acquired by the Musée National de Céramique de Sèvres in 2010.

I developed this further with a slightly larger teabowl, which was made for the *Teabowl* exhibition at the Oxford Ceramics Gallery in 2014.

I then developed a range of objects that are not only functional but are *objets d'art* in their own right. I enjoyed a more spontaneous approach to decorating, as can be seen in the *Symphony in Black and White* series, where black slip dances around pure white 'rocking' bowls, drawing the eye inside and out, over and under these rhythmic forms.

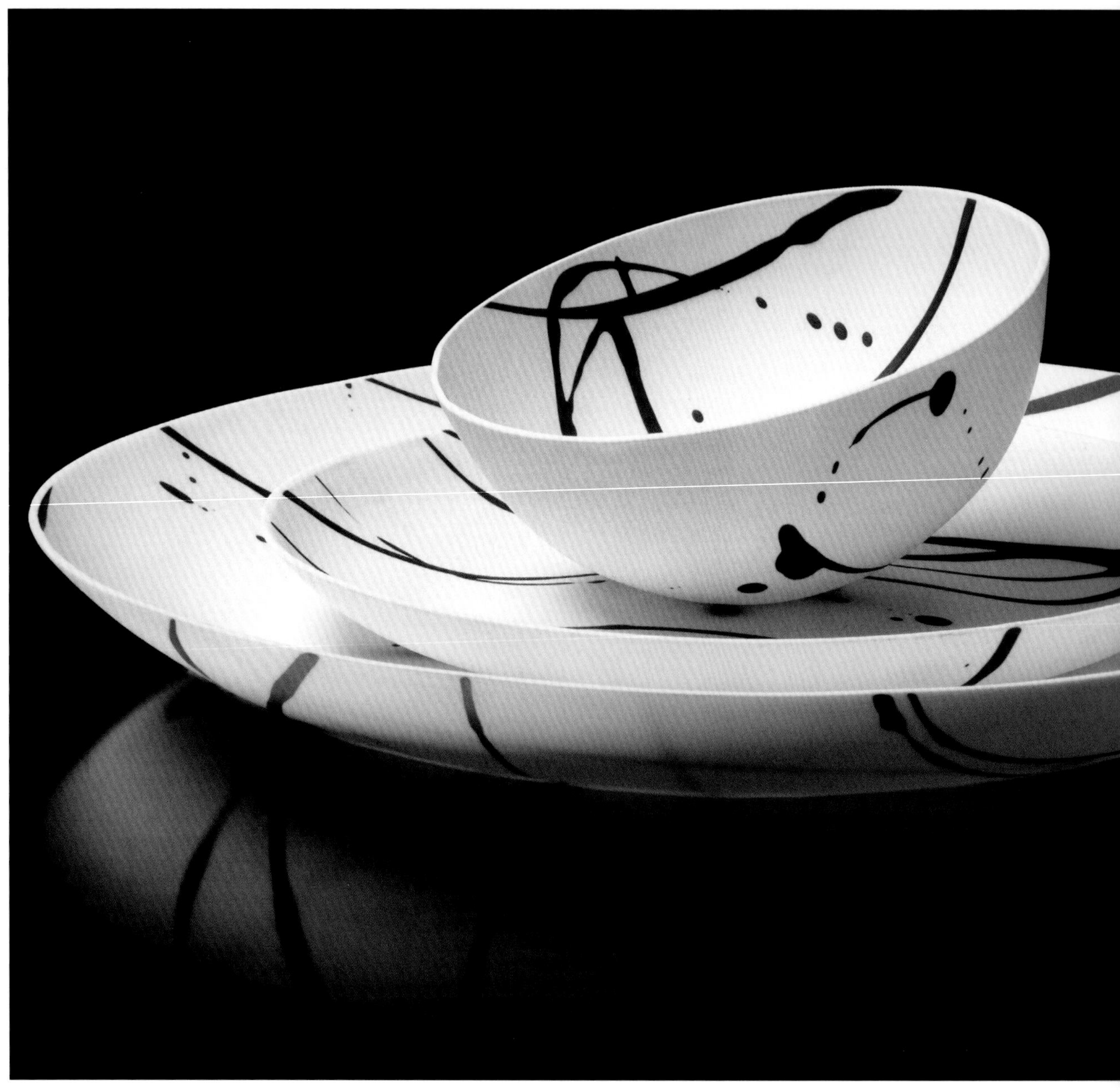

ABOVE: *Symphony in Black and White* objects, Angela Mellor, 2016. Bone china with black slip trailed into the mould. (each piece) 37.5 cm; 21.5cm; 7 x 13.5 cm (w x h). *Photo: Stephen Bond.*

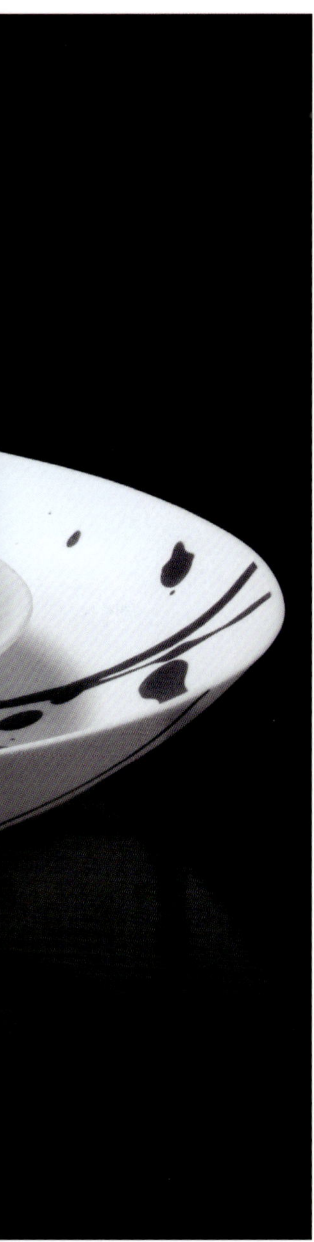

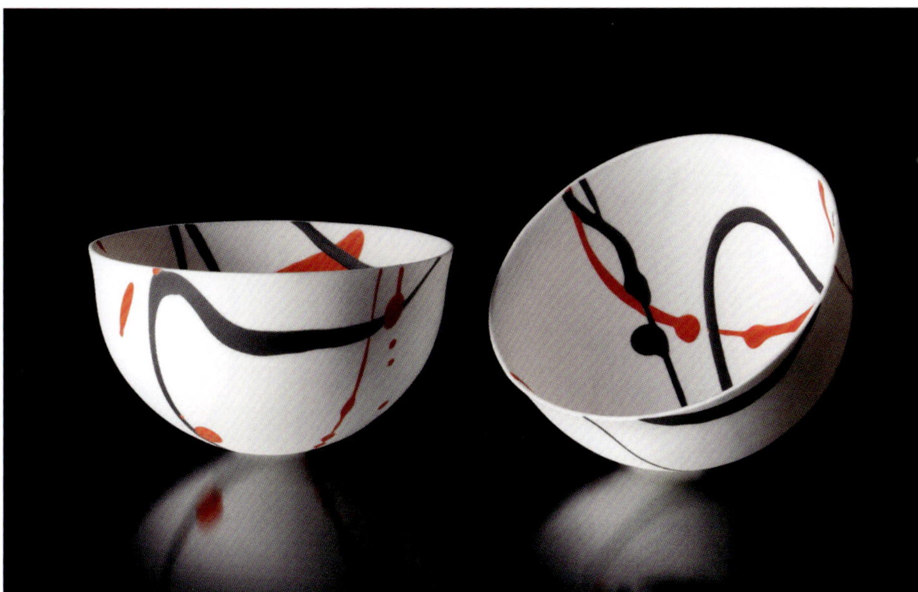

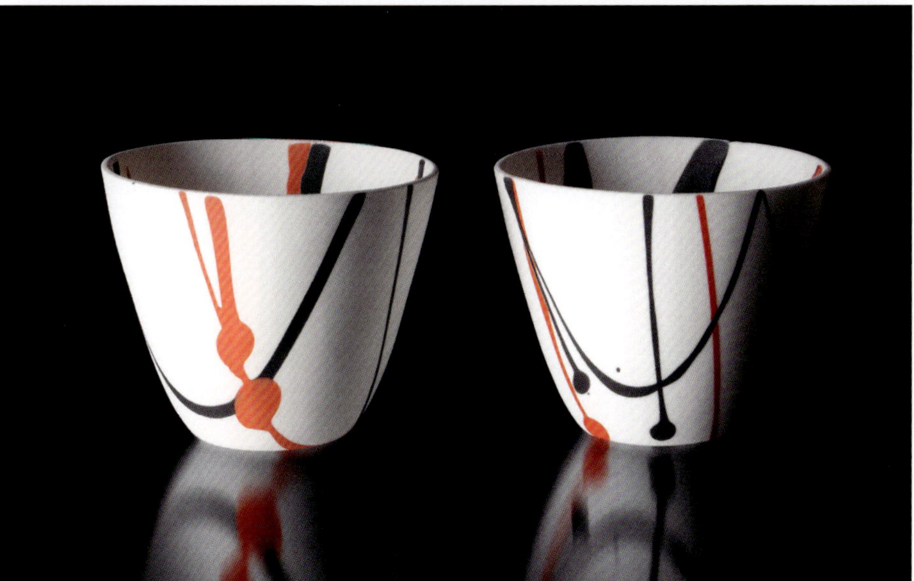

TOP: *Symphony in Black and Red* rocking bowls, Angela Mellor, 2019. Bone china with red slip trailed into the mould. 6 x 10 cm (h x w). *Photo: Stephen Bond.*

BOTTOM: *Symphony in Black and Red* cups, Angela Mellor, 2019. Bone china with red slip trailed into the mould. 6 x 5.5 cm (h x w). *Photo: Stephen Bond.*

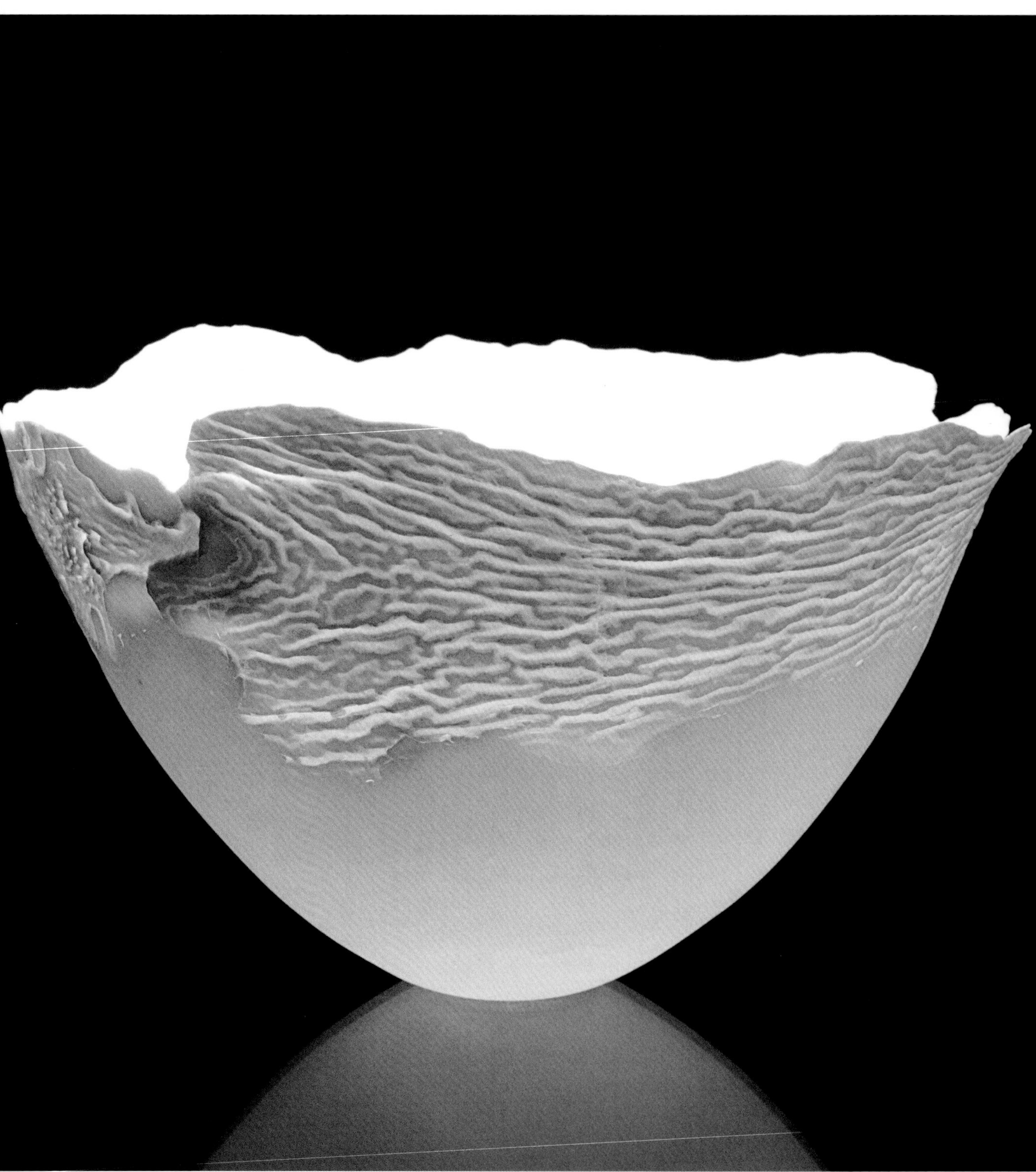

6

Lighting

LEFT: *Ocean Light* bowl, Angela Mellor, 2003. Bone china with paperclay inlay. 12 x 18 cm (h x w). *Photo: Victor France.*

Although a wide range of contemporary artists use porcelain as a means of conveying light through vessel forms, very few have used it in association with lighting. My aim in using electrical lighting has been to explore the way in which light can be transmitted through slipcast bone china, using a range of decorating techniques to optimise translucency.

The bright sunlight of Western Australia and its dramatic effect on the landscape made a great impression on me and greatly influenced the development of my work. The harsh quality of light in WA tends to separate and isolate objects, and this high contrast is what distinguishes the landscape from the British countryside, where soft light connects the elements in the landscape.

On returning to Perth after graduating in 2000, I set up my studio. I was awarded a grant from the World Crafts Council of Australia, which enabled me to collaborate with lighting designer Urs Roth, an innovative product designer for independent company Ropa Lighting. Together, we designed lights for an exhibition sponsored by Mondo Luce Lighting Solutions in Perth.

Lynda Dorrington, the executive director of Craftwest (now FORM), first suggested that I collaborate with industry to develop a range of custom-lit works that could fully realise the potential of my explorations. She saw that the partnership between Urs and me had enormous potential for the application of design in contemporary craft, and she was interested in encouraging artists to extend their practice into commercial production. Lynda introduced me to Gerry de Wind, Managing Director of Mondo Luce, who had previously designed and installed the lighting system for Craftwest's gallery. Andrew Nicholls, Artistic Programmes Director at Craftwest, was a great support throughout this project, which was developed over 18 months through an extensive study of how artificial light sources enhanced the translucency of my bone china forms. The project culminated in the *Ocean Light* exhibition at the Craftwest gallery. In a review of the exhibition for *Object* magazine, Andrew said:

> *Angela Mellor is one of Western Australia's foremost ceramicists. She is internationally acclaimed for her almost impossibly delicate minimal white vessels, vases and sake sets, and as one of the principal contemporary ceramicists exploring the incorporation of paperclay as a sculptural element with fine bone china. Since undertaking her Honours degree in 1997, the primary focus of Mellor's practice has been the exploration of light and organic form as the basis of an aesthetic in fine bone china.*

She draws her inspiration from the natural world: the sublime qualities of sea and sky, the forms and textures of subterranean environments (coral, kelp and weathered driftwood, as well as limestone cave formations) and most significantly, the play of natural light upon the landscape. Mellor attributes this fascination to her amazement at the quality of the Australian sunlight after first emigrating from the United Kingdom in 1994.

Mellor's trademark is the incorporation of paperclay inserts, often cast from natural elements—rock faces, driftwood or the shoreline—to vary the opacity of her pieces. She also employs a sponging technique to alter the thickness, and subsequent translucency, of her works. Hence, viewed in natural daylight, her vessels appear to be extremely beautiful minimal forms; but when seen under focused, electrical light they come to life, literally glowing, with intense areas of intricate detail.[1]

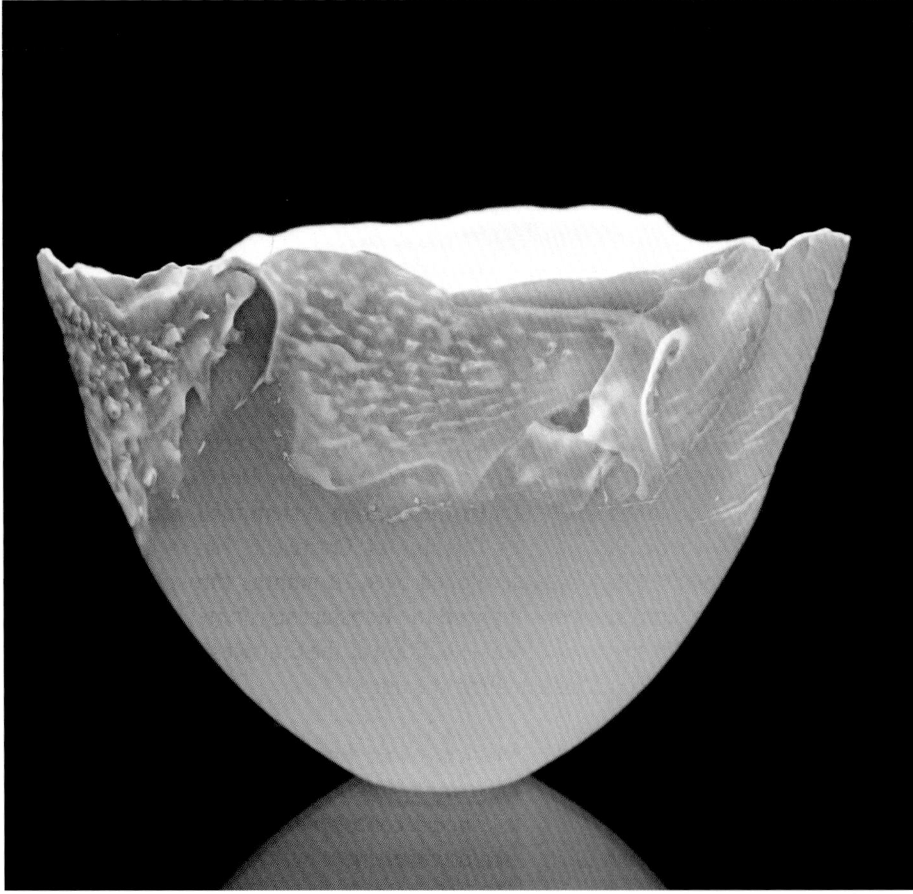

LEFT: *Ocean Light* bowl, Angela Mellor, 2003. Bone china with paperclay inlay. 9 x 13 cm (h x w). *Photo: Victor France.*

[1] Andrew Nicholls, 'Translucent Light. Ceramics and lighting combine in a collaboration between Angela Mellor and Mondoluce.', *Object*, 42, 2003, pp.36-7.

A spin-off from this came later in 2005, when I was commissioned to make 12 trophies for the 2005 State Arts Sponsorship Scheme Awards as part of the *Ocean Light* project. I made smaller bone china *Ocean Light* bowls, and had each one mounted on an aluminium base. These trophies were presented to successful businesses in Western Australia.

Penny Smith, my tutor in Tasmania, opened the exhibition and wrote an article for the catalogue entitled *Tripping the Light Fantastic: When Art and Science Converge*, in which she said:

> *In presenting* Ocean Light, *a major exhibition of Mellor's latest ceramic works, Craftwest illustrates the vital role it plays in serving its local constituency by promoting work of an international calibre. The show is also important for the collaboration it has instigated between individuals and business, which reflect a research and development culture that has been commercially sponsored rather than academically driven.*
>
> *Mellor is well known for her highly evocative work in translucent bone china that reflects a strong sense of place through her love of nature. However, it was Dorrington who recognized that one way to forward Mellor's explorations into the aesthetic and functional qualities of translucency even further, was to encourage her to collaborate with those expert in the science of mastering illumination.*
>
> *Priding itself on the promotion of unique lighting solutions for both domestic and commercial environments – from functional and aesthetic perspective – Mondo Luce responded positively to the challenge of working with Mellor on the* Ocean Light *series. Believing that light is greater than the absence of darkness and is thus an art form in its own right, de Wind was quick to see the potential of Mellor's new work and generously sponsored Urs Roth of the independent lighting company, Ropa Lighting to collaborate with Mellor.*
>
> *Roth had an established reputation in WA for finesse and originality of his lighting products and overall sense of design. The light fittings used by Roth (known as* luminaires*) are discreet and unobtrusive tools which direct and distribute light using the luminaire as an artistic decorative object in its own right. The tiny light-emitting diodes (LEDs) that Roth implanted in many of the pieces revealed the full gamut of subtle textural detail and range of glacial tones.*[2]

I had long held the idea of a large installation of spotted cones on a polished black granite base that, mirror-like, reflects the installation in reverse, giving a dramatic double image. The inspiration for this idea came from a host of translucent spotted yellow tentacles of coral, called *Dendrophyllia*, which can be seen underwater at night when illuminated by light near Perth.

[2] Penny Smith, 'Tripping the Light Fantastic: When Art and Science Converge', *Craft Arts International*, 60, 2004, pp.2-4.

I mounted multiple bone china cones of various sizes onto a polished black granite slab. A tiny LED was inset into the slab under each cone to bring a light source into the heart of each piece, creating a dramatic translucent sculpture. The technique of washing back areas of clay around the spots created differing thicknesses that affected the density of the light and tone, giving an illusion of fragility and fluidity.

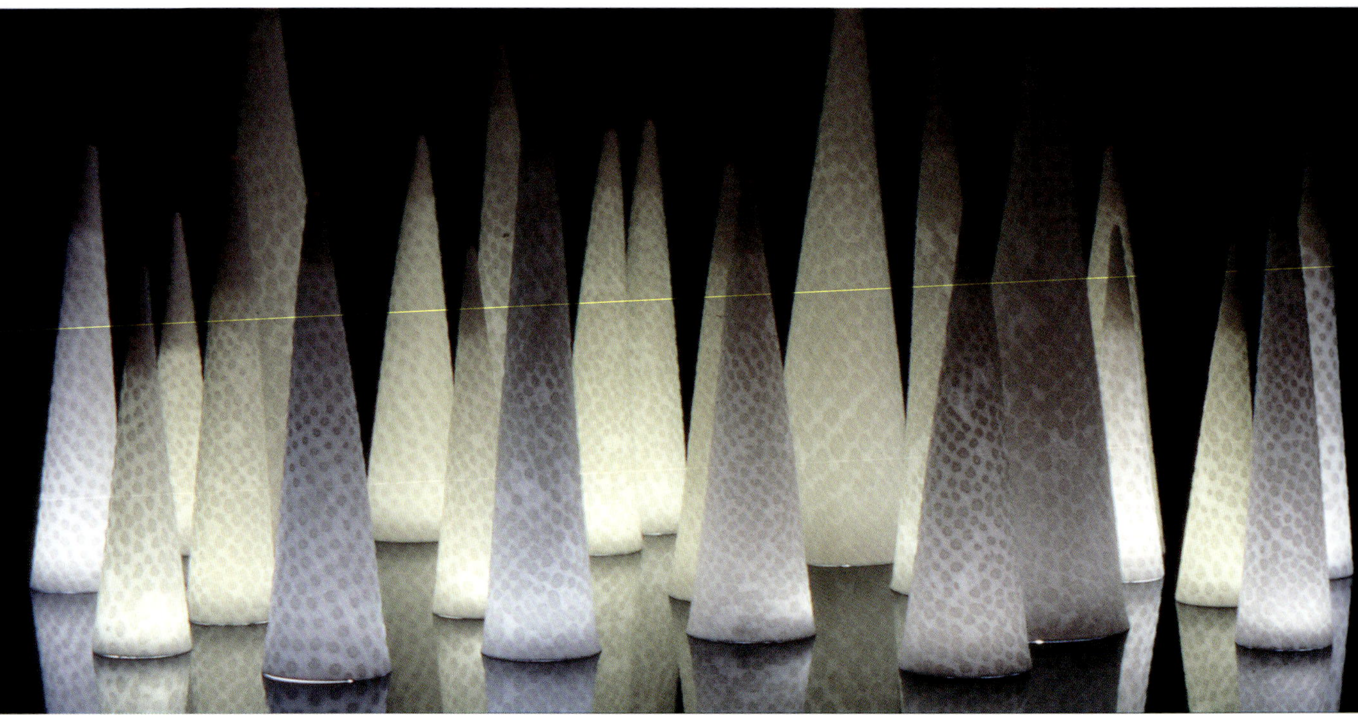

I also designed a large table light and matching wall lights that each used a single cone. Holes had to be incorporated into the unfired cone before firing to accommodate the metal fittings. The shrinkage of the holes had to be calculated to make sure that the wires or metal rods fitted properly.

> *By bringing a light source to the very heart of each ceramic element Roth and Mellor have created related groups that appear to vibrate gently with a sensuously intimate life force that unifies each grouping rather than separating their independent parts. This is particularly evident in the* Dendrophyllia *installation where a series of multiple cones of varying sizes allude to spatial depths and tones to generate a sense of gentle theatre – generating the glow, rather than trumpeting the glare.*

ABOVE: *Dendrophyllia* installation, Angela Mellor, 2003. Etched bone china, granite base, LED lights. 29 x 80 x 60 cm. *Photo: Victor France.*

OPPOSITE: *Dendrophyllia* trapeze light, Angela Mellor, 2003. Bone china, medal rods, halogen bulb. *Photo: Victor France.*

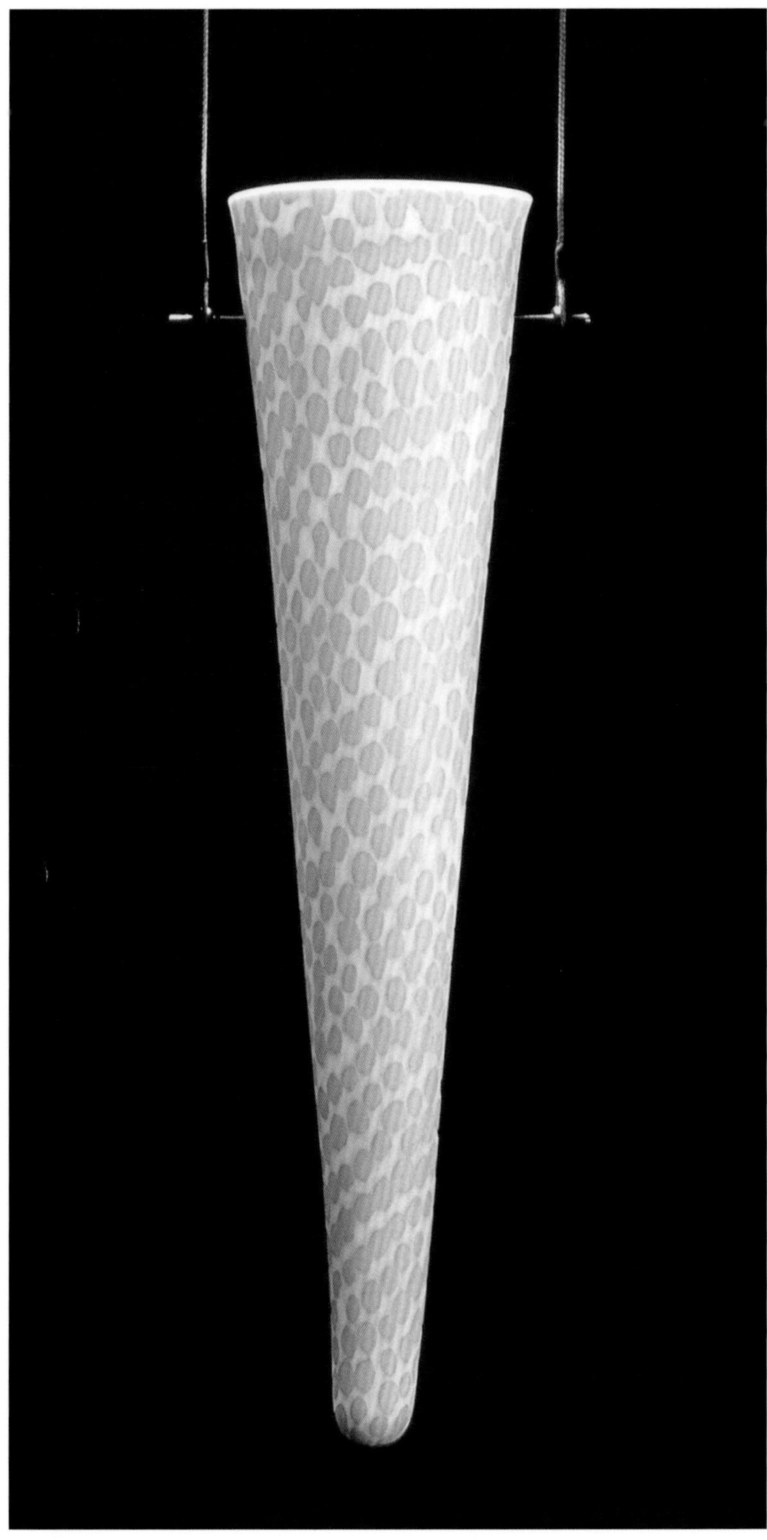

Some of the most satisfying aspects of this work are when the independent elements of both artist and designer – of art and science – converge as one.

Such examples are the floor, wall and ceiling lights, where the transition from the hard industrially turned aluminium fittings and metal rods appear to blend smoothly into the apparent fragility of each bone china bowl. By elevating these ceramic elements on their metal vertical stands – or as they undulate from the wall or cascade from the ceiling in repetitive groups – Mellor's work gains new heights of scale and grace that would have been impossible to achieve otherwise. Her technique of washing away areas of clay to produce the organic stripes and dots of different thicknesses that change the light's density and tone in each of the bowls, adds further emphasis to the paradoxical allusion to fragility and the suggestion of mobility.[3]

[3] *Ibid.*

In the *Coral Cluster* table light, five translucent coral cups are grouped together at different heights on top of slender silver-grey metal stems. I designed metal nipples to connect the cup to the rod. I also produced a *Coral Cluster* table light with three cups.

The design was based on coral, and again used acrylic resist and water etching to create the various degrees of translucency. These *Coral Cluster* table lights were a development of my *Sake Cups*, using the same form but a different surface design.

The crown of the *Coral* series and centrepiece of the *Ocean Light* exhibition was the elegant *Coral Bud Chandelier*, an exciting collaboration between Urs and me. The light fitting is a linear cascade of thin metal rods that fall in a cluster from the ceiling, emulating a splash, and gracefully separating into arcs reaching outward and upward in perfect symmetry. Each rod is crowned with a pure white etched bone china cone.

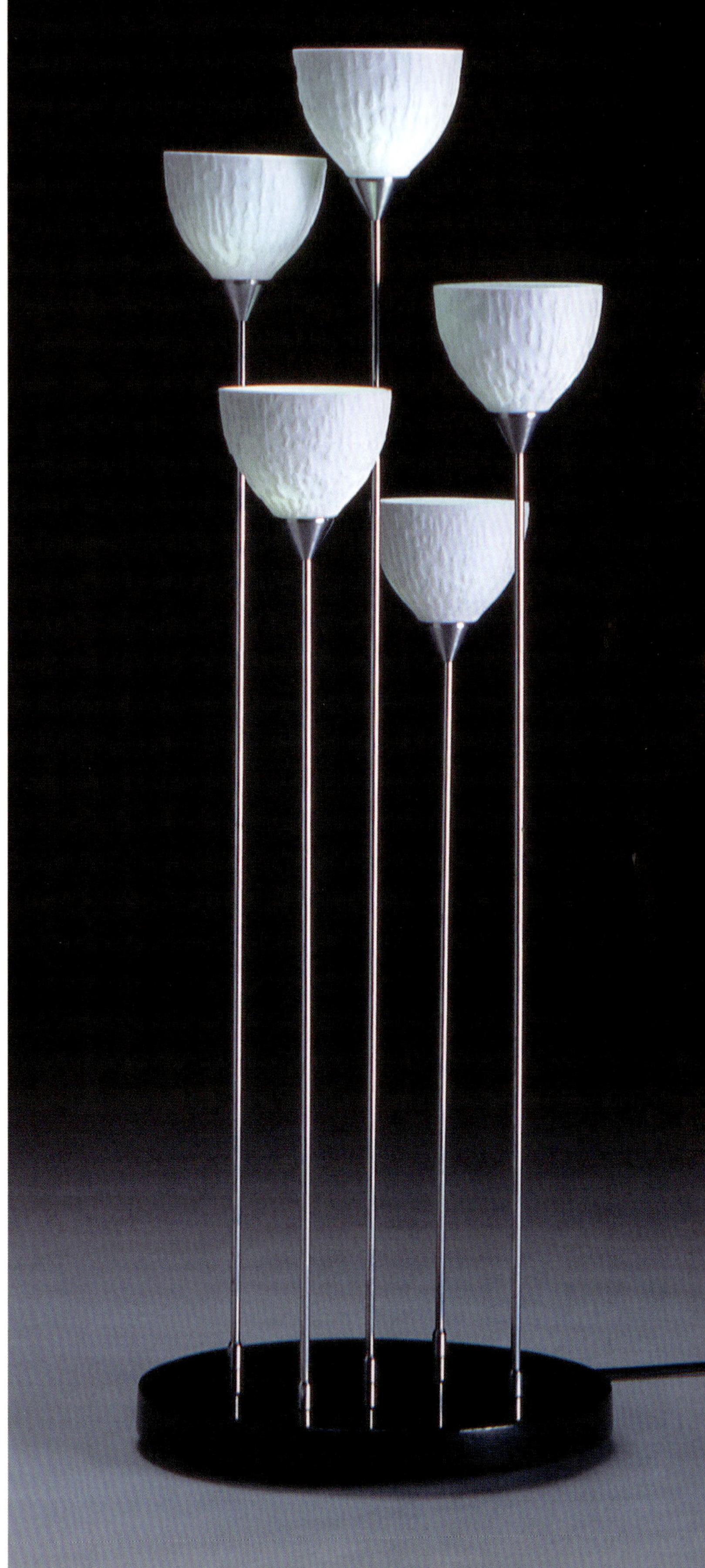

OPPOSITE: *Coral Cluster*, Angela Mellor, 2003. Etched bone china, steel, LED lights. 53 x 20 cm (h x w). *Photo: Victor France.*

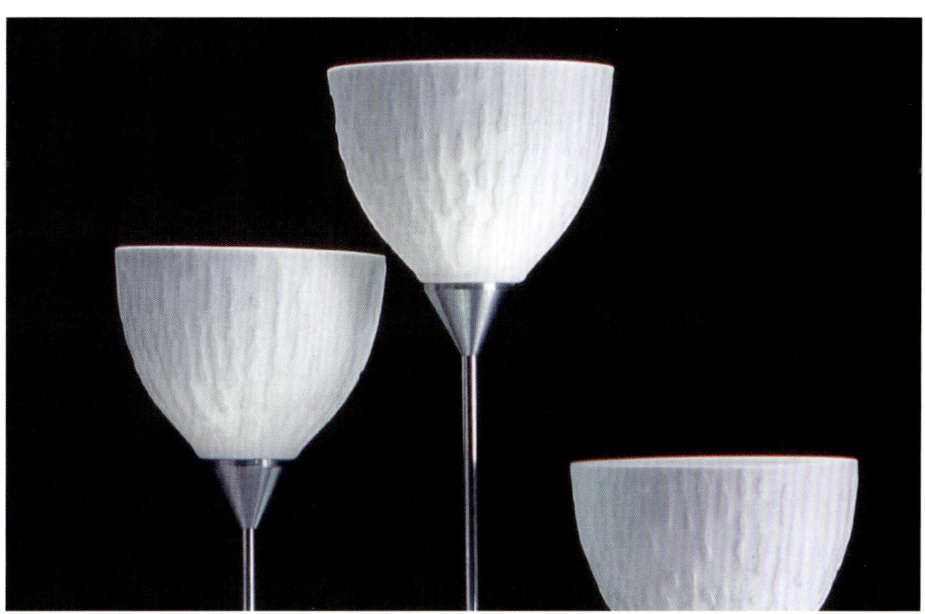

RIGHT: *Coral Cluster* (detail), Angela Mellor, 2003. Etched bone china, steel, LED lights. 53 x 20 cm (h x w). *Photo: Victor France.*

BELOW: *Coral Cluster* (interior view), Angela Mellor, 2015. Etched bone china, steel, LED lights.

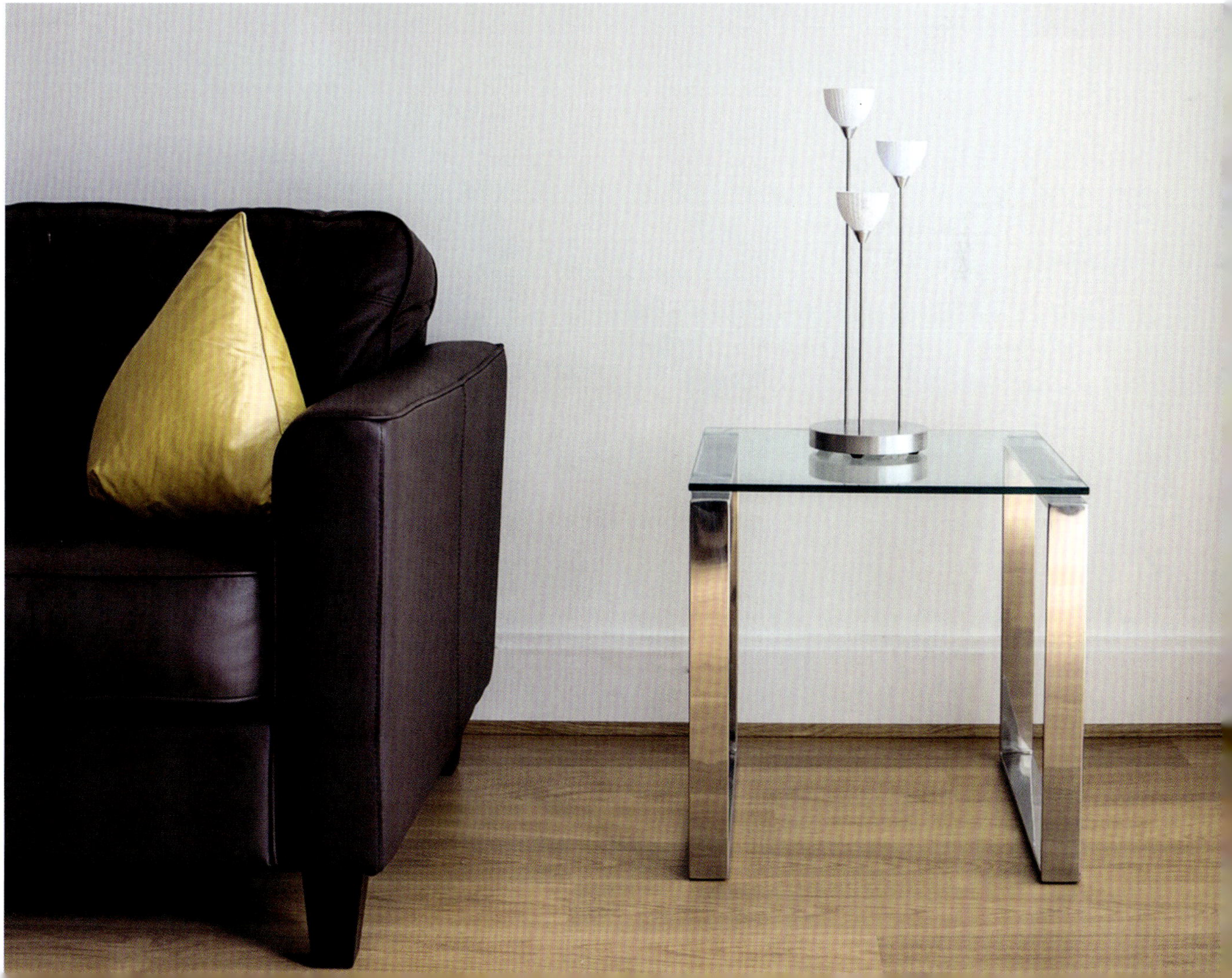

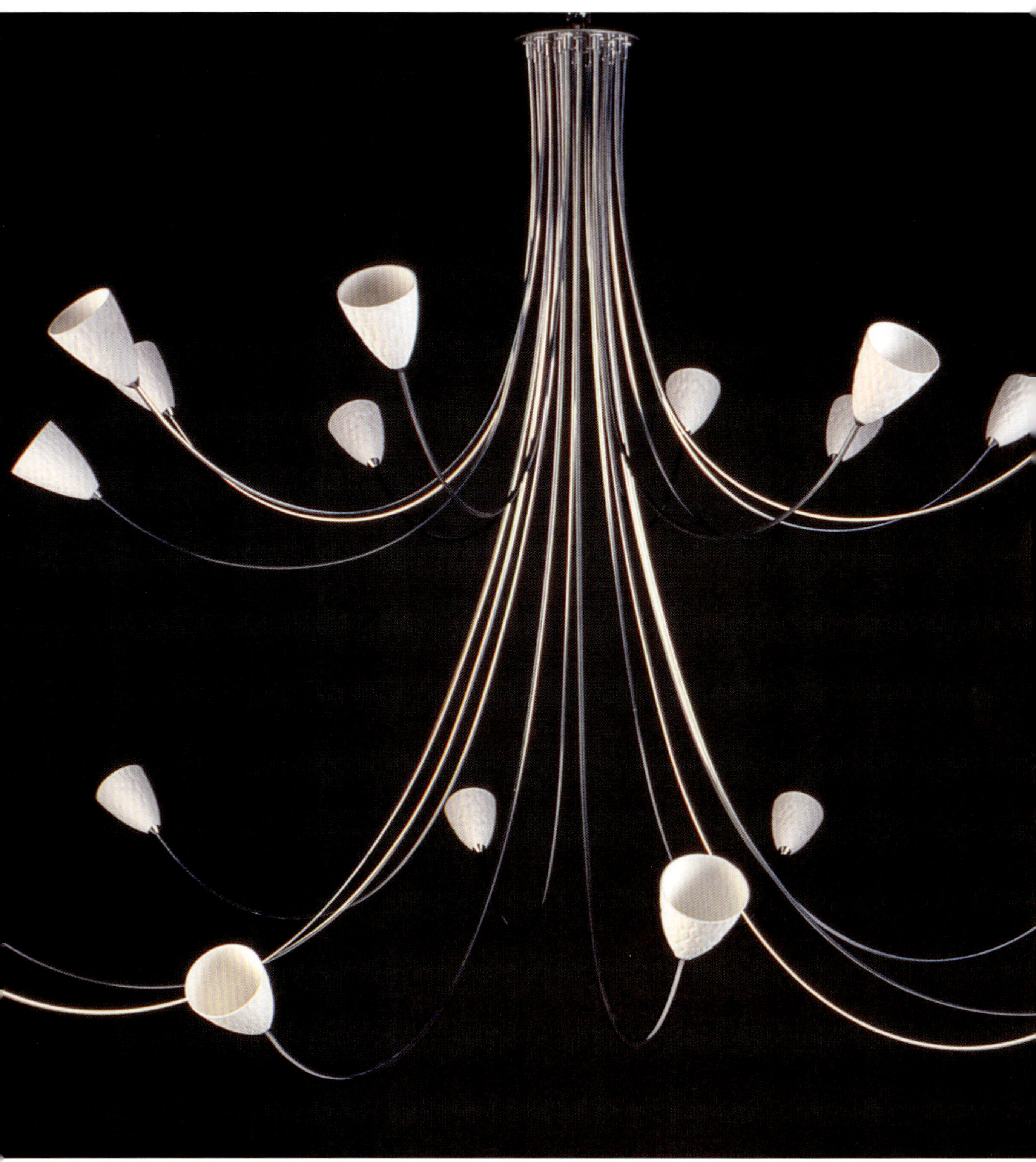

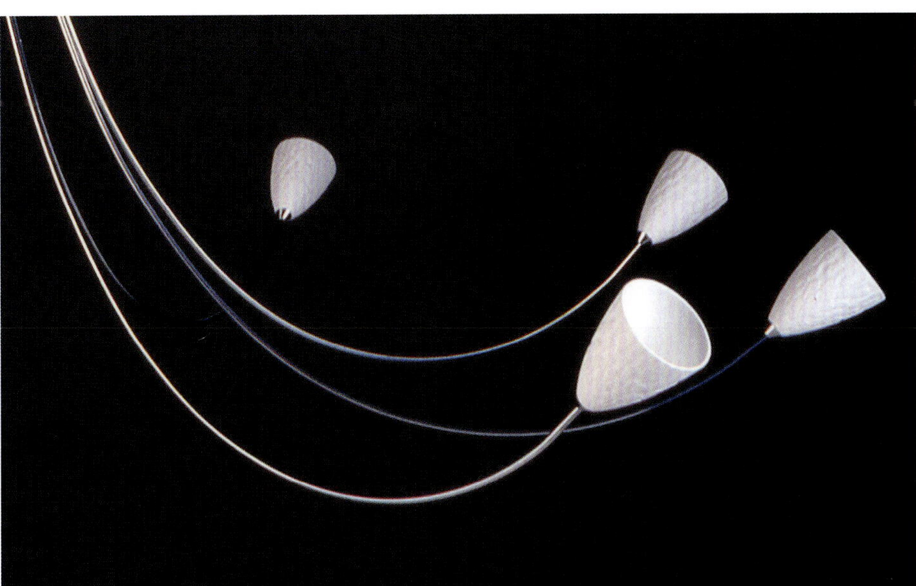

My work requires light to bring it to life. Prior to collaborating with Urs, to fully show the translucent qualities of the medium, I lit my work from above using a narrow beam of downlight. This lighting technique is still used when photographing my work.

ABOVE: *Coral Bud Chandelier* (detail), Angela Mellor, 2003. Etched bone china, steel, LED lights. 100 x 120 cm (h x w). *Photo: Victor France.*

LEFT: *Coral Bud Chandelier,* Angela Mellor, 2003. Etched bone china, steel, LED lights. 100 x 120 cm (h x w). *Photo: Victor France.*

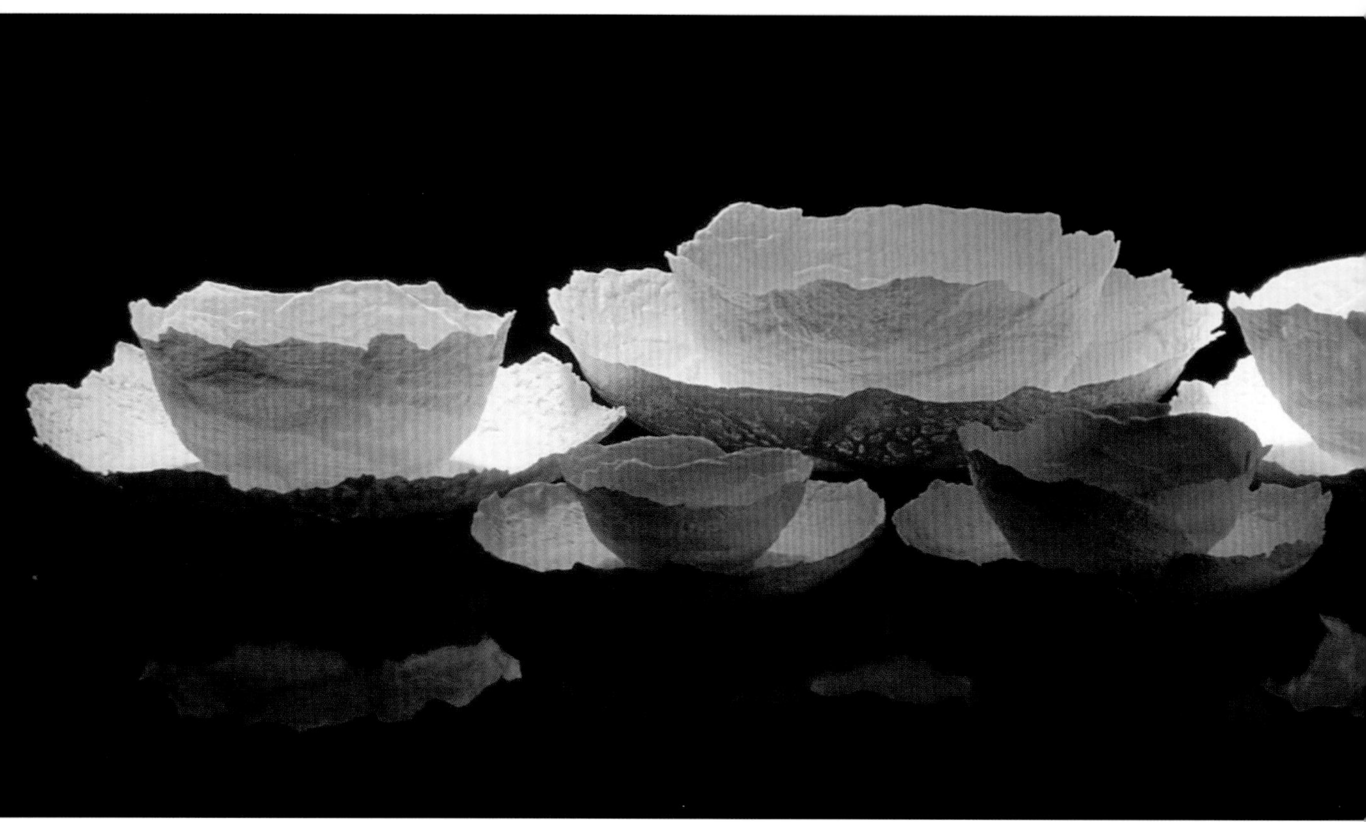

ABOVE: *Arctic Floats*, Angela Mellor, 2003. Bone china paperclay, mirror. *Photo: Victor France.*

The *Arctic Light* series was made by hand-building and manipulating torn strips of textured paperclay into a bowl and platters, which were arranged to form a group. These relied on overhead lighting and were displayed on glass for a mirror-like effect.

Using the same hand-building technique, I created another light work by manipulating the paperclay into a folded sculptural form rather like a cowrie shell, which was illuminated from within.

In the *Arctic Fold* table light, the textured bone china paperclay shell clings to the centre of a steel rod that echoes the gentle curve of the shell as it reaches upward from a polished granite base.

The *Cretaceous Light* series was inspired by a visit to Penguin Island in Western Australia, where rugged rock formations are encrusted with fossils and shells. Seeing it gave me the urge to replicate it in paperclay.

I collected fossils, corals, and shells, which I impressed into clay to create a mould, deliberately trying to emulate a seabed rich with marine life. When illuminated, the resulting texture came alive, and I decided that a cylindrical form would enhance this most effectively. I developed a cylindrical table light, wall lights, and pendant lights.

In daylight, these pieces have the subtle texture of shells and fossils, but when lit by an enclosed electric light, the fossils appear to come alive. The light within is transmitted through thin walls, illuminating these subtle images.

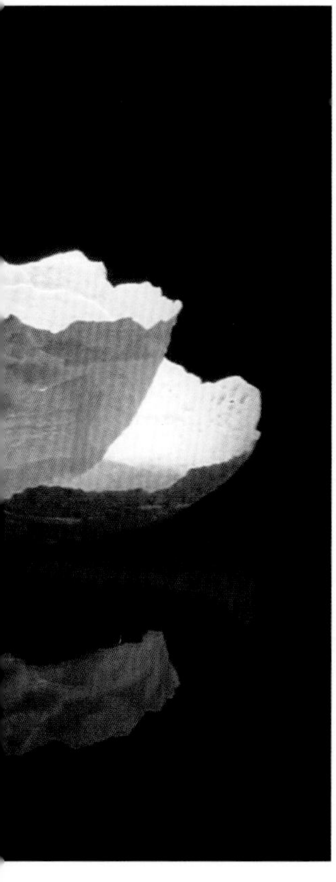

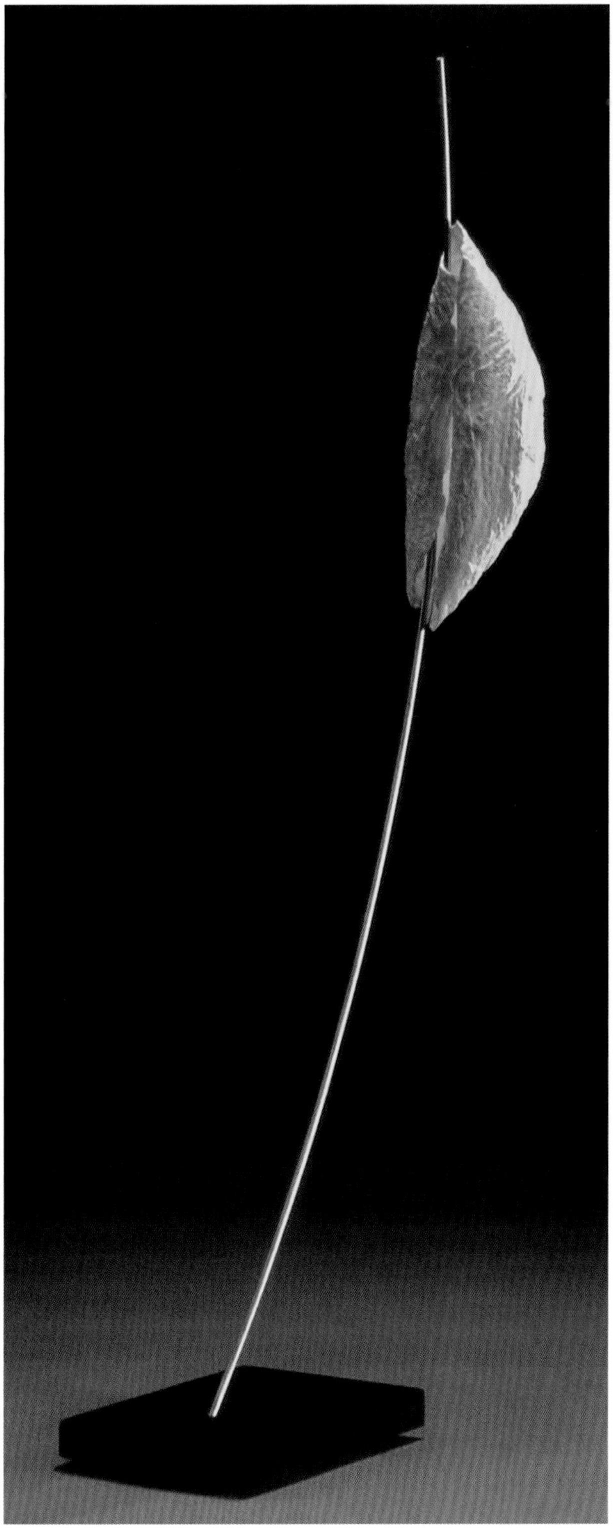

LEFT: *Arctic Fold* table light, 2003. Bone china paperclay, granite base, steel, LED light. 63 x 18 x 10 cm (h x w x d). *Photo: Victor France.*

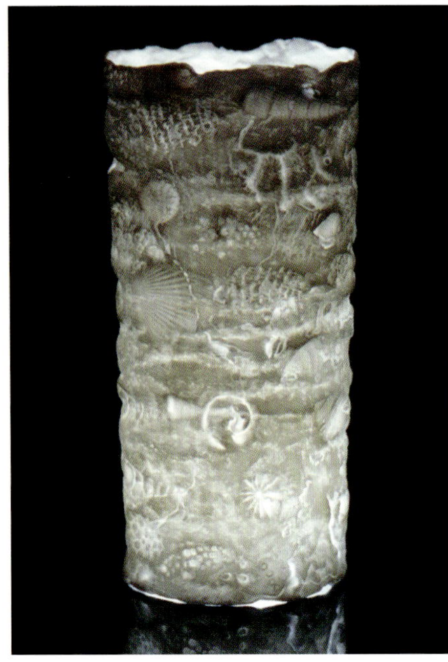

Pleuractis is the name of the coral that inspired the next design. The pendant lights were lit by incorporating a small halogen bulb at the end of a metal rod, which was inserted into the top of each shade; these bulbs gave a soft, warm light.

Working on this project allowed me to forge new ground in my practice through innovative collaboration with skilled industry professionals.

The collaboration also helped me to see my work not just as works of art or sculpture, but additionally as objects of light, both decorative and functional. I also began to view my work from a commercial perspective, which proved to be an exciting challenge. The *Ocean Light* project enabled me to bring ten years of extensive research to a resolution.

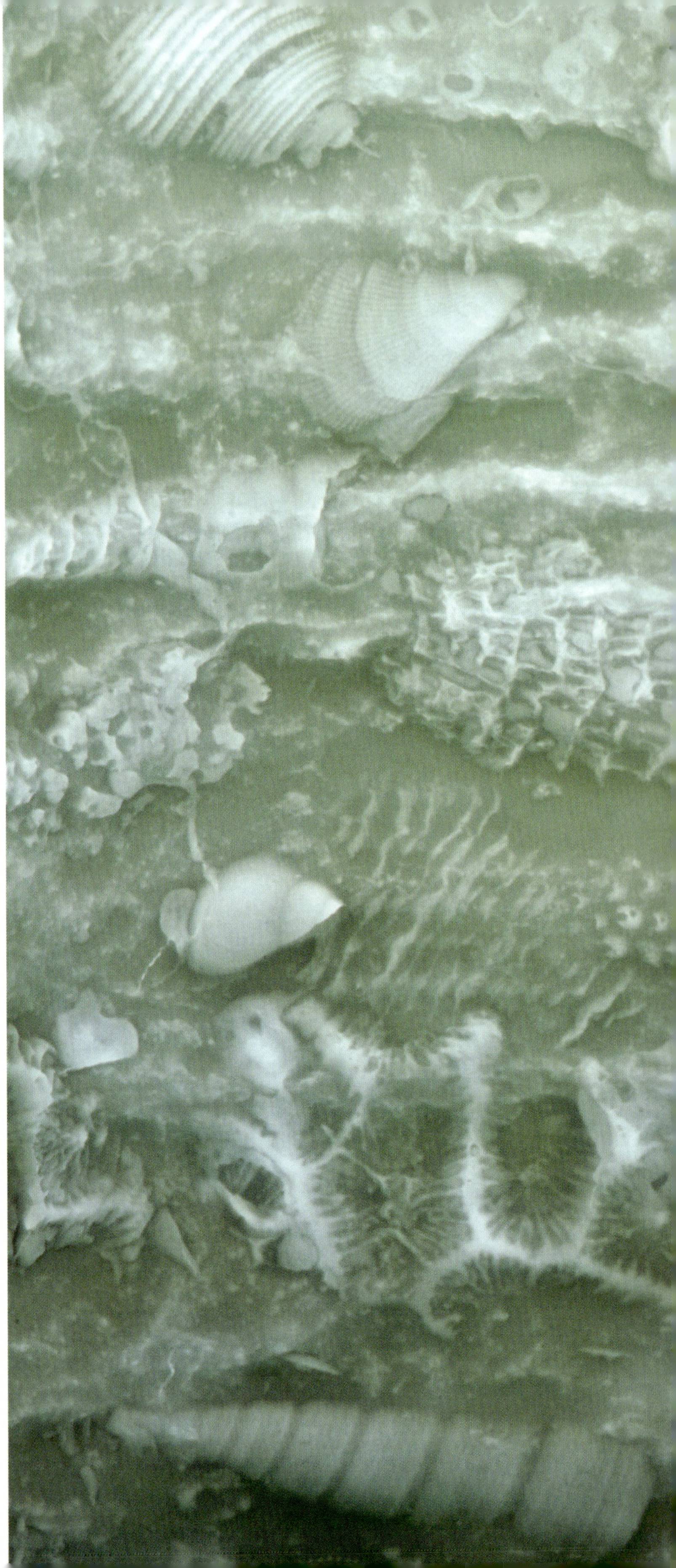

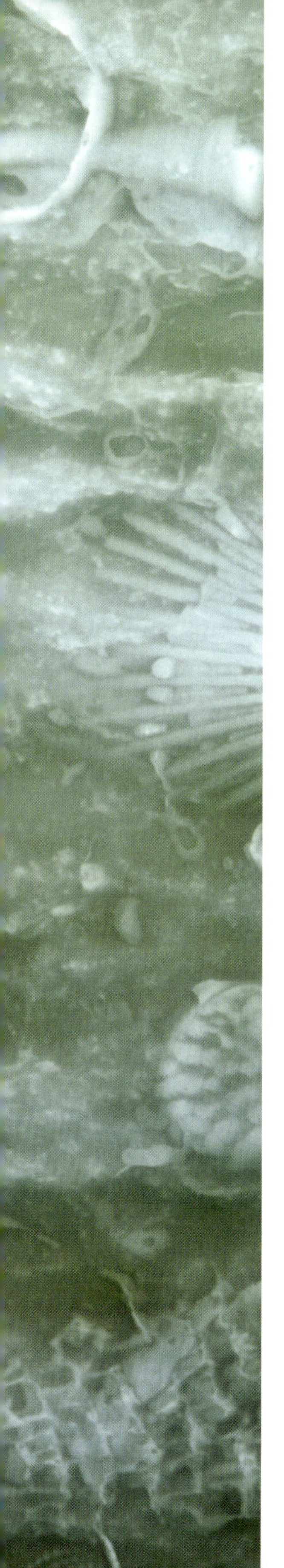

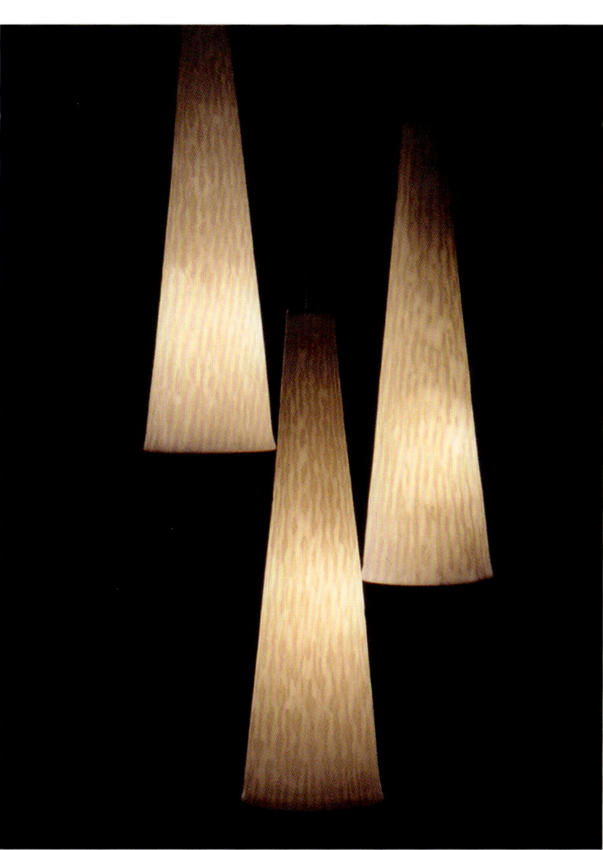

OPPOSITE FAR LEFT: *Cretaceous Light*, Angela Mellor, 2003. Bone china paperclay, LED light. 17 x 6.5 cm (h x w). *Photo: Victor France.*

CENTRE: *Cretaceous Light* (detail), Angela Mellor, 2003. Bone china paperclay, LED light. 17 x 6.5 cm (h x w). *Photo: Victor France.*

LEFT: *Pleuractis* pendant light, Angela Mellor, 2003. Etched bone china, halogen lights. 37.5 x 10.5 cm (h x w). *Photo: Victor France.*

International ceramists exploring paperclay and lighting

Anne Türn (Estonia)

Anne, when describing her work, says:

> *For me, being a ceramist is a way of living along with love for nature and wilderness. I have been experimenting with traditional materials – clay, glass and light – for more than 20 years, trying to break the rules, trying to make things that, at first, seemed impossible. With ceramics it is as with nature: you have to listen. Never go against. The mountains tell a mountaineer what is permitted; it is the same for ceramics. I love to let the materials play, let things happen, let the materials create miracles. I love to expand borders, to walk on the edge.*
>
> *Inspiration comes from the glaciers of Alaska and the Himalayas, the endless world of snow, clouds and whiteness, the power of the mountains.*
>
> *I have been making 'crazy' surfaces for quite a time. First, building 'hair' drop by drop, now letting the 'hair' form on its own. The materials and I work; I give them possibility, like rain on a stormy sea. I use paperclay and let the mix of glazes and glass melt as they wish. I give them the freedom, they create the miracles. I just sit back and wait. It is about the fragility of the nature, the fragility of life.*

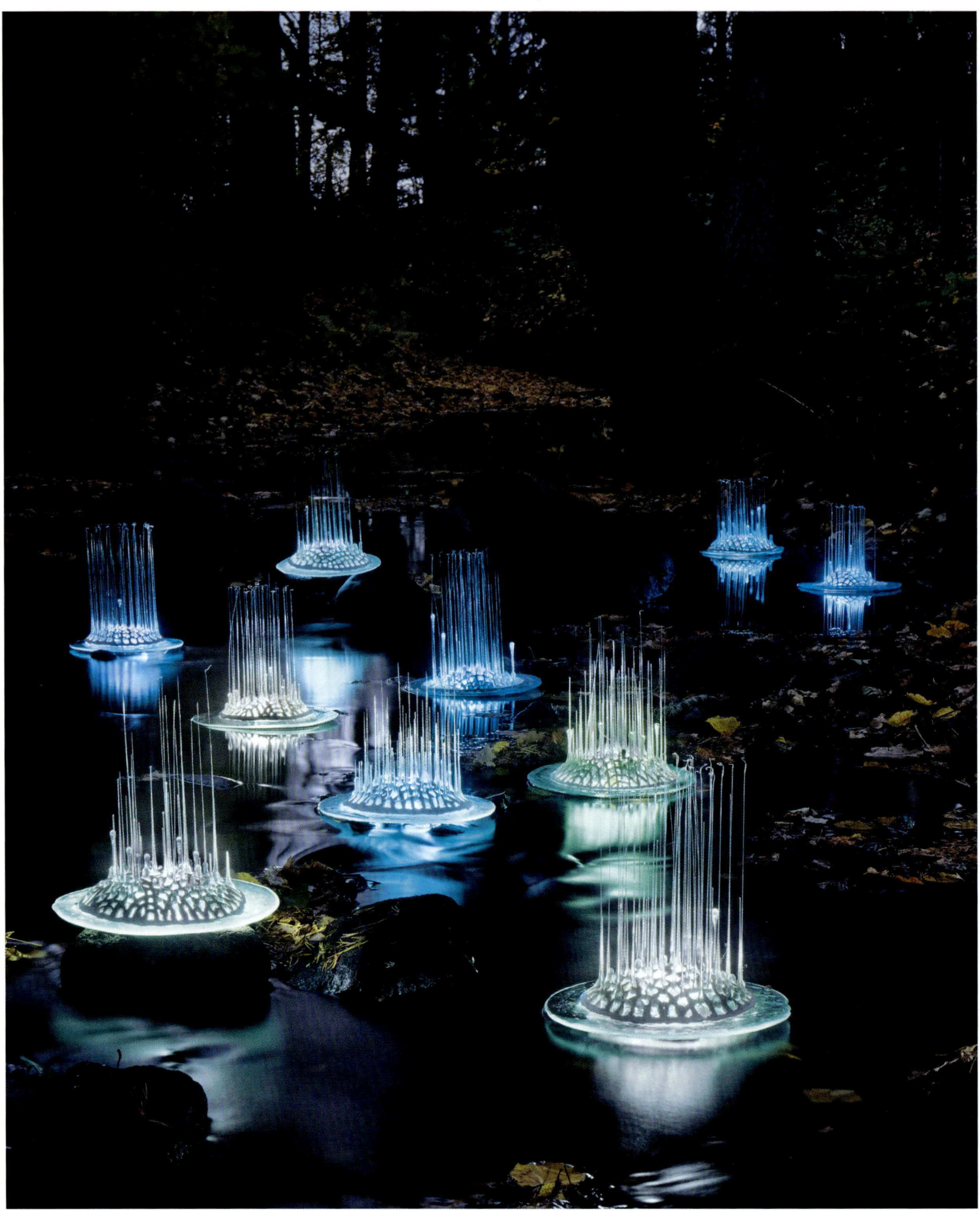

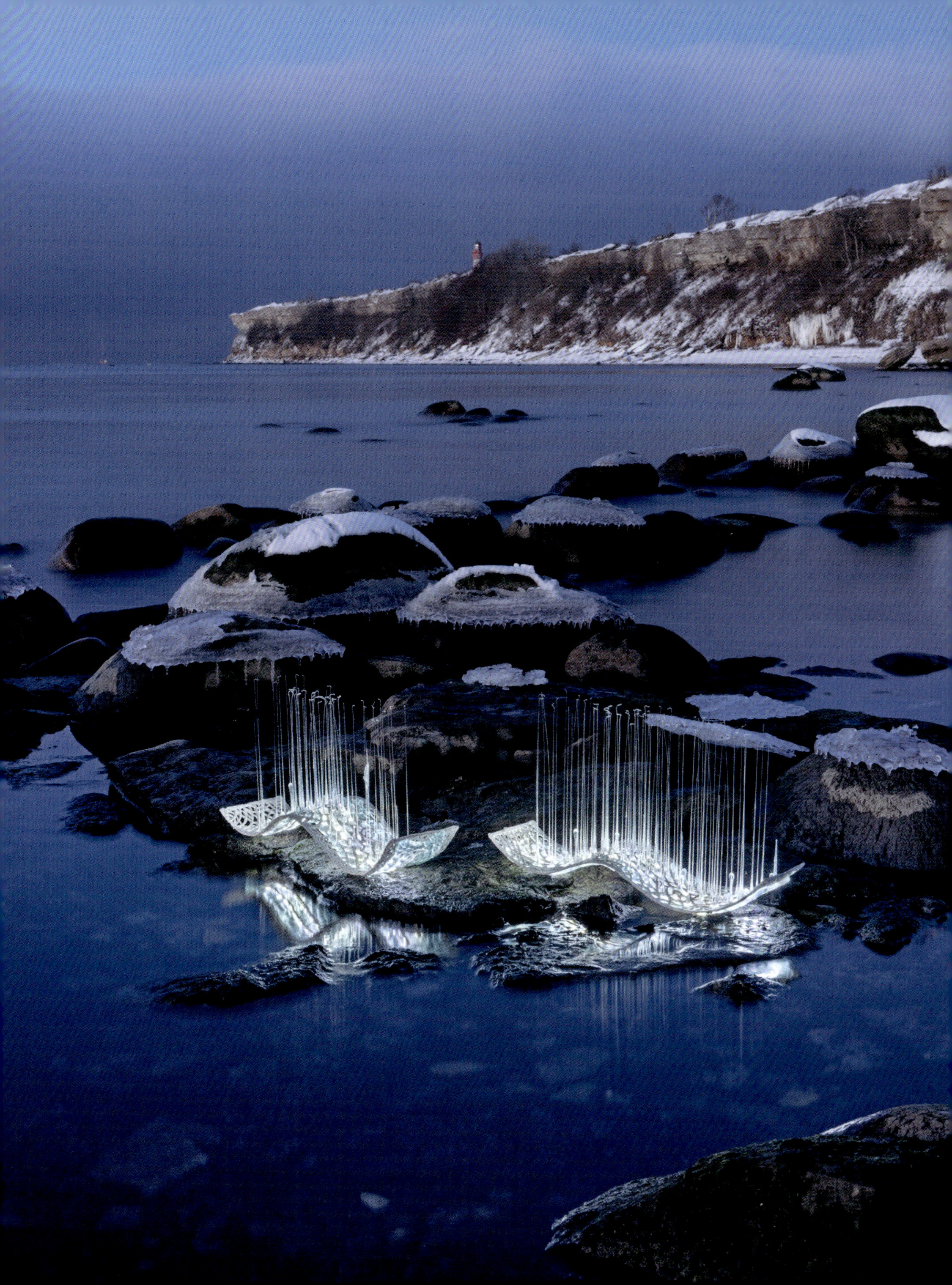

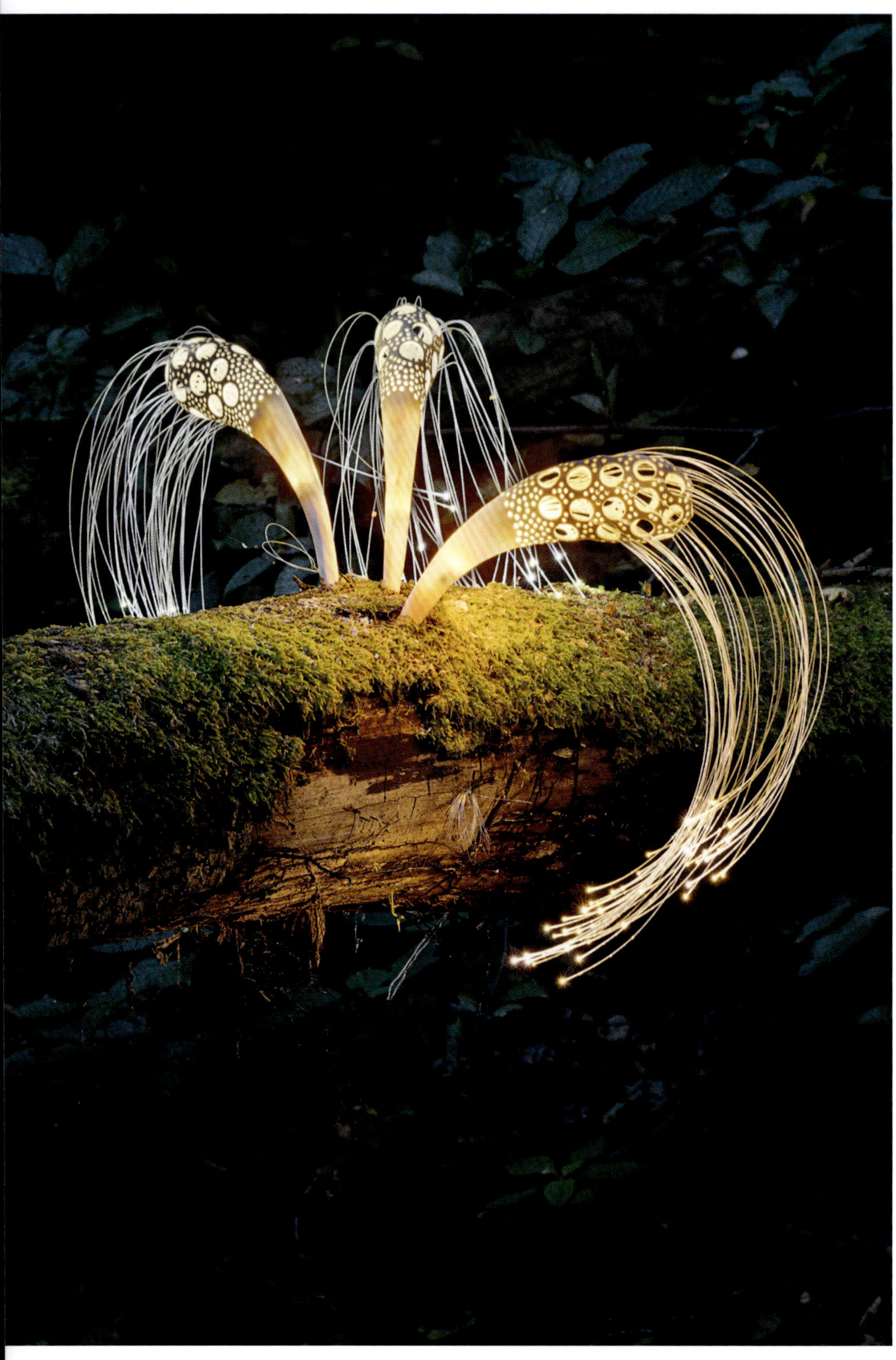

PREVIOUS PAGE LEFT: *Reflections*, Anne Türn, 2021. Paper porcelain fired at 1200°C, a mix of glass and glaze fired at 869°C, LED light. (each piece) 30 x 30 x 40 cm. *Photo: Toomas Tuul.*

PREVIOUS PAGE RIGHT: *Waves*, Anne Türn, 2022. Paper porcelain fired at 1200°C, a mix of glass and glaze fired at 869°C, LED light. (each piece) 45 x 40 x 10 cm. *Photo: Toomas Tuul.*

LEFT: *Drops*, Anne Türn, 2022. Porcelain fired at 1250°C, fibre optics, LED. (installation of three pieces) 50 x 50 x 25 cm. *Photo: Toomas Tuul.*

RIGHT: *Shower-flowers*, Anne Türn, 2022. Porcelain fired at 1250°C, fibre optics and LED. (each piece) 12 x 30 x 60 cm. *Photo: Toomas Tuul.*

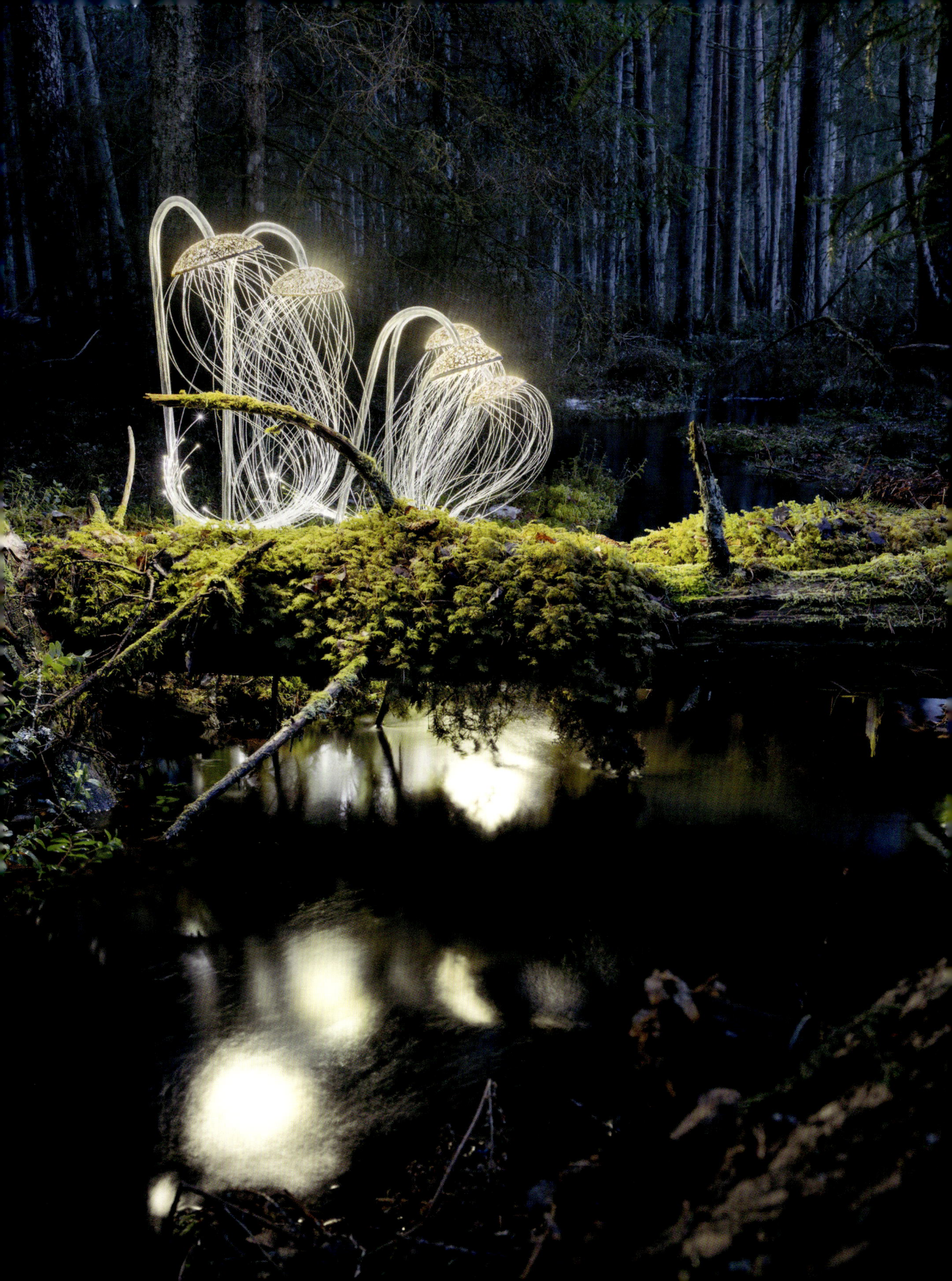

Antonella Cimatti (Italy)

Antonella Cimatti was a student of Carlo Zauli at the Institute d'Arte in Faenza, receiving a degree with distinction from the Fine Arts Academy in Bologna. She taught Design at the Liceo Faenza from 1979 until 2017, and is a member of the IAC, UNESCO, Geneva.

Antonella is a distinguished ceramic artist. Her work blends design, technology, and clay to produce innovative sculptures that vividly express her poetic and creative philosophy. Of her process, Antonella says:

> *My design is born from a re-reading of the past artistic production through a filter of formal personal sensibility directed towards the making of a functional or sculptural object. The forms generated are aesthetically accurate and display a strong sense of the real feminine character, of grace, of elegance, and of attention to detail.*
>
> *My design starts from a reinterpretation of the artistic production of the past: my way of working is not traditional but starts from tradition; my aim is to create lightness in ceramics, not only in terms of weight, but also visually.*
>
> *For this, I started from the majolica models of the* crespine, *finely modelled bowls made in Faenza in the sixteenth century and used in European royal courts as luxury objects. In 2005, I started revisiting these bowls, wanting to create a contemporary porcelain version, modifying both the material and the technique in which decoration and structure are one. They are almost immaterial artefacts: impalpable threads, ornaments, openwork lace, and filigrees of white lace.*

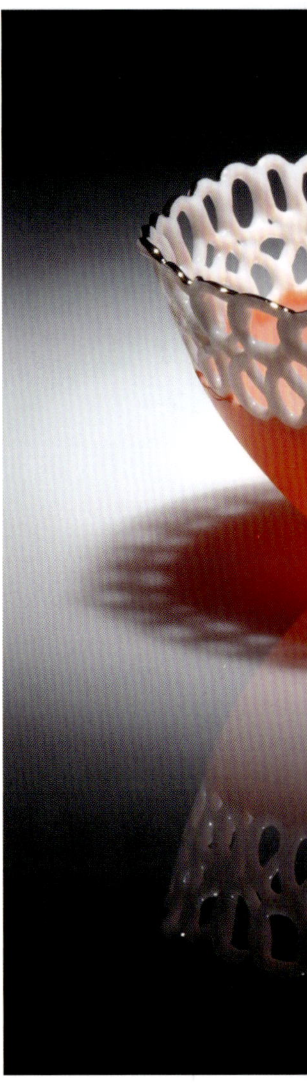

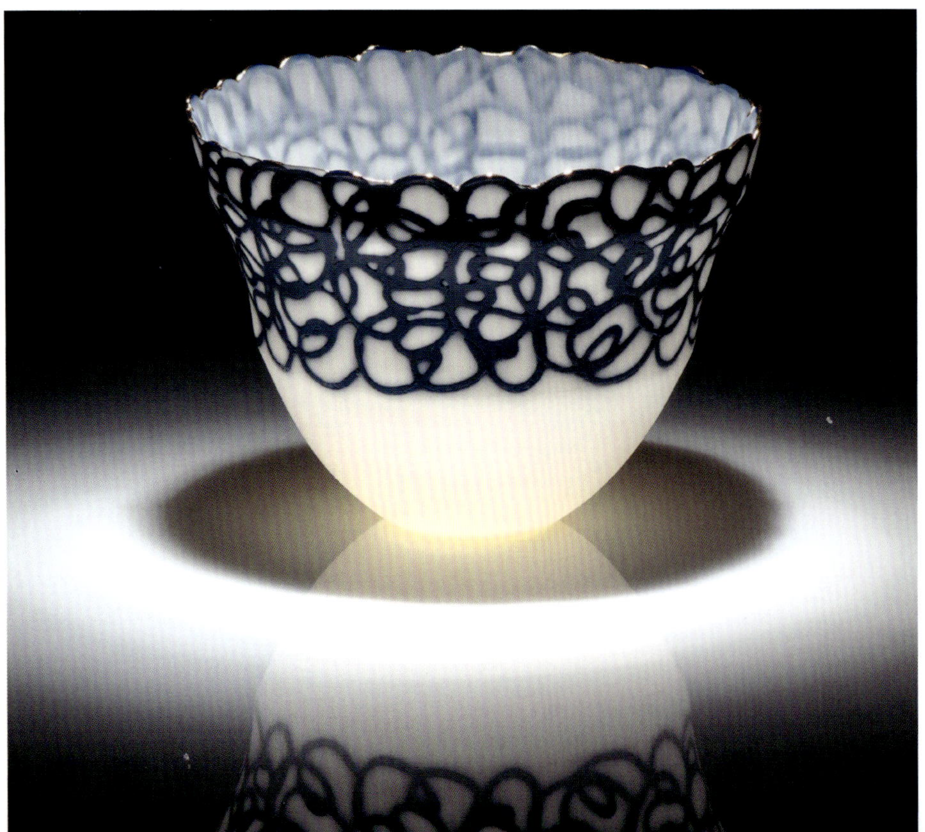

LEFT: *Crespina*, Antonella Cimatti, 2020. White porcelain slip casting and coloured slip trailing porcelain paperclay. *Photo: Raffaele Tassinari.*

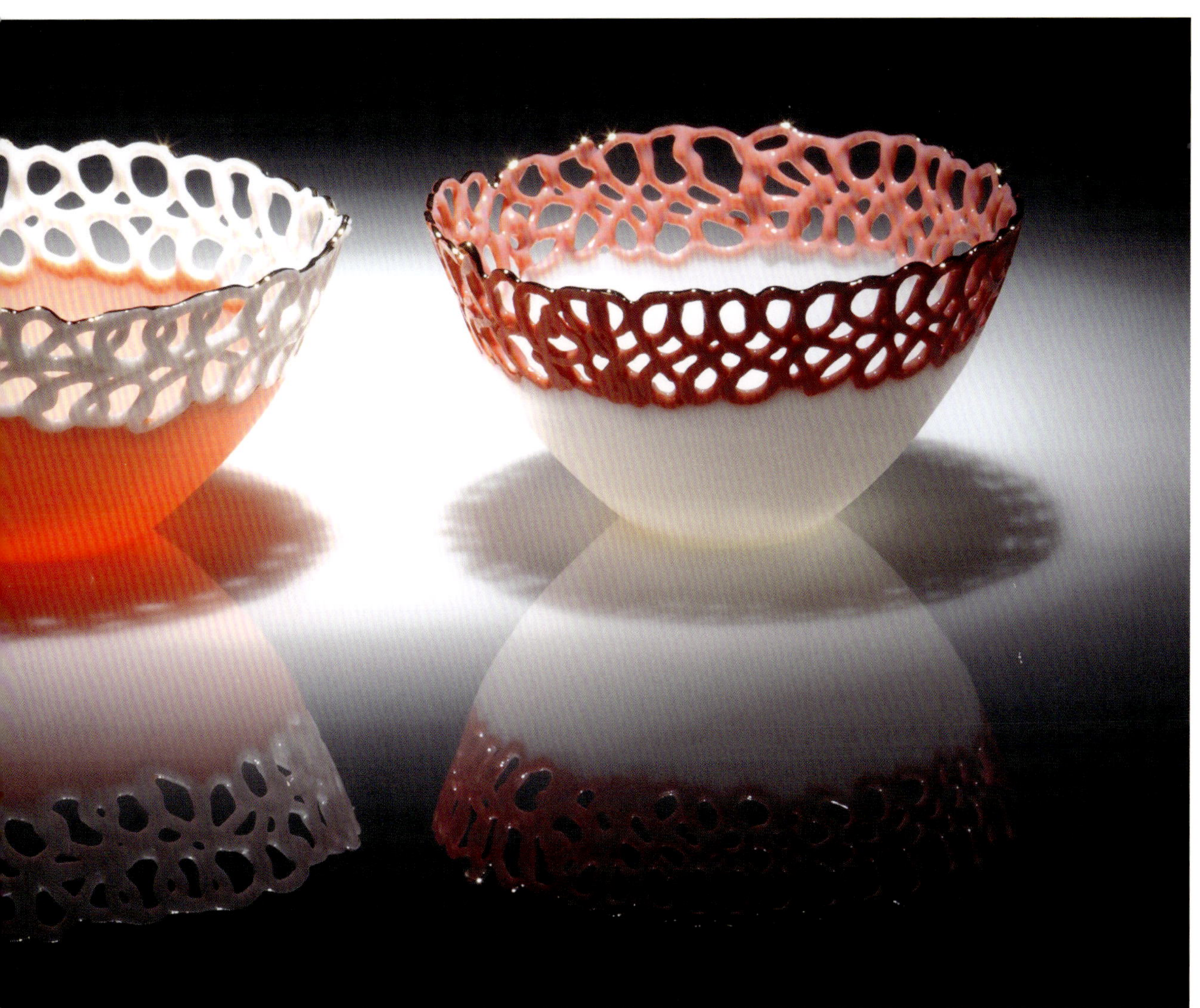

ABOVE: *Crespina*, Antonella Cimatti, 2016. White porcelain slip casting and coloured slip trailing porcelain paperclay. *Photo: Raffaele Tassinari.*

My first trip to Japan was in 1981 at the age of 25, when I participated with the Faenza delegation in a cultural exchange with the twin city of Toki-shi. I immediately fell in love with Japan and was impressed by their ancient culture. More recently, I have travelled several times to China and South Korea, and every time, I bring home some little technical secret about the porcelain so little known in Faenza.

More recently, I have started working with shadows, which have always fascinated me, and an advanced ceramic material (sintered alumina substrates) with which I create silhouettes that reproduce through shadows. Such objects were made in Faenza in the Renaissance, and are now preserved in many important museums around the world. I thus create a sort of virtual museum where the models of the past – objects of everyday life – made in Faenza are all together again.

> *More generally, I am very inspired by fashion and design. In fact, in recent years, there has been a major trend in these areas of Italian excellence towards lightness and attention to detail, which is so incredibly in line with my way of being and working.*

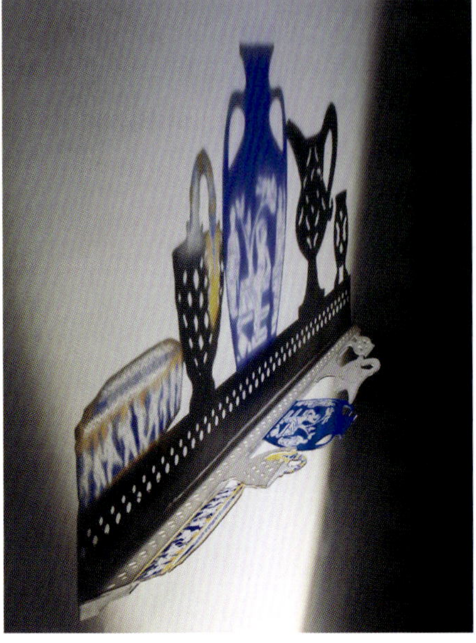

LEFT: *Virtual Museum #4*, 2020. Translucent laser cut-out of sintered alumina-advanced ceramic attached to the wall at 90° and under-lit, projecting its shadow. Coloured film over plexiglass. *Photo: Raffaele Tassinari.*

Her most extravagant work is a light installation entitled *Trame di Luce* (Plots of Light).

> *It is three metres high, where optic fibres and porcelain shapes intertwine and dialogue. Vibrant light sheets fill the space and intersect in performed bodies of highly translucent paper porcelain. The fibres take on a 'bringing together' function, enveloping and permeating the whole, placing the material in space, lighting and lifting the work as it rises and hovers in the void.*

To install it, she spent three days climbing on a scaffold.

> *My way of working is not traditional. My objective is to create lightness in ceramics, not only in terms of weight, but also visually.*
>
> *I've been working with porcelain to make large installations. They are environmental sculptures that produce light, thanks to optical fibre technology: the light plays with the porcelain and, getting into its gaps and holes, creates new shapes. Thus, the ceramic loses its body and weight to become flexible and dynamic. The space comes to life and a closely woven set of optical illusions surfaces from the dark.*

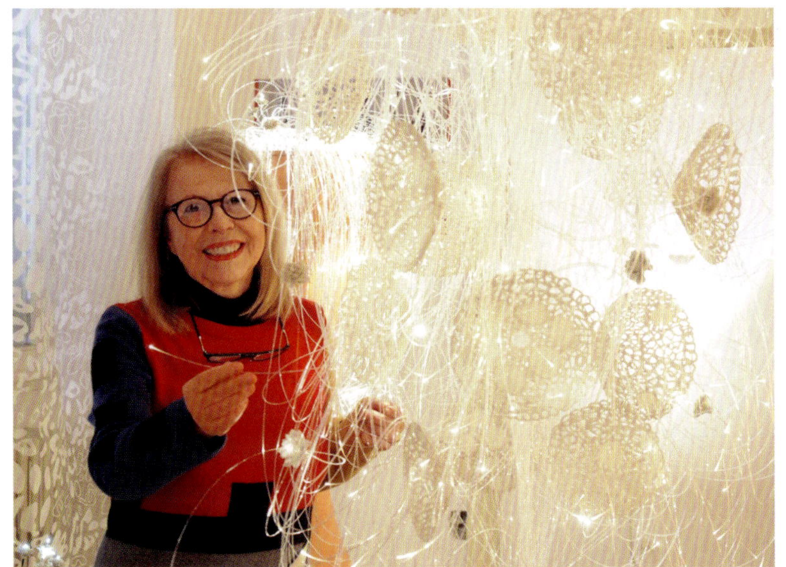

ABOVE: *Trame di Luce* (detail), Antonella Cimatti, 2008 (photographed 2024). Translucent porcelain paperclay, fibre optics, and hand-moulded porcelain flowers. *Photo: Raffaele Tassinari.*

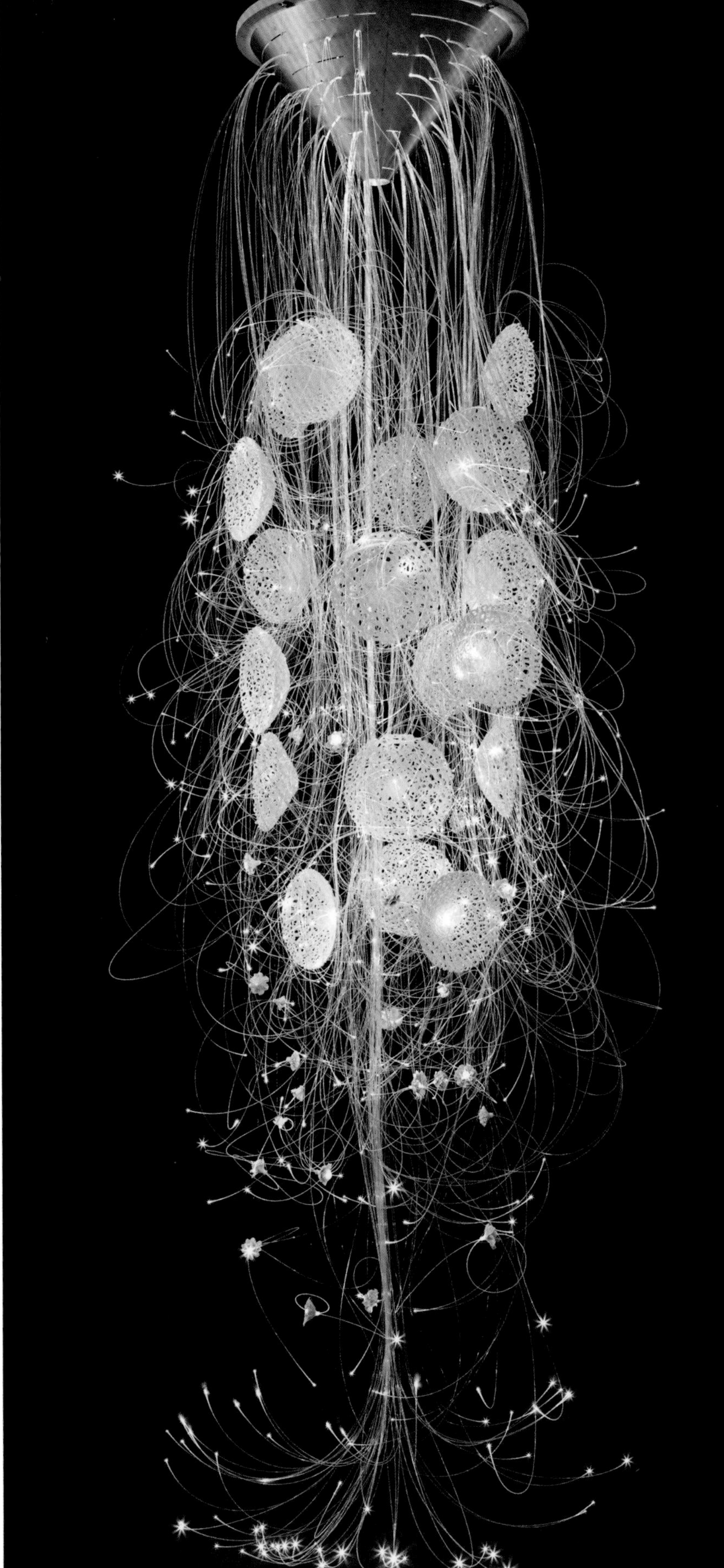

LEFT: *Trame di Luce*, Antonella Cimatti, 2008. Translucent porcelain paperclay, fibre optics, and hand-moulded porcelain flowers. *Photo: Bernardo Ricci.*

Patty Wouters (Belgium)

Vulnerability is a recurring theme in Patty's work. She works mostly with porcelain, which radiates fragility through its essential qualities of whiteness, thinness, and translucency. Patty says:

> *These light pillars have a cracked and broken surface. The fractured skin can suggest wounds and suffering. The metallic structures are a support during the making but become very weak and brittle after the firing. The porcelain cylinders enter the kiln vertical and straight; during the high-temperature firing, the clay melts – it warps and distorts, and also becomes translucent. Light emphasises the frailness of the porcelain and of human beings, but also symbolises the power of hope.*

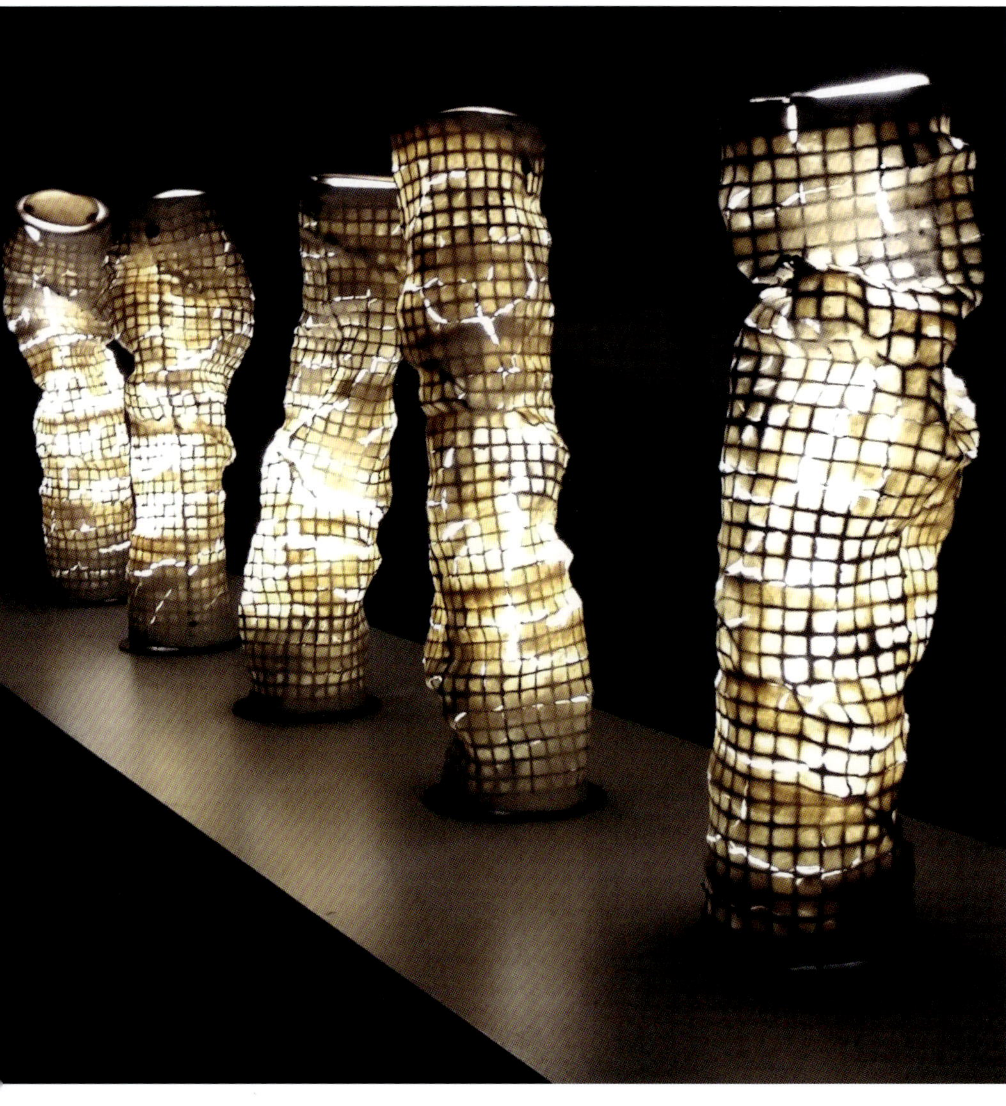

LEFT: *Fractured Light Pillars*, Patty Wouters, 2016. Paper porcelain, metal, LED lights. (each piece) *c.*50 x 12 cm. Fired in oxidation at 1250°C. *Photo: Patty Wouters.*

Patty sees *Cocoons* as vulnerable packages in transition to a new existence, saying:

> *These closed forms are thin and fragile and therefore imply vulnerability. The wrapped works resemble mummies or cocoons. The theme of containment is further investigated through the creation of assemblages of the elemental forms in gauzy bag made of silk. The sacks are hung with the elements forcing themselves against the fabric skin, suggesting attempts to escape.*
>
> *Closed forms have always played an important role in my ceramic design. A closed form such as a box has an aura of something intriguing, something fascinating, and something mysterious. Such a container can be a protection for a precious object or a wrapping for a valuable present.*
>
> *Clay is in itself a symbol of transformation. Clay is the result of millions of years of rock erosion. Through the hands of an artist or a crafter, it is shaped into a new form. The fire converts it into a new substance: rock-hard ceramic. Therefore, I deliberately chose to make these cocoons from paper porcelain to emphasise the concept of vulnerability and transformation.*
>
> *Cocoons and mummies have a loaded symbolic and metaphysical meaning. According to Jung and his contemporary followers, all artistic creations contain forms and images which have symbolic meanings, which are archetypal, and in this way communicable to others. The mummy as well as the cocoon, symbolises an intermediate stage: both are a metaphor for an emerging new life.*

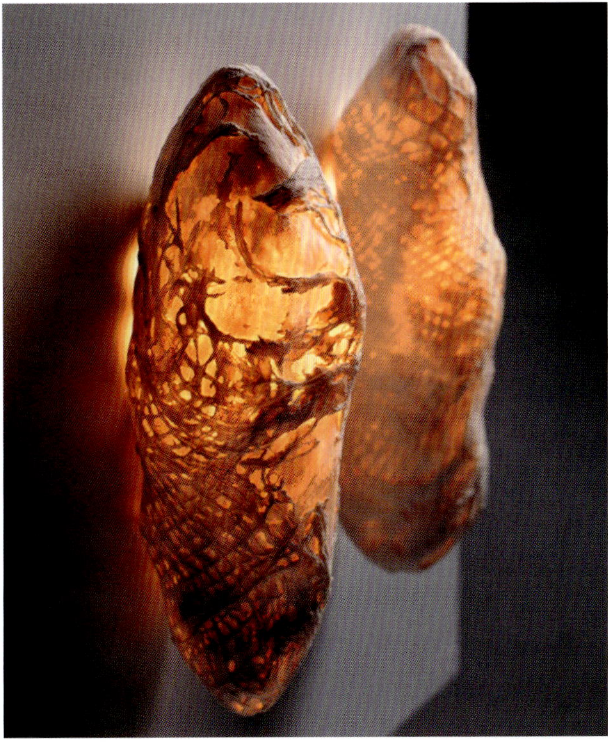

LEFT: *Cocoons*, Patty Wouters, 2018. Paper porcelain, LED lights. (each piece) *c.*22 x 10 x 7 cm. Fired in oxidation at 1250°C. *Photo: Dries Van Den Brande.*

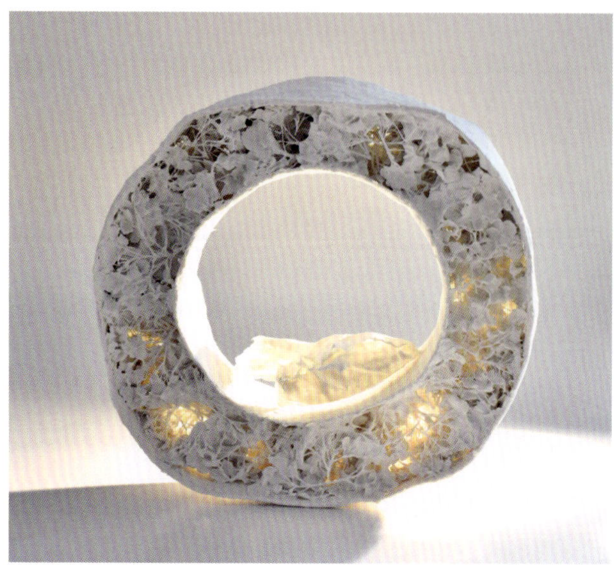

RIGHT: *Vulnerable I*, Patty Wouters, 2020. Paper porcelain and organic material. 35 x 10 cm. Fired in oxidation at 1250°C. *Photo: Patty Wouters.*

Patty describes *Vulnerable I* as:

> *A circular shape made in porcelain and organic materials, flowers from nature. The wafer-thin leaf in the opening is a metaphor for nature in general and how sensitive and vulnerable nature is.*
>
> *For centuries, humans have been more concerned with economic growth and have neglected their enormous impact on nature. Nowadays, we experience the consequences: the fires of the woods in Montiferru, just a stone's throw away from Bosa, are a result of the global warming. 20,000 hectares of forest have been destroyed. 1,500 people needed to leave their houses and all their belongings, not knowing if or when they would be able to return. A terrible tragedy! In my own country, and in Germany, we recently experienced severe floods, never seen before. More than 200 people died and hundreds of houses were damaged. All of this is a result of global warming. We urgently need to be more conscious about nature and preserve our ecosystem.*

The intention of Patty's porcelain work is to command the attention of spectators and confront us with the vulnerability of nature. She asks us all to be more conscious and listen to the messages of people like Greta Thunberg, and to cherish nature so that we can save it from destruction.

Vulnerable I was awarded the Fratelli Melis prize at the International Ceramic Biennale in 2021, held in Bosa, Sardinia; *Vulnerable II* was awarded the first prize at the International Festival of Contemporary Ceramics, in Sofia, Bulgaria, in 2021.

RIGHT: *Vulnerable II*, Patty Wouters, 2021. Paper porcelain and organic material. 25 x 25 x 8 cm. Fired in oxidation at 1250°C. *Photo: Frederik Hamelynck.*

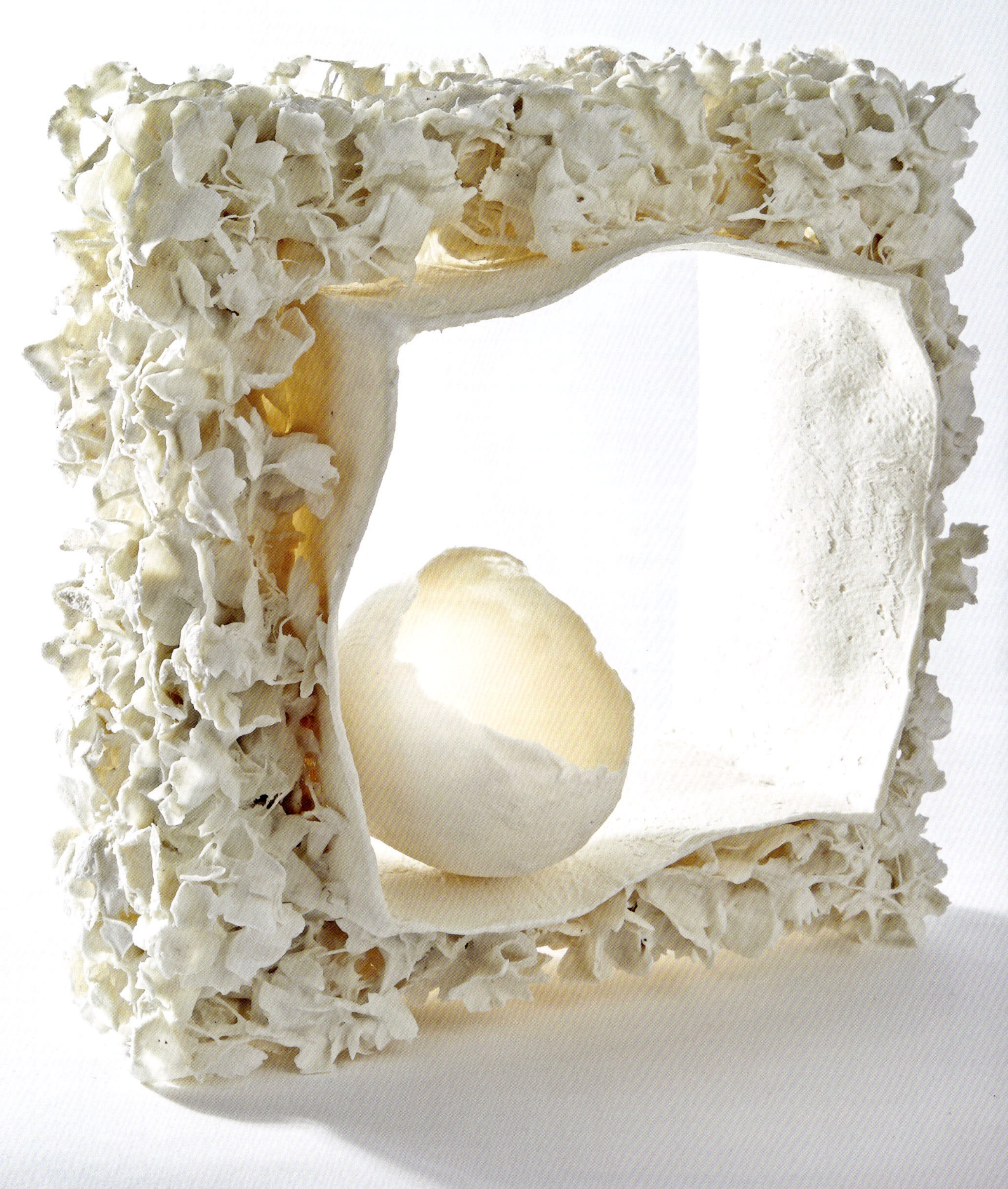

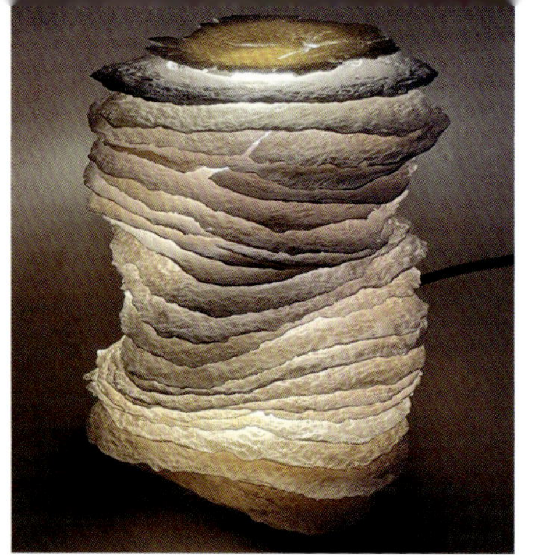

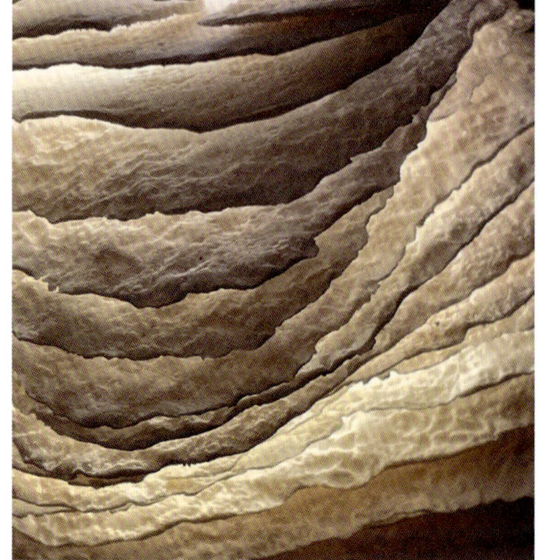

FAR LEFT: *Ethereal Strata*, Ingrid Bakker, 2024. Valentine's flax paperclay ES600, cut glass beads and silicone glue. 14 x 20 cm. Fired at 1250°C. *Photo: Ingrid Bakker.*

LEFT: *Ethereal Strata* (detail), Ingrid Bakker, 2024. Valentine's flax paperclay ES600, cut glass beads and silicone glue. 14 x 20 cm. Fired at 1250°C. *Photo: Ingrid Bakker.*

Ingrid Bakker (Holland)

Ingrid Bakker is an anthropologist from Amsterdam who fell in love with clay over 30 years ago. Ingrid is largely self-taught but has honed her craft through workshops focusing on throwing, raku, and most recently, 'paper porcelain light objects', under the guidance of Patty Wouters in Belgium. The light object *Ethereal Strata* was inspired by the works of Claire Muckian and Amina Rose, whose art was showcased in Patty's presentation while she was attending Patty's workshop.

Discussing her work, Ingrid states:

> *Throughout the years, I've created various lamps using traditional clay and materials like driftwood. However, the opacity of regular clay always posed limitations, requiring openings or holes to let light pass through. Discovering the delicate, luminous qualities of paper porcelain translucency has been a true game-changer for me. Ever since, my mind has been brimming with ideas, eager to explore the endless possibilities this material offers.*
>
> *The* Ethereal Strata *lamp is composed of 25 rounded triangles and a smaller top piece with cut-glass beads in between, supported by porcelain feet. The triangles were rolled out to a thickness of 2 mm, and the edges thinned using a roller, which also gave them a slightly irregular shape. Surface texture was added by pressing the surfaces with lava stone. The triangles were allowed to dry naturally, resulting in an organic, irregular form.*
>
> *Once dried, the triangles were fired at 1250°C in sand and then assembled into the final shape. Small cut-glass teardrop beads were used as spacers between the triangles, and these were secured with silicon adhesive. The triangles were twisted slightly during assembly to create a sense of movement and to produce interesting shadow effects.*
>
> *The overall structure nearly forms a full circle, with a smaller triangle on top that has a small crack. A paper column placed in the hollow centre of the lamp hides the bulb's warm light. Because the white paper column is open at the top, the light there is warmer than on the sides, making the top layer look more yellow.*

Nemanja Nikolić Prika (Serbia)

> *Nemanja Nikolić Prika is an artist who lives and works in Belgrade. He studied Ceramics at the Faculty of Applied Arts at the University of Belgrade in Serbia, and is a member of the Serbian Applied Arts Association ULUPUDS. He uses different forms and shapes to express his art – from sculpture to illuminating objects – and employs a wide range of techniques and materials. His work has been exhibited both nationally and internationally.*[4]

His lights, when lit, show detailed drawings on paper porcelain, creating exquisite illuminating effects. When unlit, the wall pieces and lamps become relief sculptures and paintings.

Circular Variations of Circumstance (see p.143) shows one segment from a composition of round porcelain shapes of different dimensions. In this artwork, Nemanja explored the possibilities of the material when he discovered how translucent it was. The translucency inspired him to make more layers, between 1 and 15 mm thick. He tried to capture the moist, soft porcelain texture in this artwork, which was made from the finest Dehua porcelain at the Blanc de Chine art residency in China.

His fully handmade ceramic lamp received an award at the 15th International Ceramic Competition, Carouge, Switzerland. The porcelain panels include exclusive artwork on the surface, the interventions made on it creating extraordinary dance and light shadows.

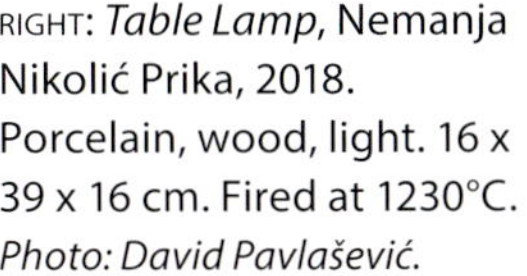

RIGHT: *Table Lamp,* Nemanja Nikolić Prika, 2018. Porcelain, wood, light. 16 x 39 x 16 cm. Fired at 1230°C. *Photo: David Pavlašević.*

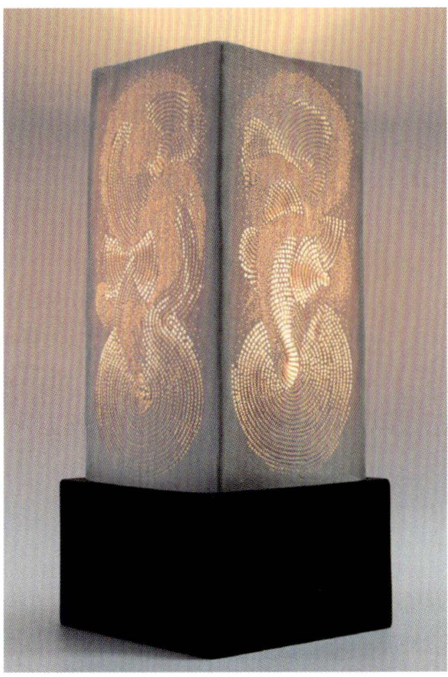

[4] Nemanja Nikolić Prika, nemaniko.com/about (accessed 25 March 2025).

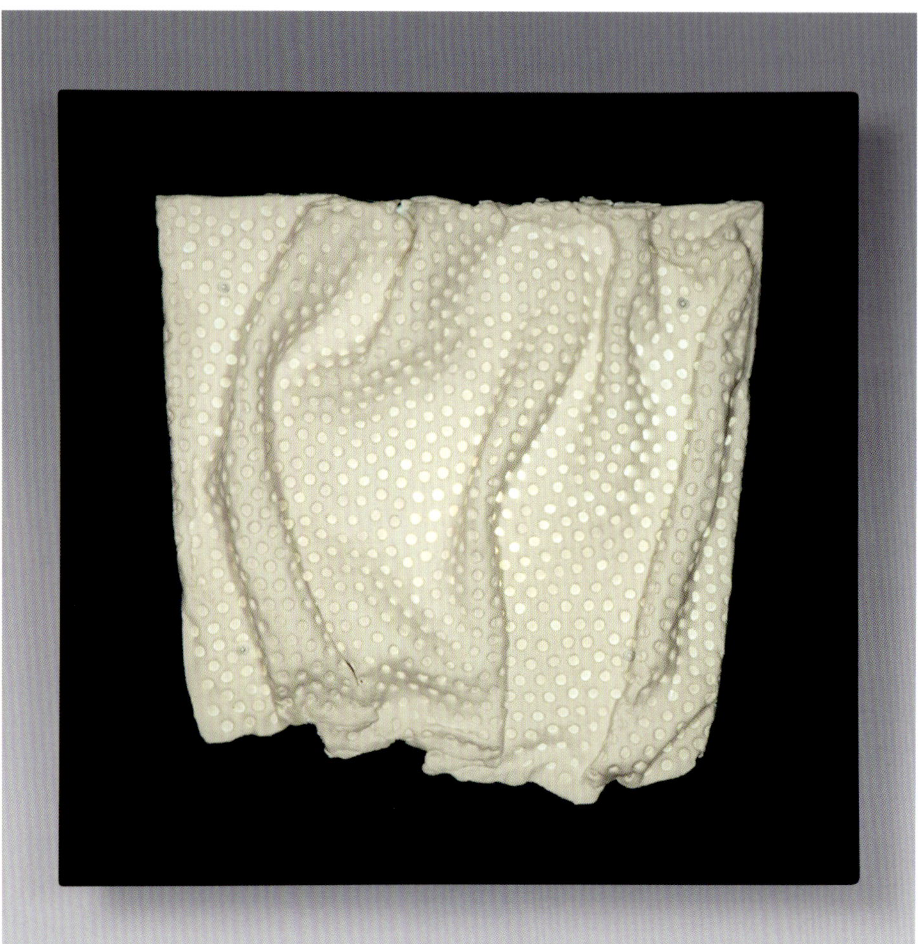

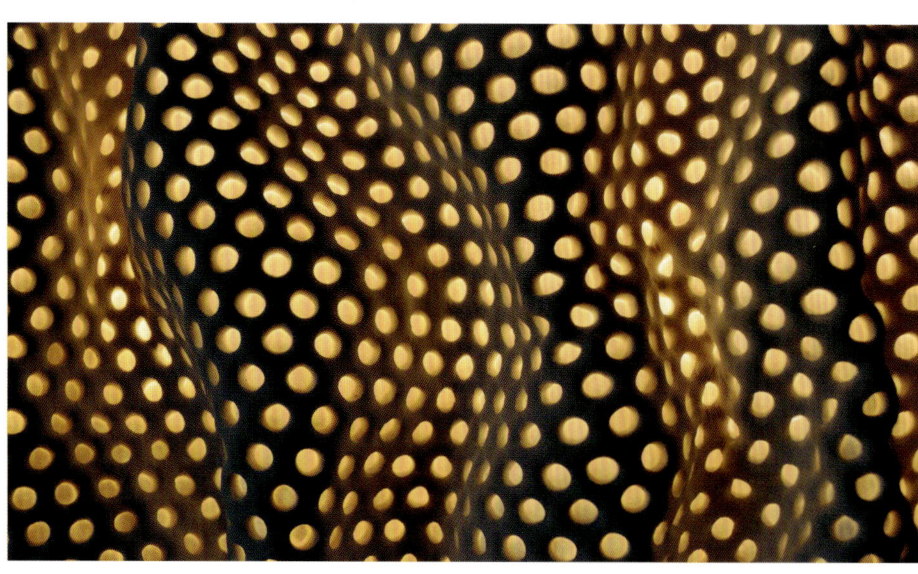

LEFT: *Whispering Wave*, Nemanja Nikolić Prika, 2009. Porcelain, wood, light. 45 x 45 x 9 cm. Fired at 1250°C. *Photo: Milena Rakočević.*

BOTTOM LEFT: *Whispering Wave* (detail), Nemanja Nikolić Prika, 2009. Porcelain, wood, light. 45 x 45 x 9 cm. Fired at 1250°C. *Photo: Milena Rakočević.*

RIGHT: *Circular Variation of Circumstances*, Nemanja Nikolić Prika, 2023. Porcelain, light. 30 cm. Fired at 1300°C. *Photo: Liang Ying.*

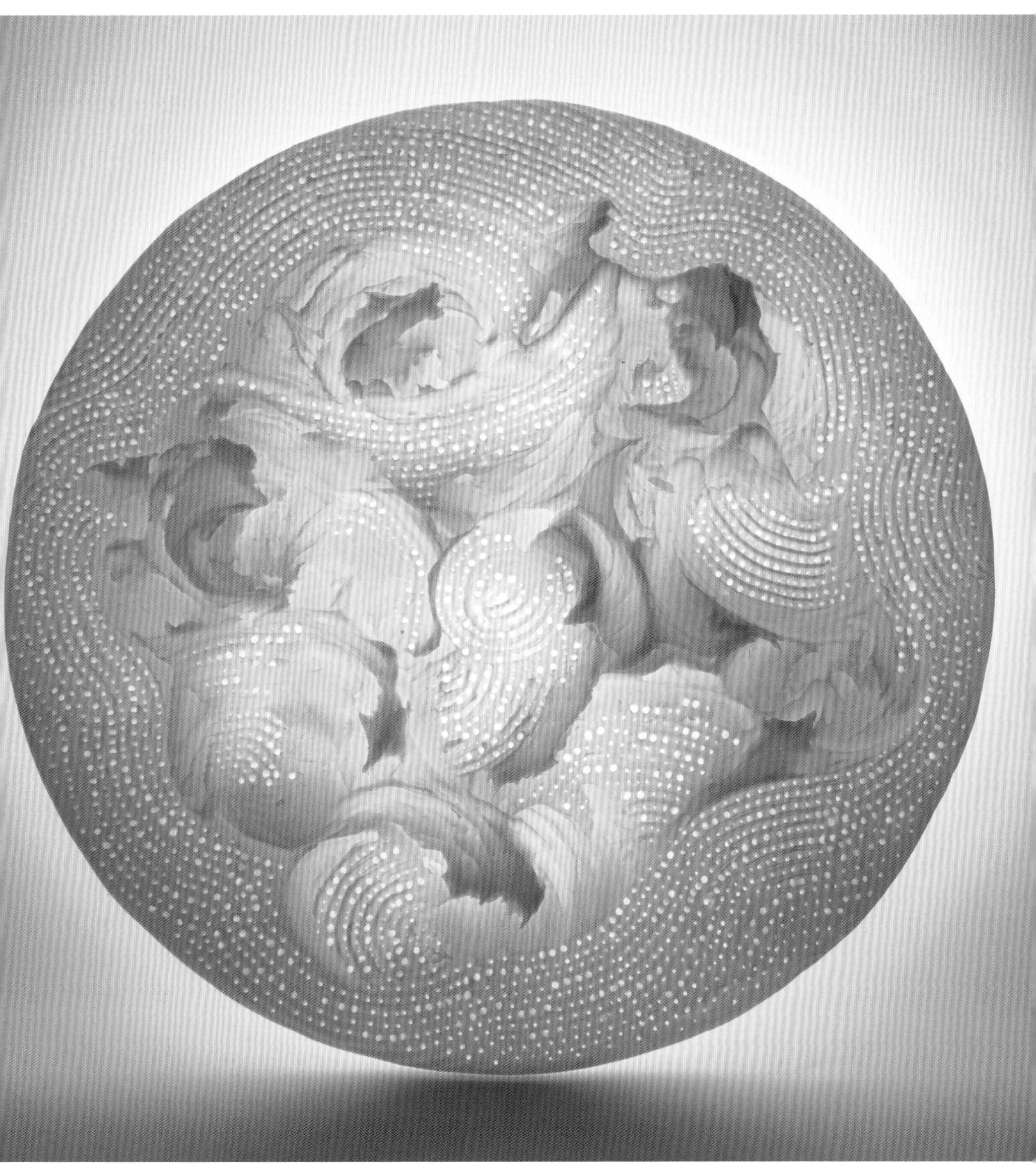

Ilona Romule (Latvia)

Ilona Romule graduated from the Latvian Art Academy in 1985 (MFA in Ceramics) and has been a Member of the IAC since 2001.

Ilona is a freelance artist, working with high-temperature porcelain. She has held over 20 solo shows and participated in over 50 international juried exhibitions and over 30 group exhibitions, winning eight awards. Her work is in the collections of many international institutions, including the International Ceramics Studio (ICS) Kecskemét, Hungary; Taipei County Ceramics Museum; FLICAM, Fuping, China; and the Kennedy Museum of Art, Athens, OH, USA.

Ilona has received international recognition for her slipcast porcelain figures. She examines the figure as a hollow vessel. Her human or animal figures are meticulously drawn on the sides of her pots, at the same time partially emerging from them as three-dimensional forms.

She is known for her use of ironic and erotic imagery, both in the form of her fine porcelain pieces and in the surface decoration with onglaze china paints. Ilona is interested in the opportunity for three-dimensional expression, using the scope of graphics and painting.

Lithophanes are fine sheets of translucent white porcelain with a decorative motif etched in negative. When lit from behind, these sheets show the scene in positive, creating a three-dimensional effect. They were produced from about 1830 to about 1900, mostly in Germany, by the Royal Porcelain Factory in Berlin and by Meissen.

Ilona is one of the few artists making three-dimensional lithophane objects. Of her process, she says:

> *I refer to my Art as, three-dimensional stories in colours.*
>
> *Drawing continues the form and the form continues the drawing, while together they make a story.*
>
> *Both the events of my daily life and my imagination are reflected in my work. As well as a game with symbols and mystical fantasy creatures …*
>
> *Images: a man and a woman, sometimes represented as animals.*
>
> *So … the story …*
>
> *Everybody has their own story. This is a deeply individual matter.*
>
> *I make one-of-the-kind artworks, using industrial processes. The material and technique should not be the primary focus. They should simply be professional enough not to surpass the idea.*
>
> *Porcelain is the language of my 'story', and this language should be fluent and flawless.*
>
> *It is not important in what language you keep silent,*
> *The most important is in what language you speak …*[5]

OPPOSITE RIGHT: Light object, Ilona Romule. Porcelain slipcast, assembled. *Photo: Ilona Romule.*

OPPOSITE FAR RIGHT: Light object, Ilona Romule. Porcelain slipcast, pierced, assembled. *Photo: Ilona Romule.*

OPPOSITE BOTTOM RIGHT: 3D lithophane objects, from the artist's solo exhibition *Breakout Light* at the Martinsons House in Daugavpils, Ilona Romule, 2023. Porcelain, slipcast. *Photo: Ilona Romule.*

[5] Ilona Romule, www.ilonaromuleporcelain.com/artist-statement (accessed 25 March 2025).

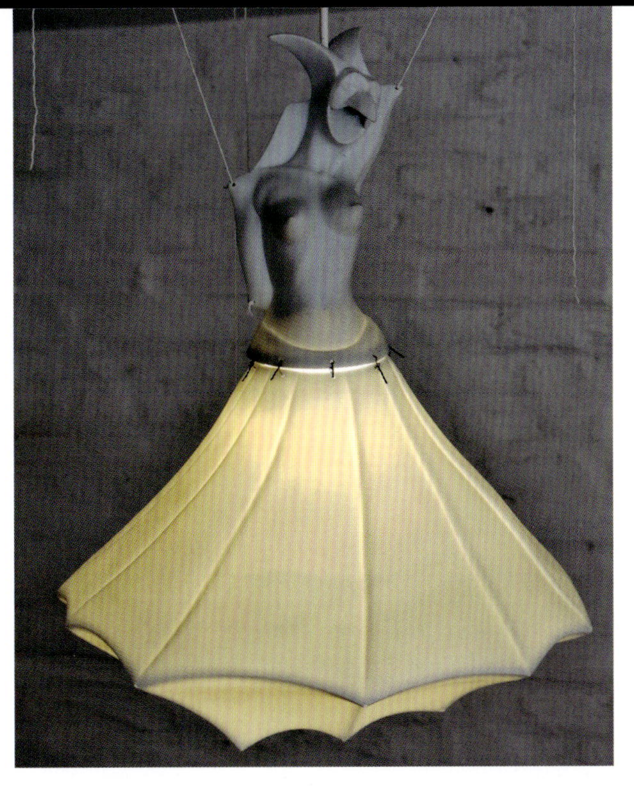

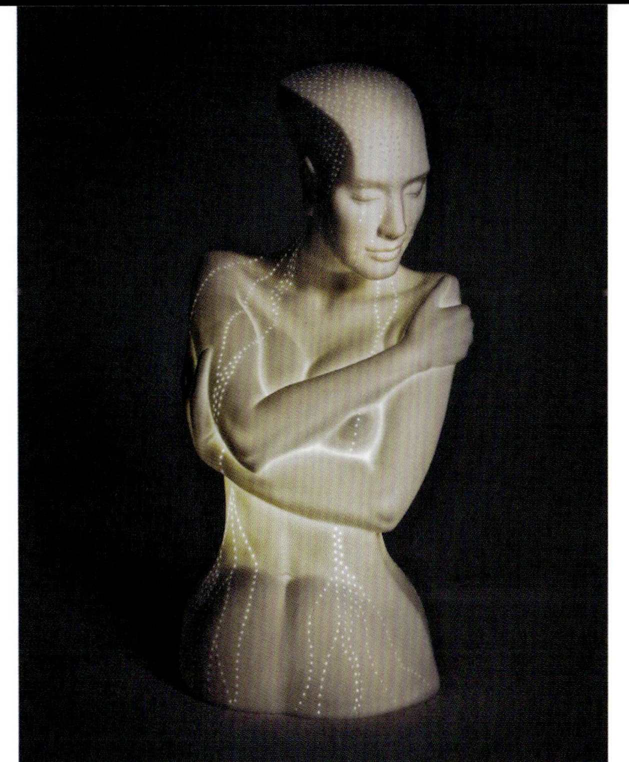

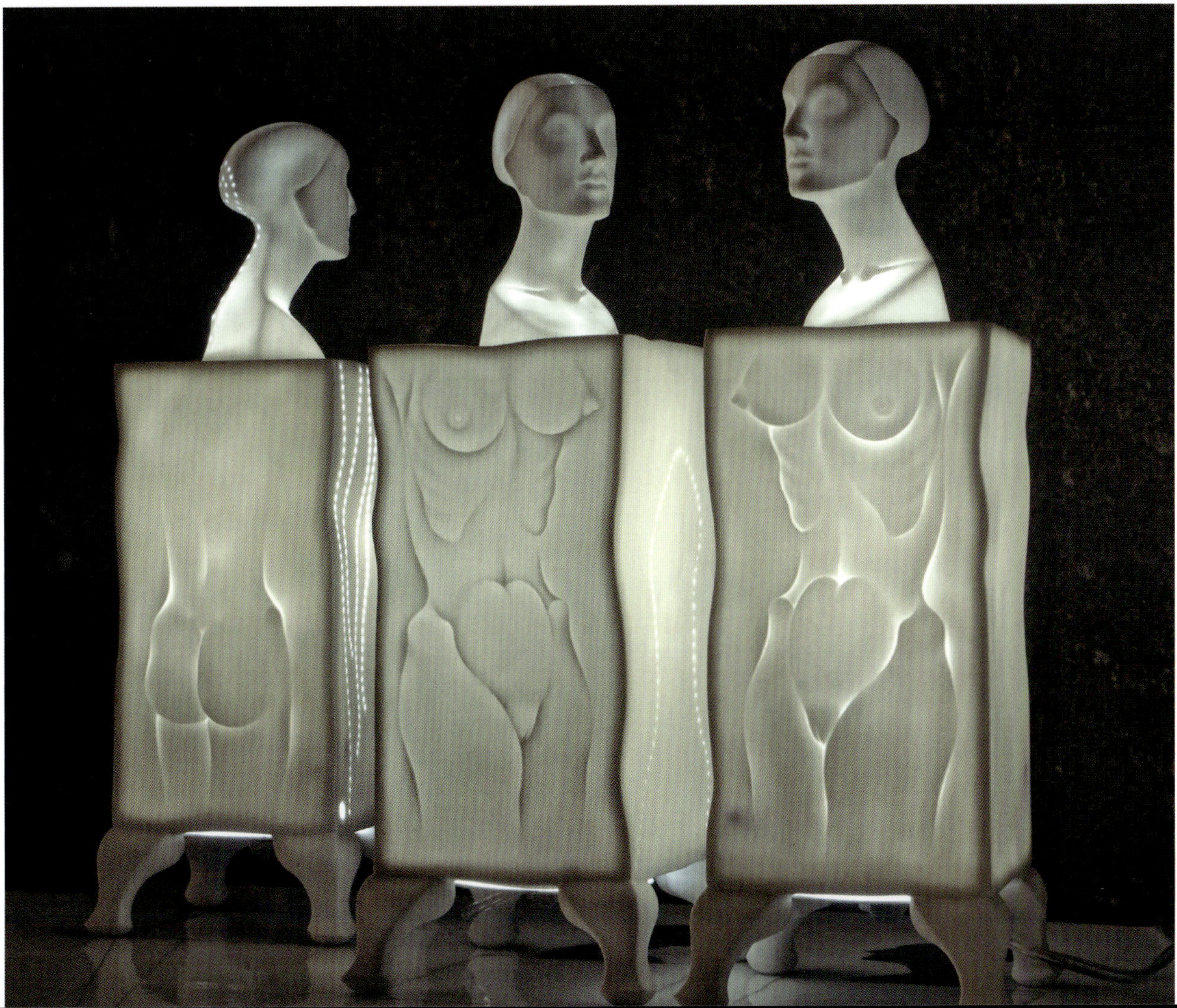

Chris Wight (UK)

Chris Wight was born in Glasgow in 1967 and grew up by the sea on the west coast of Scotland. He has been working with bone china for over 25 years, and was trained by designers and modellers from Wedgwood and Royal Dalton during his MA in Industrial Ceramics at Staffordshire University.

Here, Chris describes the challenges and rewards of using bone china as a medium:

> *A highly alluring yet notoriously 'stubborn' clay body, bone china presents many technical challenges – an inflexible 'body' prone to crumbling when worked by hand, an inability to be wheel-thrown and a propensity for pyroplastic deformation (loss of shape during firing) – characteristics I have grown to relish over time. To create works, often I adopt high-risk unconventional techniques to push bone china to its limits, sometimes utilising modern technologies as I seek to redefine conventional notions and develop new modes of expression with this most traditional of materials.*
>
> *With works spanning domestic scale vessel forms and sculptures to architectural scale panels and screens my output ranges greatly – each piece however shares a common drive to harness and exploit the unique qualities of bone china – its pristine whiteness, ability to take on crisply defined texture and carving and of course its astonishing light responsive properties. Its capacity to switch between translucent and opaque states – instantaneously or gradually – with changes in light is as magical to me today as it was when I first encountered bone china as a young student many years ago.*[6]

Chris's work has been shown in exhibitions at venues in Europe including Sotheby's, the Victoria and Albert Museum and the Saatchi Gallery in London. He has also shown in a number of group and solo exhibitions in Japan and Korea. He recently showed work in the USA with renowned gallerist and scholar Garth Clarke.

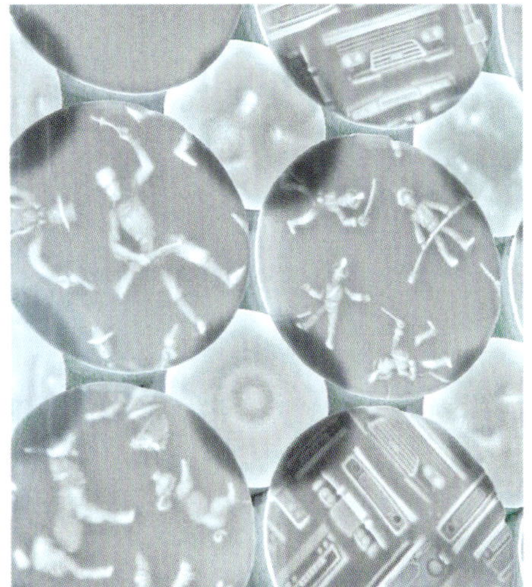

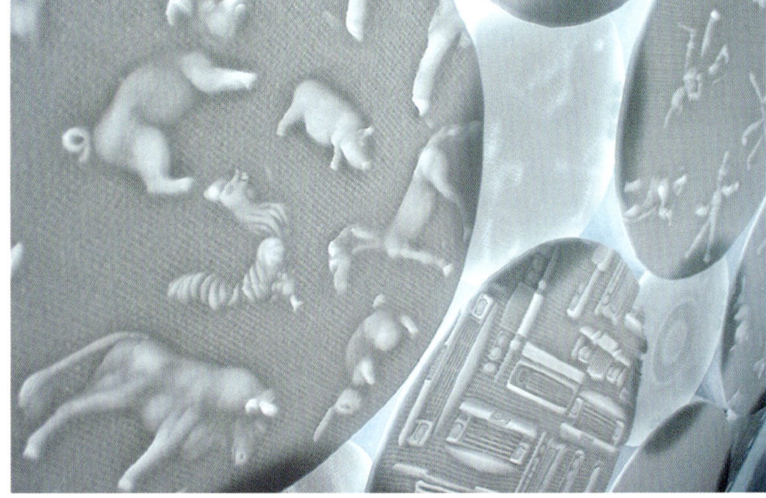

[6] Chris Wight, cone8.co.uk/bio-statement (accessed 25 March 2025).

OPPOSITE BOTTOM FAR LEFT: *'Toys' Panel* (detail), Chris Wight. *Photo: Chris Wight.*

OPPOSITE BOTTOM LEFT: *'Toys' Panel* (detail), Chris Wight. *Photo: Chris Wight.*

RIGHT: *'Toys' Panel*, Chris Wight. Glass-encased handmade bone china discs. Bone china, toughened glass, perspex, stainless steel; (base cladding) rubber *'bleu pâle'*. 58 x 27 x 3 cm (h x w x d); (base depth) 14 cm; (large discs) 7 cm; (small discs) 4.5 cm. *Photo: Chris Wight.*

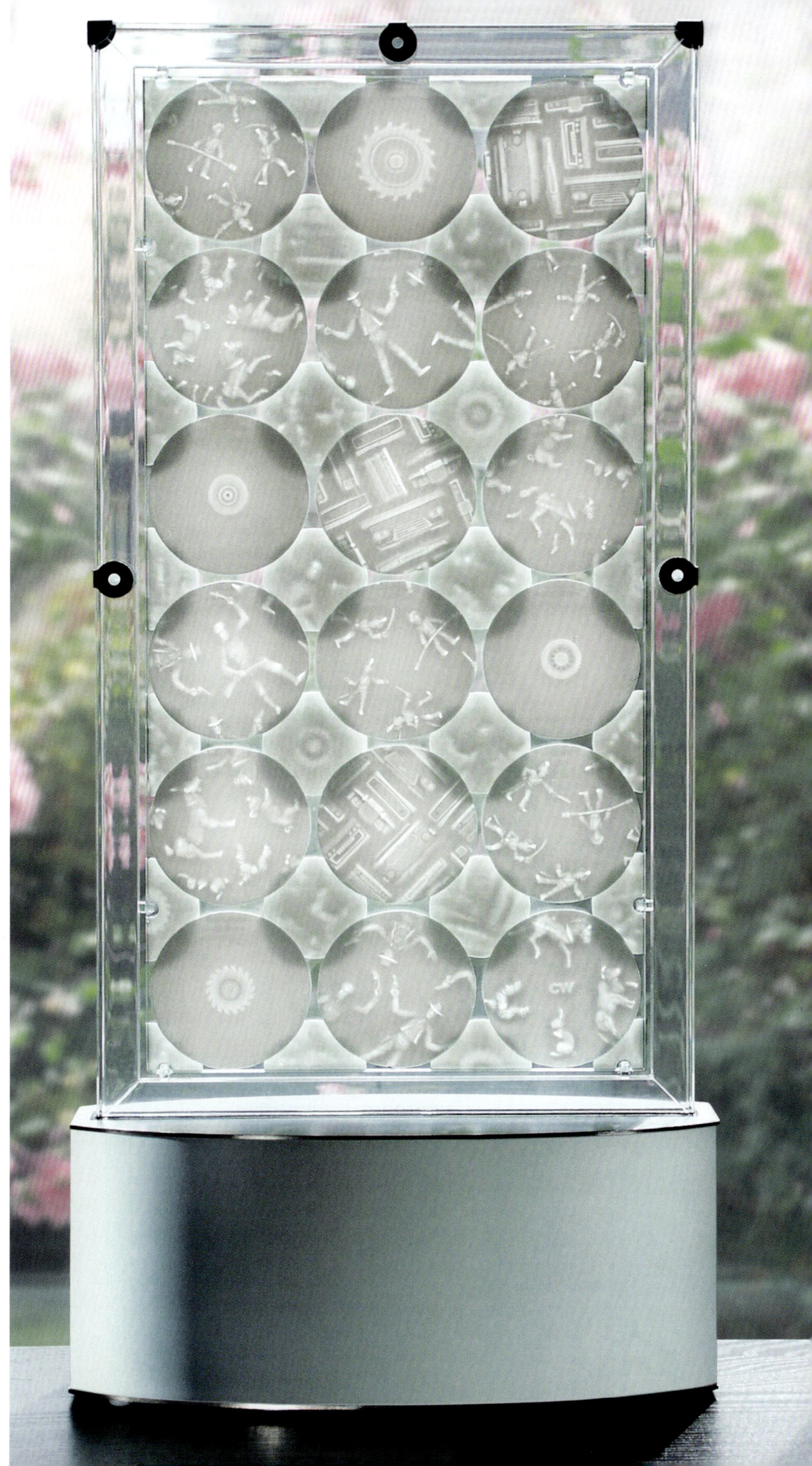

Describing the *'Toys' Panel*, Chris relates how he was fascinated by the markings that his toys left in plasticine.

> *The* 'Toys' Panel *revisits this simple, very direct and unsophisticated technique , though now as an adult it is the imperfect impressions it creates that appeals – clearly defined elements blending with vague distortions echoing the memories of my own childhood.*[7]

Organic Modular Construction is slipcast, hand-carved, and glazed bone china.

> *Inspiration for a work often comes to me via a dream, where upon waking I make a frantic sketch to capture the essence of what I had seen. This particular structure seemed to flow from my observations and meditations on 'the architecture of the natural world.'*[8]

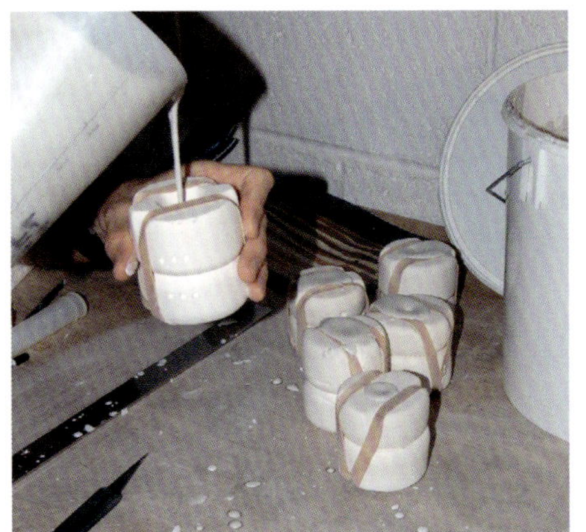
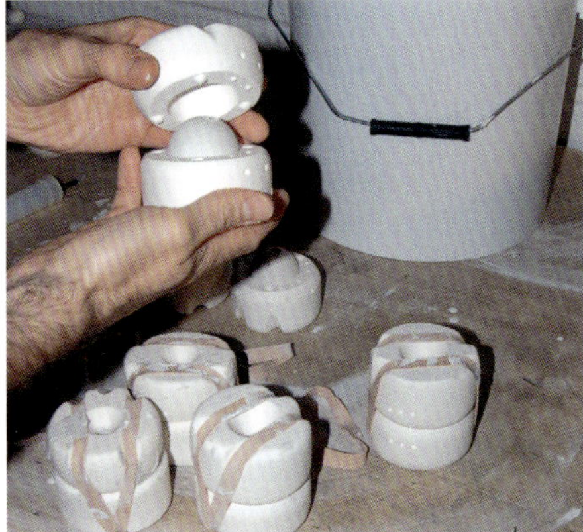
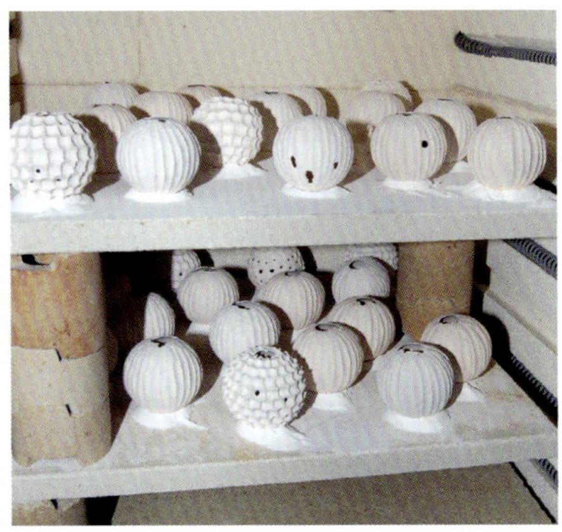
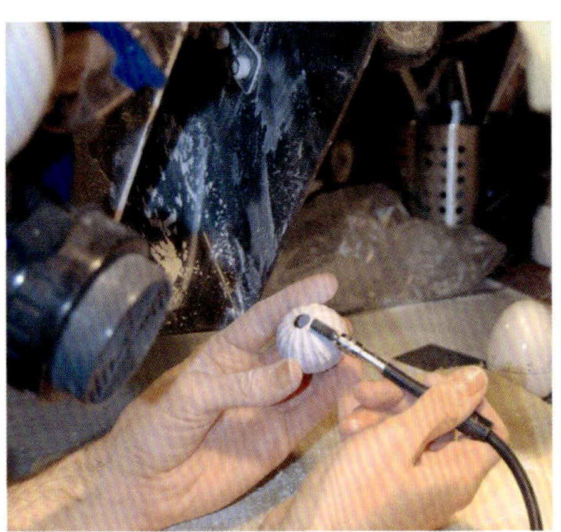

LEFT: Stages in making *Organic Modular Construction*, Chris Wight: (clockwise from top left) slipcasting; releasing from moulds; carving; packed in the kiln. *Photo: Chris Wight.*

[7] Chris Wight, cone8.co.uk/toys-panel (accessed 24 January 2025).

[8] Chris Wight, cone8.co.uk/project/organic-modular-construction (accessed 25 March 2025).

RIGHT (TOP TWO ROWS): Stages in making *Organic Modular Construction*, Chris Wight: (clockwise from top left) applying a small bead of glaze at each point of contact to create a joint; close-up of glazed joints; kiln brick support to hold work while firing; the completed structure after firing. Glaze bonding. *Photo: Chris Wight.*

BOTTOM LEFT: *Organic Modular Construction* (detail), Chris Wight. Slipcast hand-carved and glaze-bonded bone china. Bone china, sterling silver (wire support), concrete (base). (various) 21 x 8 x 20 cm (w x d x h). *Photo: Chris Wight.*

BOTTOM RIGHT: *Organic Modular Construction* (detail), Chris Wight. *Photo: Chris Wight.*

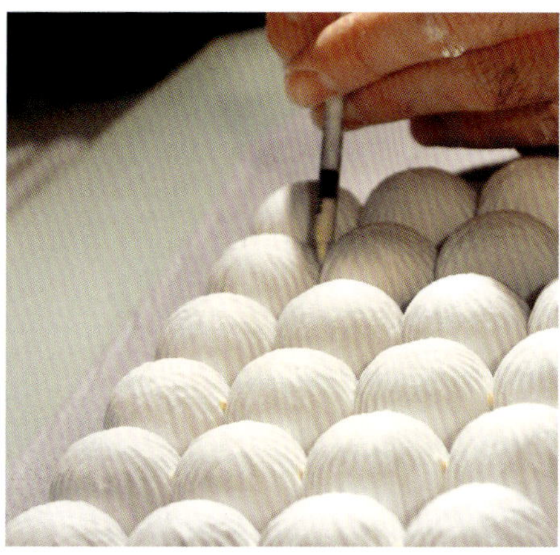

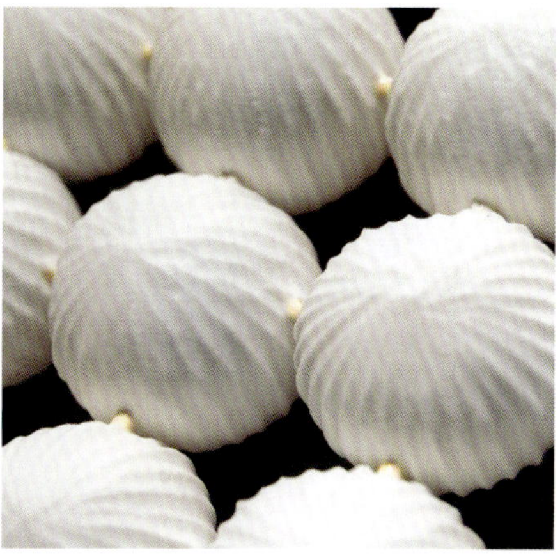

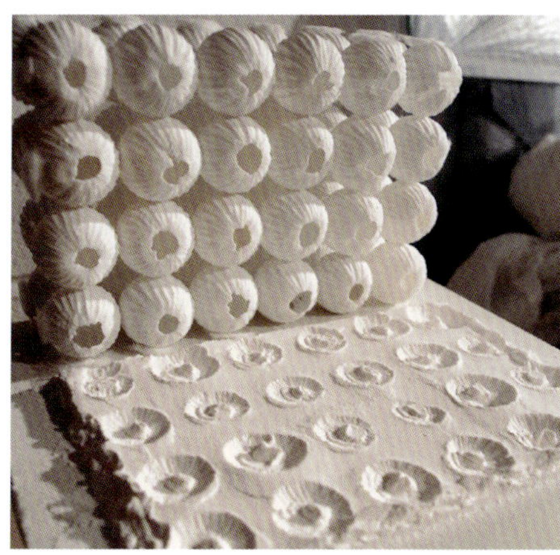

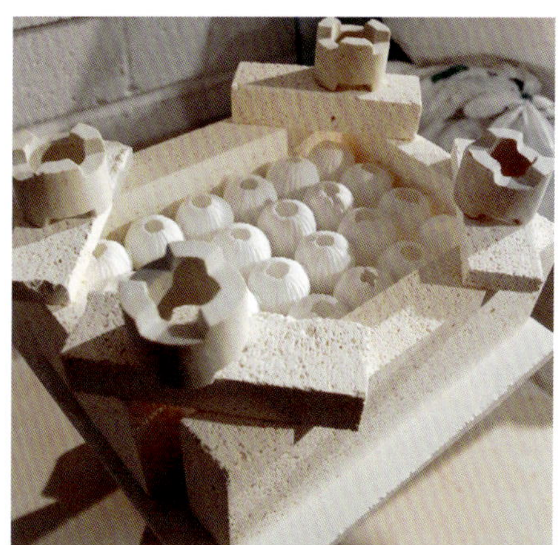

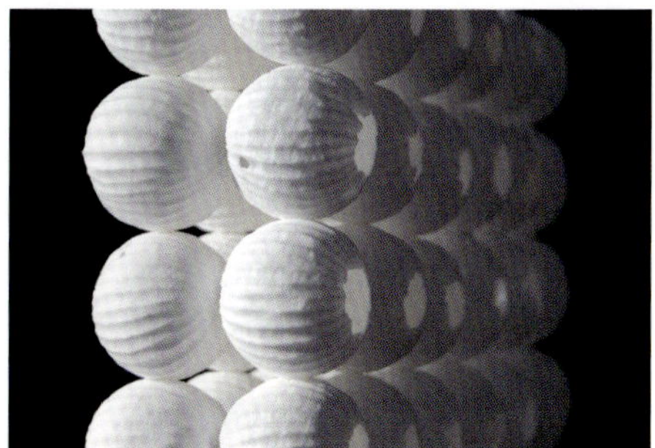

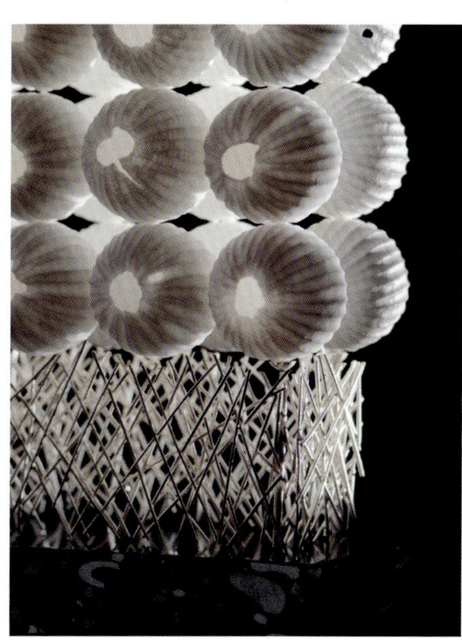

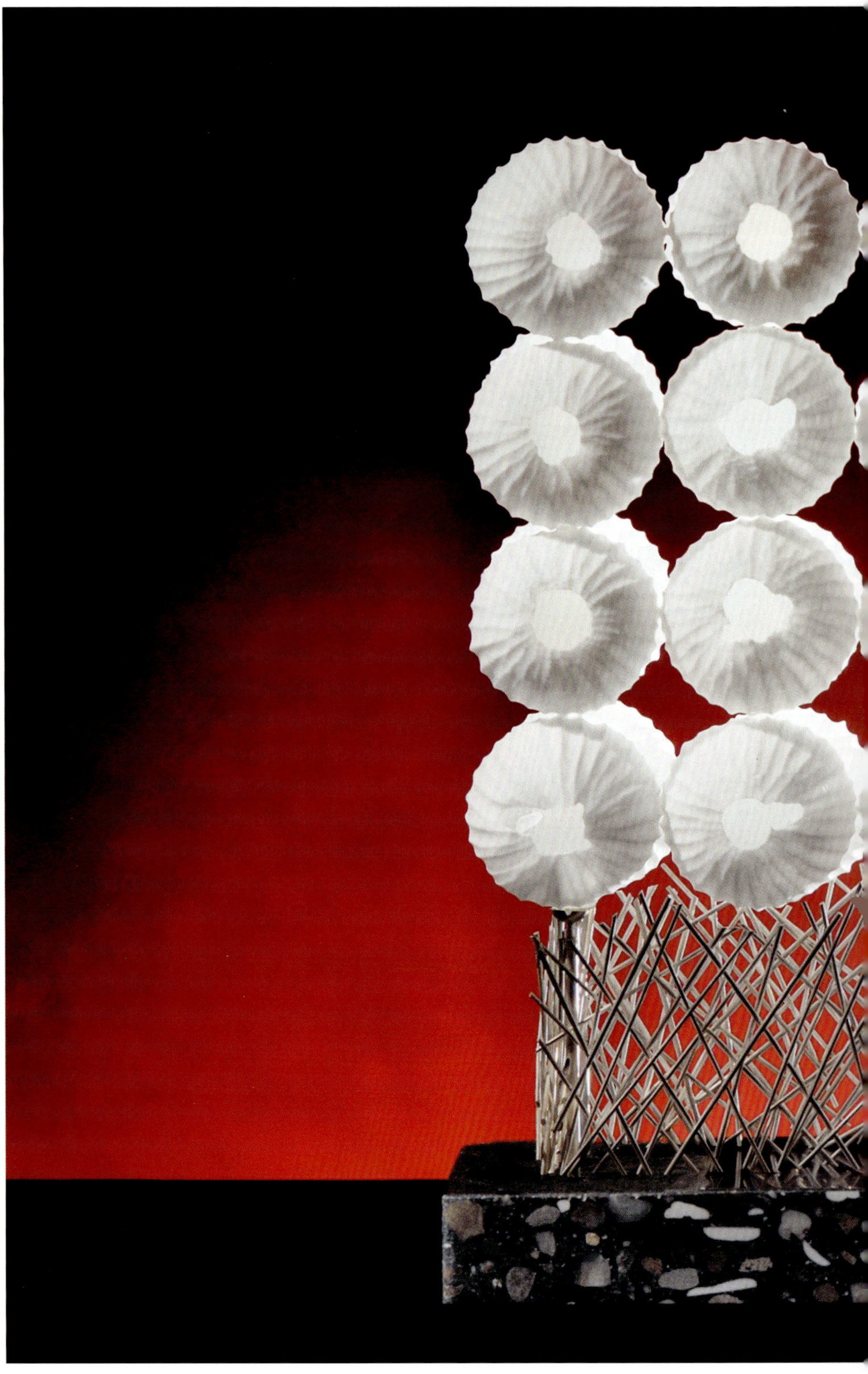

RIGHT: *Organic Modular Construction*, Chris Wight. Slipcast hand-carved and glaze-bonded bone china. Bone china, sterling silver (wire support), concrete (base). (various) 21 x 8 x 20 cm (w x d x h). *Photo: Chris Wight.*

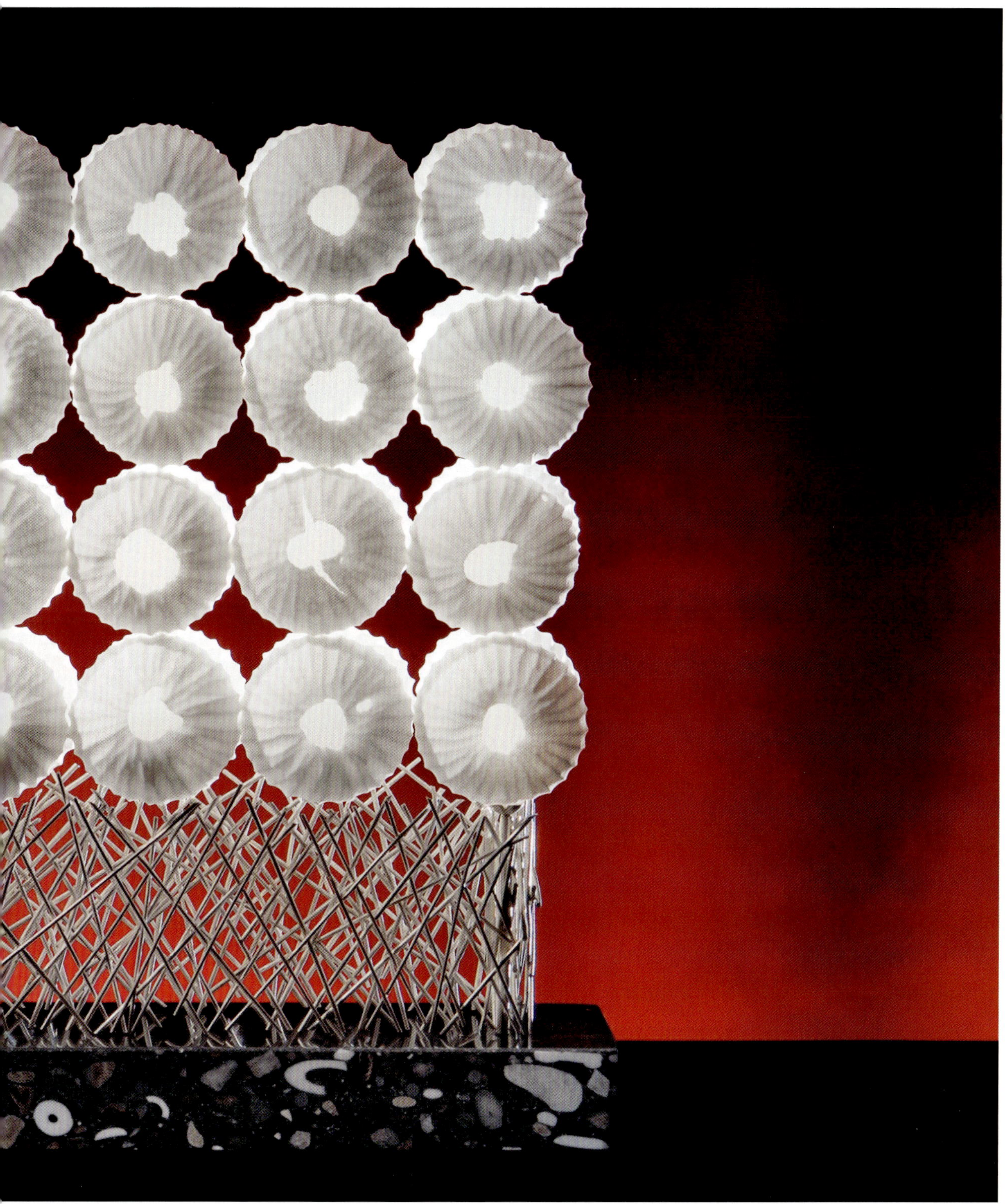

Margaret O'Rorke (UK)

Margaret, when describing her work, says:

> *For over 40 years, the translucency of fine high-fired porcelain has led me to create thrown porcelain forms incorporating light.*
>
> *Inspiration can come from anywhere. A generous invitation from Ryoji Koie to go to Japan for three months to make my work in his studio was the first of many residencies spent in other countries, including Australia, USA, China, Denmark, and Finland. All have contributed to the development of my works that begins on the potter's wheel. I re-form the thrown porcelain, then, using a blowtorch to firm the soft clay, I can handle and join it. The works are fired to 1300°C in a gas kiln, mostly in a reduced oxygen atmosphere.*
>
> *Engineering and incorporating the electrical fittings has been a long and arduous process; I always want the fittings not to be seen.*
>
> *I also create thrown and re-formed porcelain models that are manufactured. I work in the ceramics industry in Stoke-on-Trent, the Royal Copenhagen Porcelain factory in Denmark, and in Jingdezhen and Xian in China to develop lively thrown manufactured models.*
>
> *Collaborating with the Finnish art weaver Sirkka Paikkari has resulted in our creating a number of lit woven steel, fibre optics, and porcelain installations and chandeliers.*

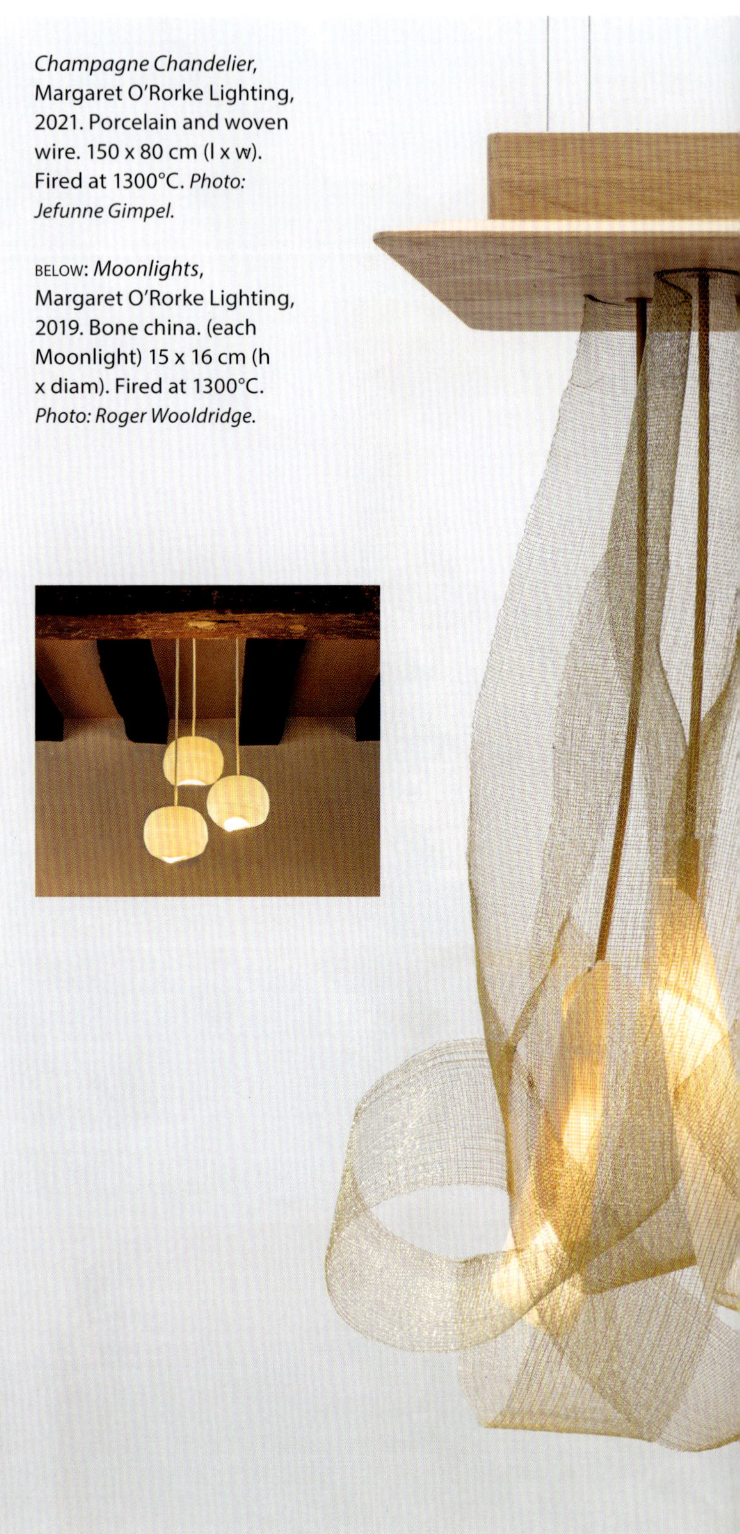

Champagne Chandelier, Margaret O'Rorke Lighting, 2021. Porcelain and woven wire. 150 x 80 cm (l x w). Fired at 1300°C. *Photo: Jefunne Gimpel.*

BELOW: *Moonlights*, Margaret O'Rorke Lighting, 2019. Bone china. (each Moonlight) 15 x 16 cm (h x diam). Fired at 1300°C. *Photo: Roger Wooldridge.*

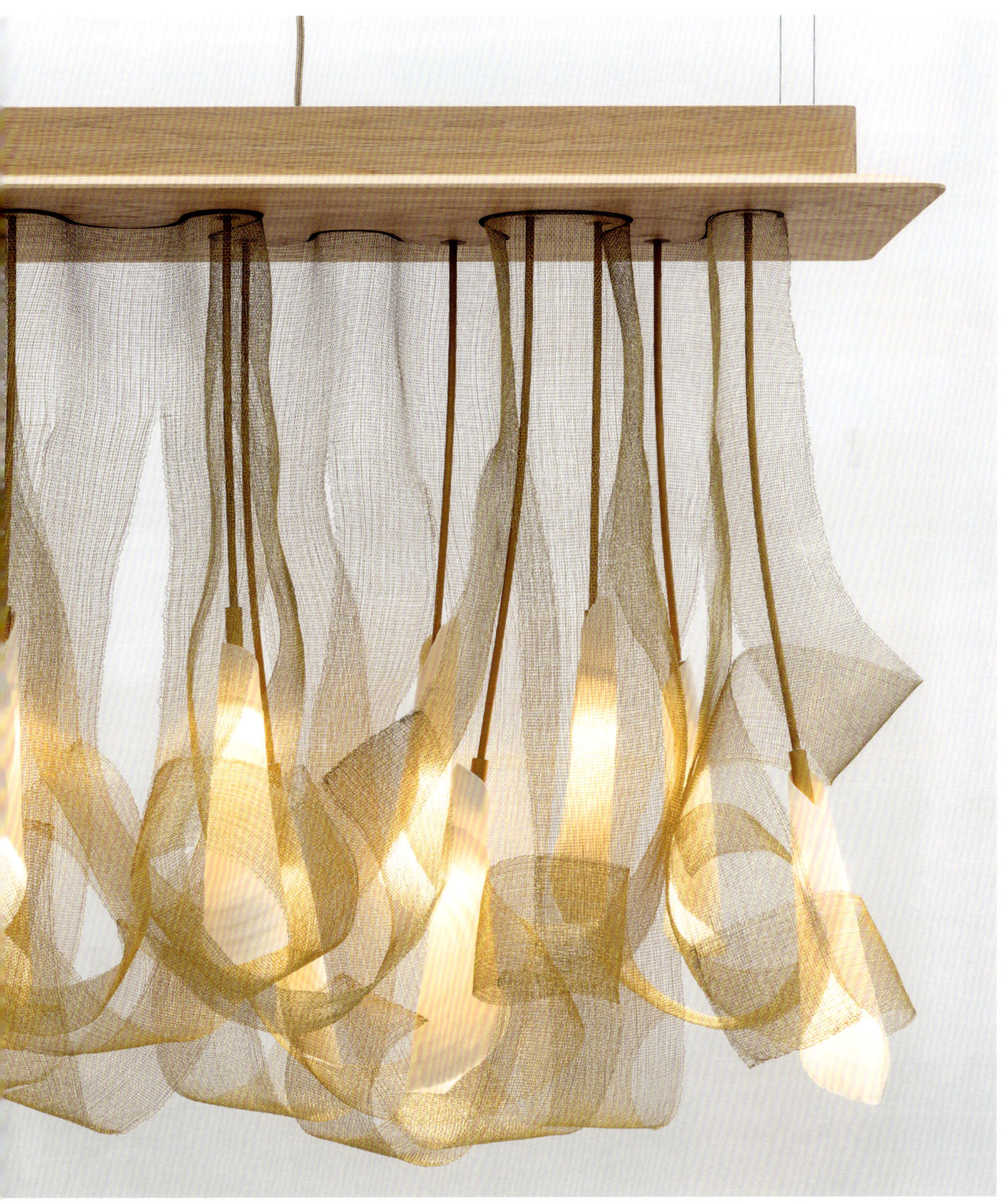

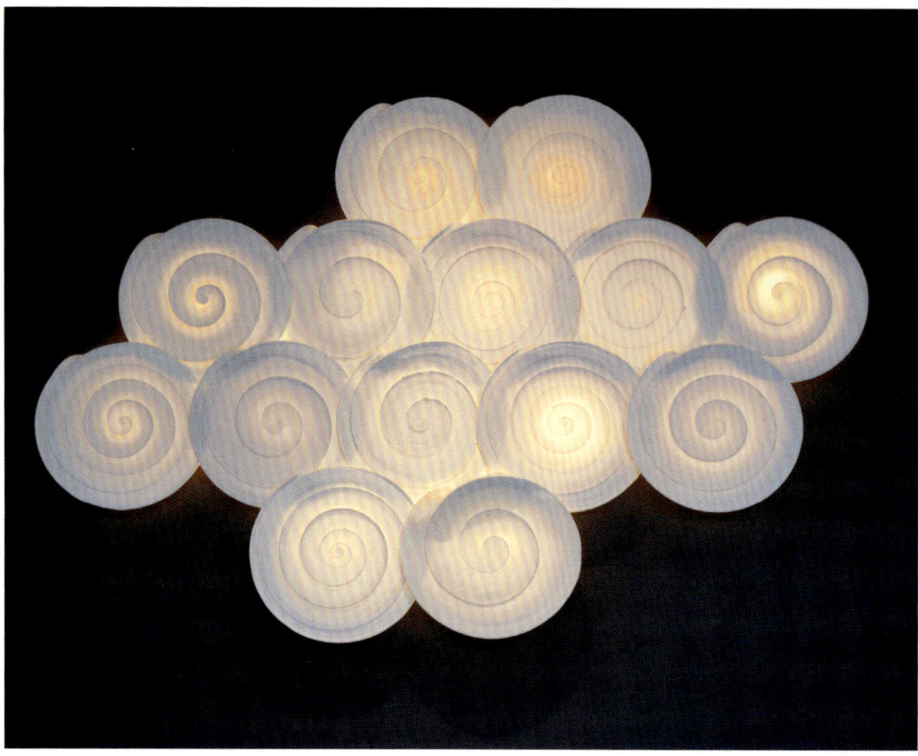

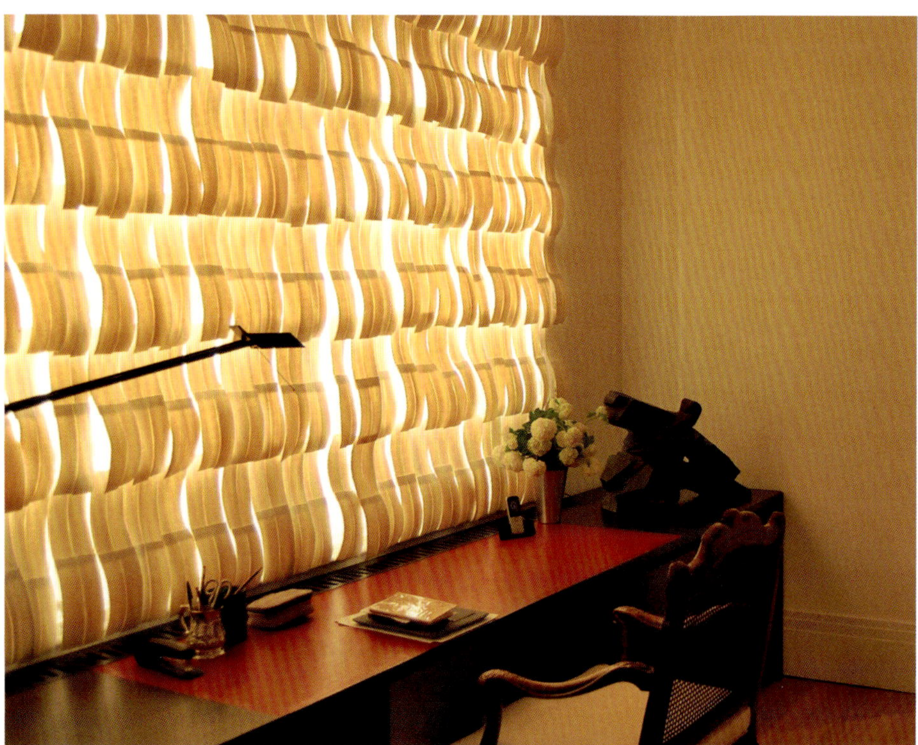

LEFT: *Cloud Discs*, Margaret O'Rorke Lighting, 2017. Porcelain. (complete work) 85 x 128 cm (h x w). Fired at 1300°C. *Photo: Kasia Konairek.*

BOTTOM LEFT: *Porcelain Curtain* (illuminated), Margaret O'Rorke Lighting, 2016. Porcelain. 183 x 222 cm (h x w). Fired at 1300°C. *Photo: James Crabbe.*

OPPOSITE RIGHT: *Vertical Rods*, Margaret O'Rorke Lighting, 2019. Porcelain. (each pod) 19 x 14 cm (diam x d). Fired at 1300°C. *Photo: Roger Wooldridge.*

OPPOSITE FAR RIGHT: *Woven Light* and *Porcelain Curtain*, Margaret O'Rorke Lighting, 2009. Porcelain and light fibres. 300 x 80 x 16 cm (h x w x d). Fired at 1300°C. *Photo: Rauno Träskelin.*

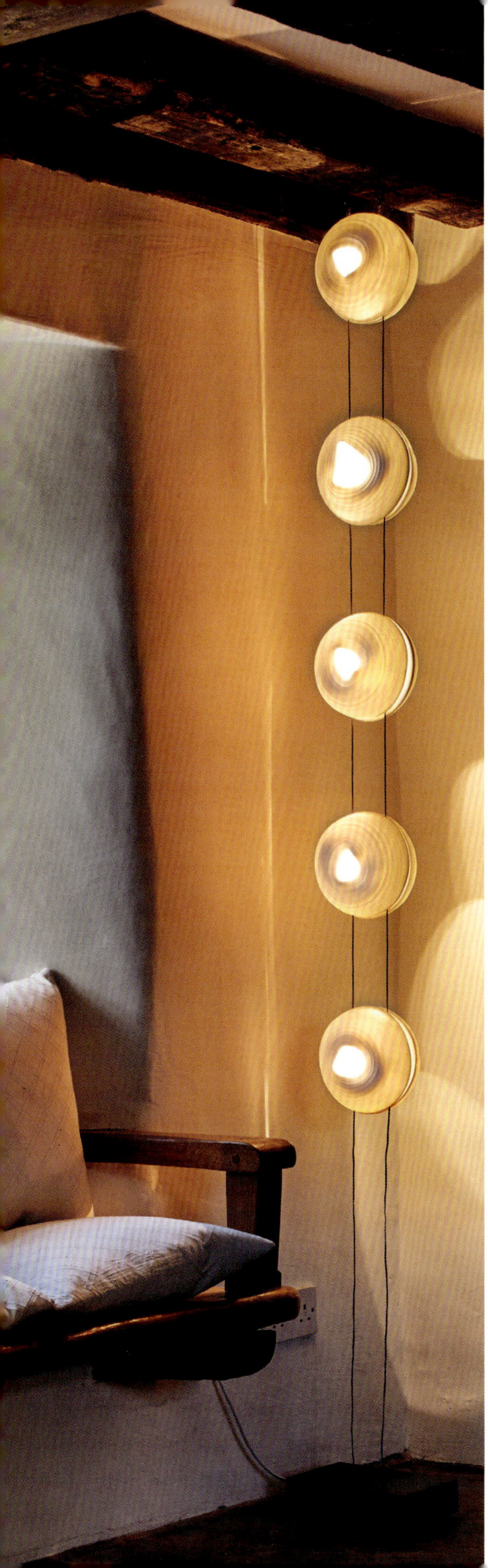

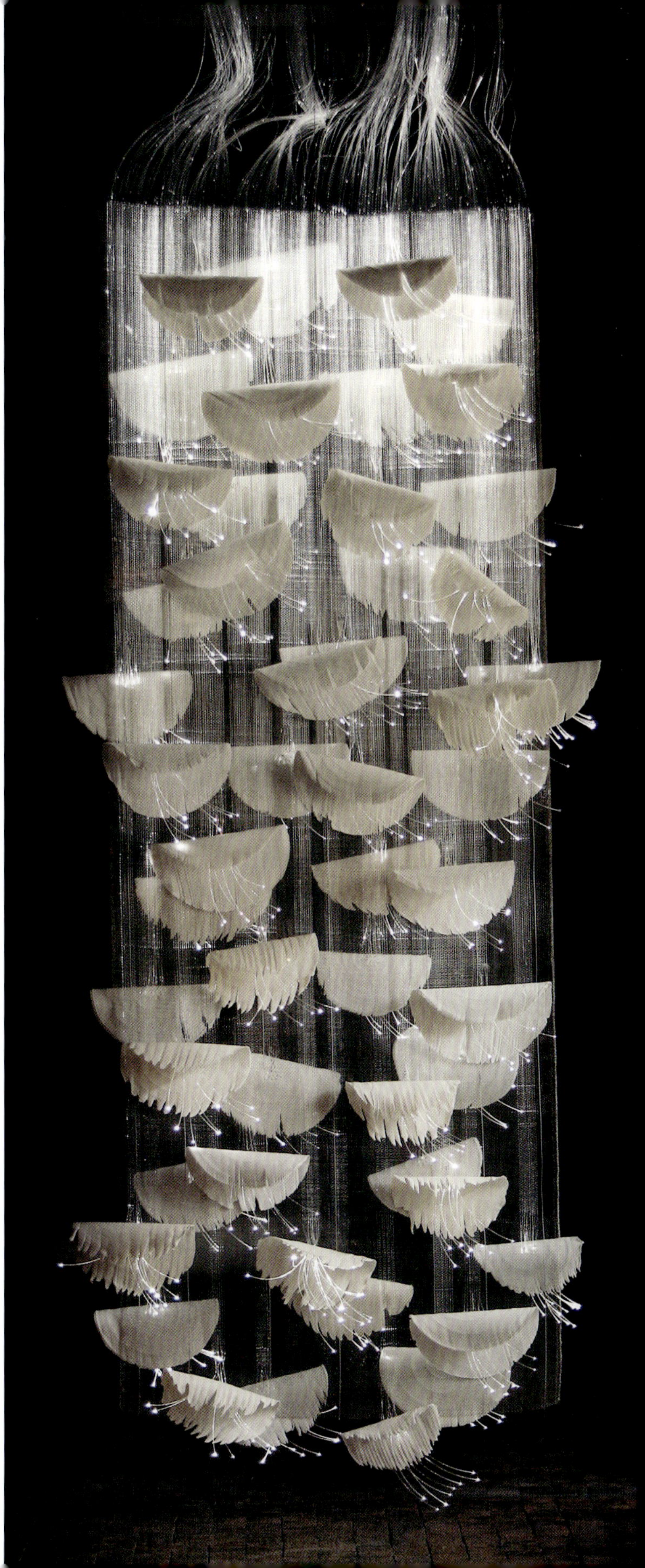

Sasha Wardell (UK)

Sasha describes the processes she uses to exploit the translucency of the bone china:

> *The first focuses on specific points of translucency by using a 'slicing and layering' technique. Subtle facets on the curved outer surface of the pieces involve slicing through three or four coats of different coloured slips to reveal underlying and increasingly transparent layers. Similar techniques exist in glass production, which serve as the inspiration and starting point for this body of work.*

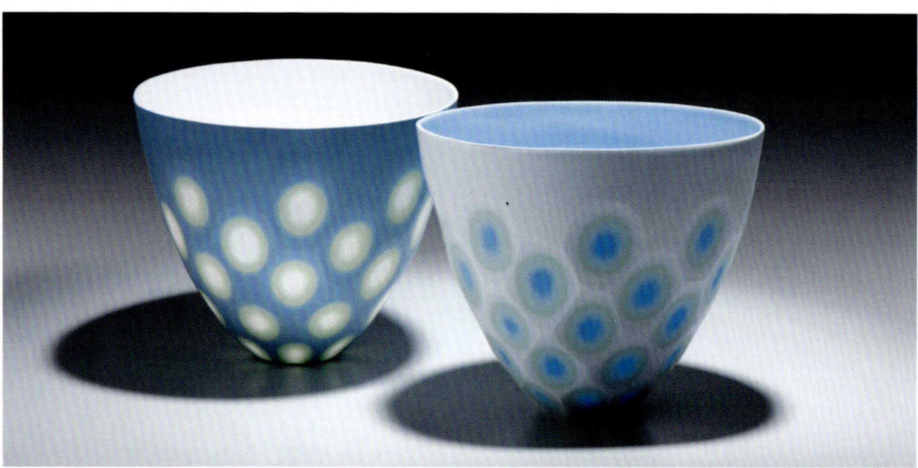

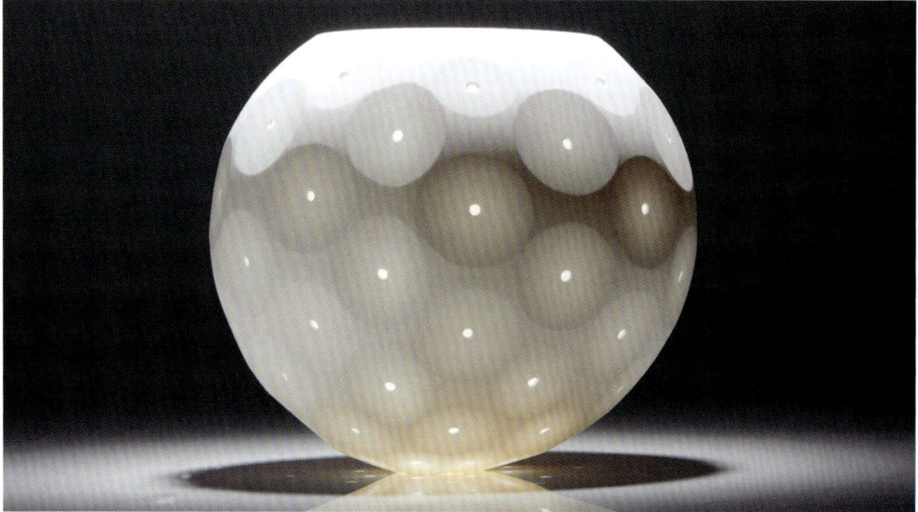

LEFT: *Pair of Galaxy Bowls*, Sasha Wardell, 2024. Slipcast bone china, usng a layered and sliced technique. 23 cm (h). Fired at 1280°C. *Photo: Mark Lawrence.*

BOTTOM LEFT: *Sparkle Globe*, Sasha Wardell, 2024. Slipcast bone china, using a sliced and pierced technique. 16 cm (diam). Fired at 1280°C. *Photo: Mark Lawrence.*

OPPOSITE: *Shoal Pendant Light*, Sasha Wardell, 2024. Slipcast bone china, using a layered and incised technique. 35 cm (h). Fired at 1280°C. *Photo: Mark Lawrence.*

> *The second technique uses the same layering process as above; however, marks are made with a loop tool to incise or gouge through the layers. This has the same effect of revealing the underlying colours until the interior layer is exposed. Marine references inform the pattern making in this collection.*

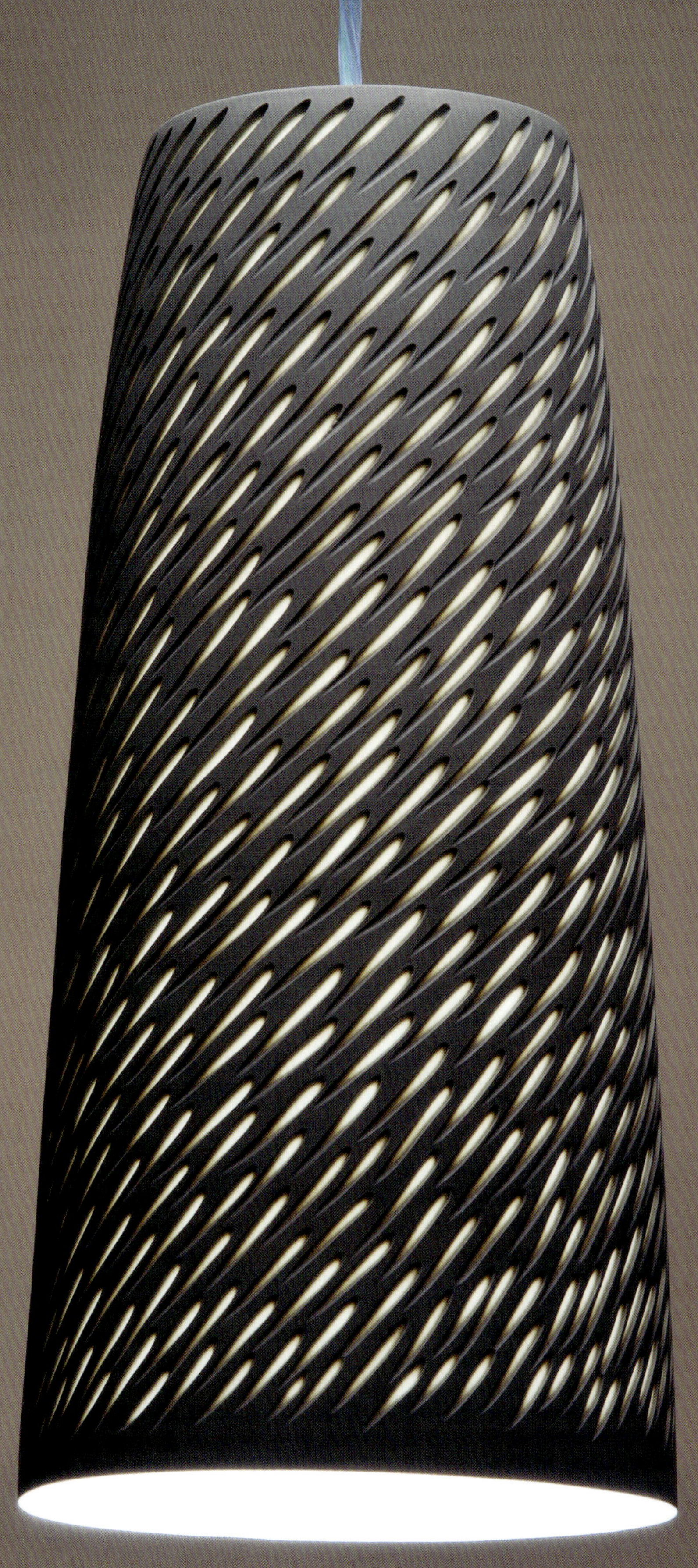

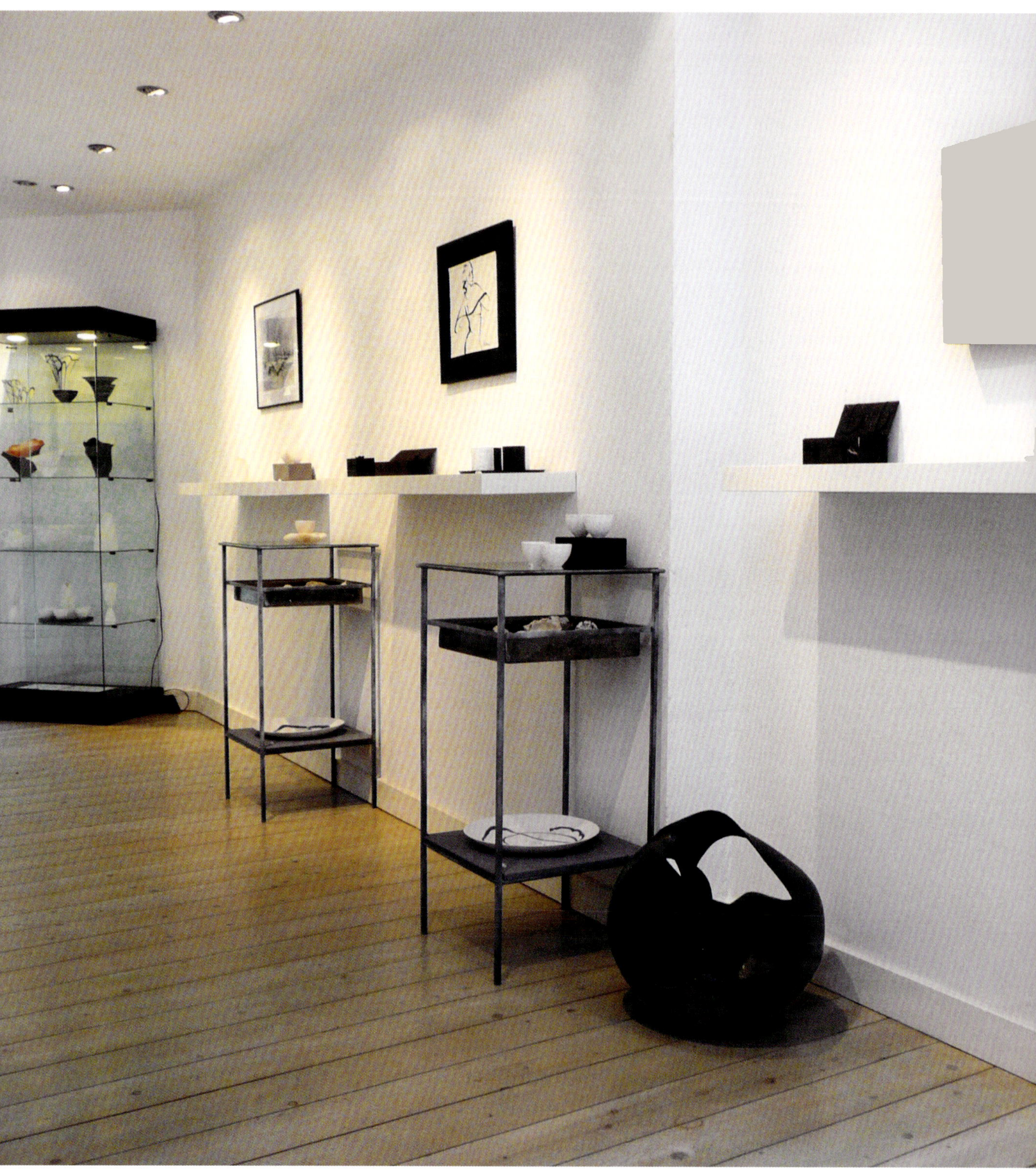

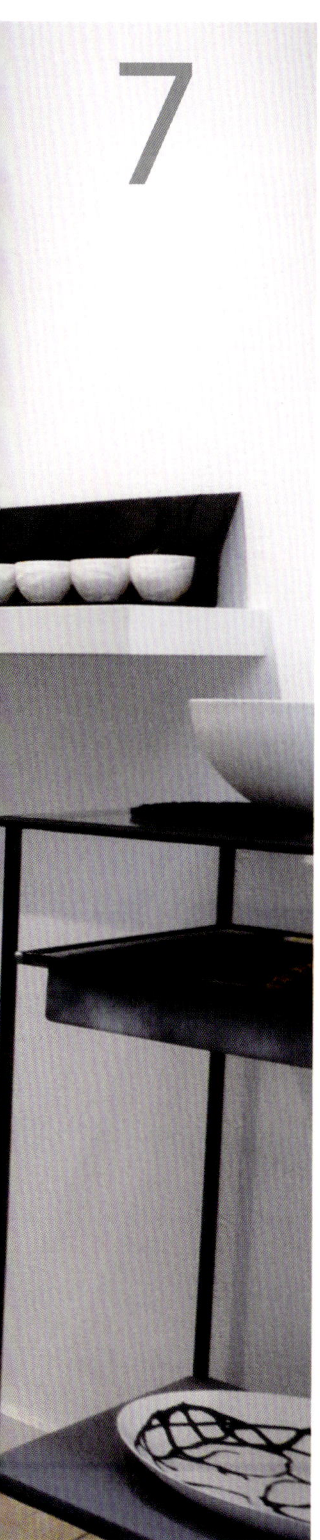

7 My work since 2006

It had always been a dream of mine to own my own gallery, so after ten years of living in Australia and the loss of my husband, I returned to England in 2006 to be near my sons. Ronald Pile, owner of Primavera in Cambridge until 1999, had decided to sell his gallery in the historic city of Ely. The gallery is situated in a Grade II listed building, dating from 1658, opposite Ely Cathedral, and overlooking St Mary's Church and Oliver Cromwell's House.

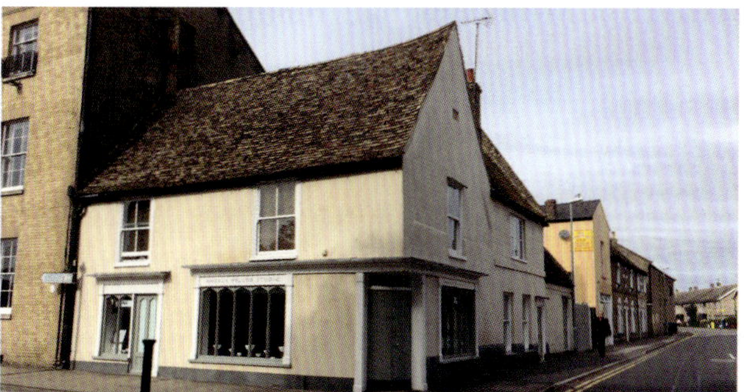

LEFT: The Angela Mellor Gallery, exterior, 2019.

FAR LEFT: The Angela Mellor Gallery, upper gallery, 2010.

BELOW: The Angela Mellor Gallery, 2019. *Photo: Layton Thompson for* Ceramic Review.

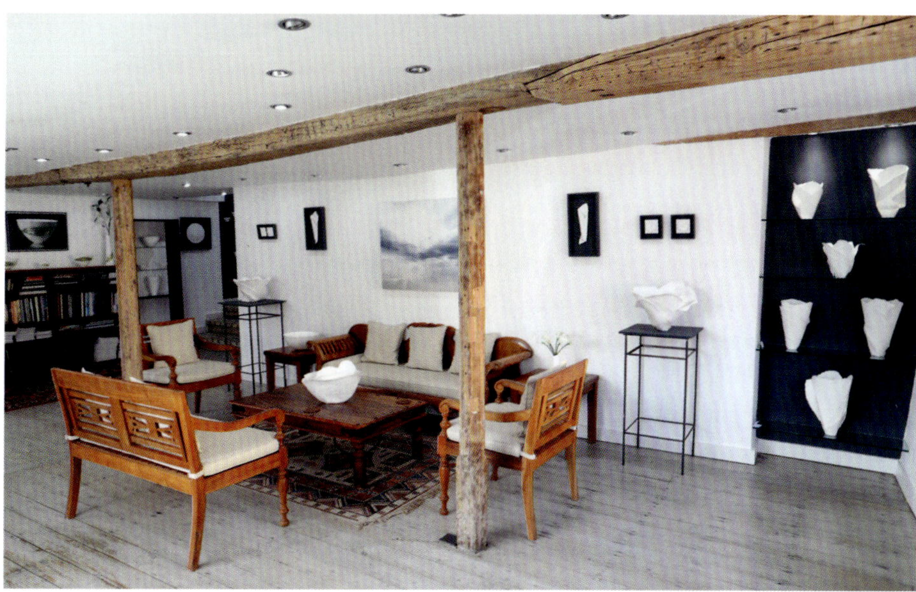

I lived above the gallery and, while waiting for my kiln and studio equipment to be shipped back from Perth, I began organising exhibitions for the gallery. I planned four mixed exhibitions a year, to include ceramics, sculpture, painting, textiles, and jewellery. The first exhibition was at Christmas 2006, and as I hadn't had time to source many UK artists, I decided to invite international artists, including some Australian ceramists who I knew were working in bone china and porcelain. It was to be an all-white show for Christmas, entitled *I'm Dreaming of...* I had missed seeing a white Christmas while in Australia, and I thought it would be nice to experience one in my new gallery.

I exhibited works by Perth ceramist Pippin Drysdale, who has since received several major accolades, including being awarded an Honourable Doctorate of Arts by Curtin University, and being named a Living Treasure of Western Australia. She exhibited a beautiful collection of vessels in the subtlest shades of white. Also included were Sandra Black's intricate pieced white bowls, and a group of stunning white translucent bowls by Switzerland's Arnold Annen.

Contrast was provided by Jenny Beavan's highly textured porcelain wall panels, which incorporated combustible material with china clay, glaze, glass, sand, and pebbles found on the beach close to her home in Cornwall. Exquisite gold and silver jewellery by Perth's Dorothy Erickson adorned the gallery, next to jewellery by local artists.

Subsequent exhibitions were based on the four seasons, as I had also missed the changing light and weather patterns that the English seasons bring.

I was very proud and honoured to be elected a Member of the IAC in 2005, and I exhibited a piece at *World Clay*, the IAC members' exhibition held at the Latvian National Museum of Art, Riga, in 2006.

In 2006, my work was also shown at *COLLECT* at the Victoria & Albert Museum in London, represented by the Raglan Gallery. I was pleased to keep ties with Australia, exhibiting in *And there was Light* at Cudgegong Gallery in New South Wales. In 2008, I took part in an exhibition called *Porcelain Bowls from the UK* at St Joseph Galerie in Leeuwarden, Netherlands.

In the 2008 IAC members' exhibition in Xian, China, I exhibited a large *Sea Bowl* painted with soluble salts.

It was quite an experience seeing contemporary ceramics in a place where esteemed traditional ceramics from China were everywhere. Special new museums had just been built to house diverse new ceramics from different countries, and these have been extended over the years. It was wonderful to visit Jingdezhen and see the huge, diligently and skilfully painted thrown porcelain pots. I also visited a studio where tiny eggshell-thin porcelain bowls are thrown, some with pierced decorative holes filled with glaze, others painted with the finest cobalt blue brushstrokes. I was mesmerised!

The *Coastal Light* series

Following on from my *Ocean Light* series in Australia, I was inspired by the coastline around the UK, often visiting the Norfolk and Suffolk coasts near my home in Ely. I was very aware of the difference in light in the UK; it is much more subdued, with cloudy skies – a far cry from the bright sunlight of Western Australia.

I spent some time developing my inspiration from the UK coastlines into the *Coastal Light* series. *Coastal Light* bowl II is a kinetic bowl inspired by the gentle rhythmic movements of deep-sea marine life. This piece was exhibited in the 58th International Competition of Contemporary Ceramic (Premio Faenza) in Italy in 2013.

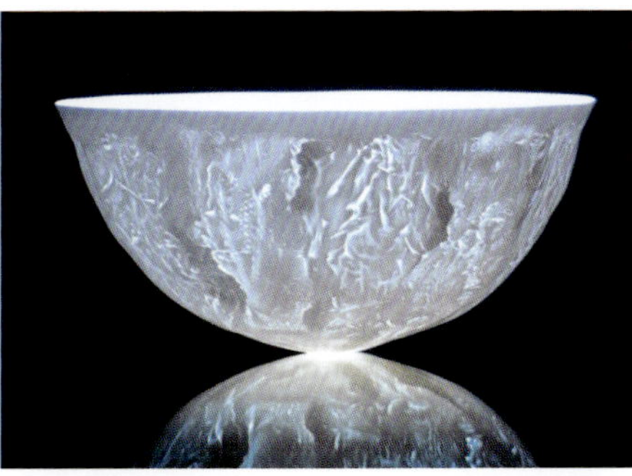

TOP: *Coastal Light* bowl II (kinetic), Angela Mellor, 2012. Bone china with paperclay inlay. 12 x 26 cm (h x w). *Photo: Stephen Bond.*

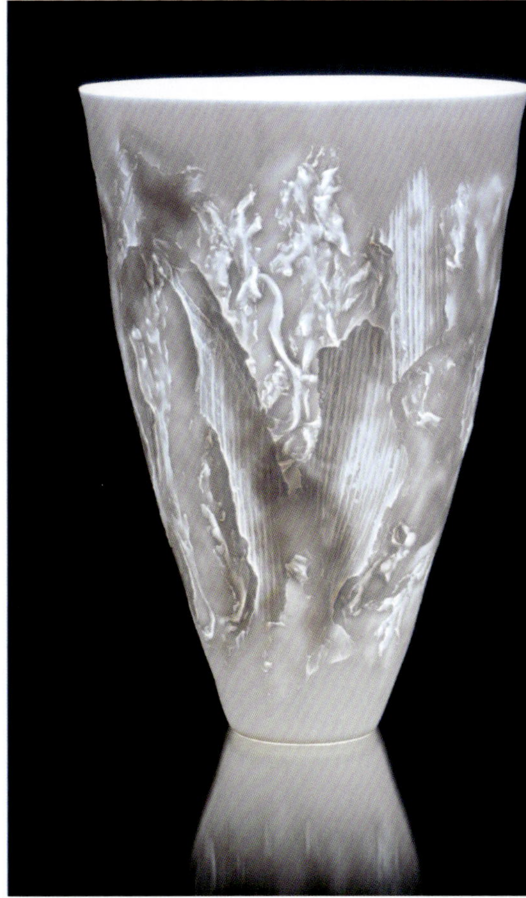

FAR RIGHT: *Coastal Light* vessel, Angela Mellor, 2012. Bone china with paperclay inlay. 40 x 11 cm (h x w). *Photo: Stephen Bond.*

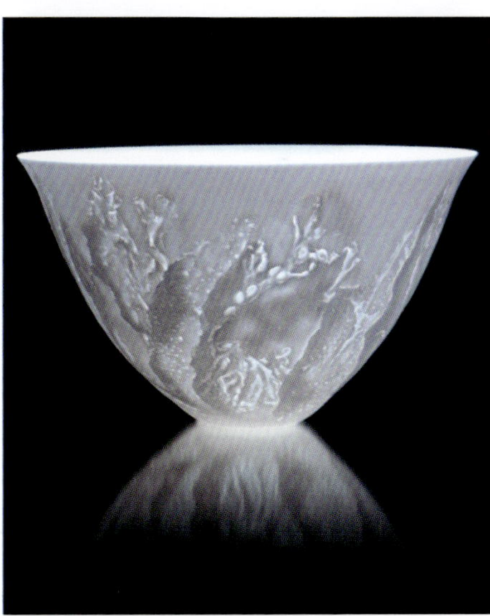

BOTTOM: *Coastal Light* bowl, Angela Mellor, 2012. Bone china with paperclay inlay. 13.5 x 22 cm (h x w). *Photo: Stephen Bond.*

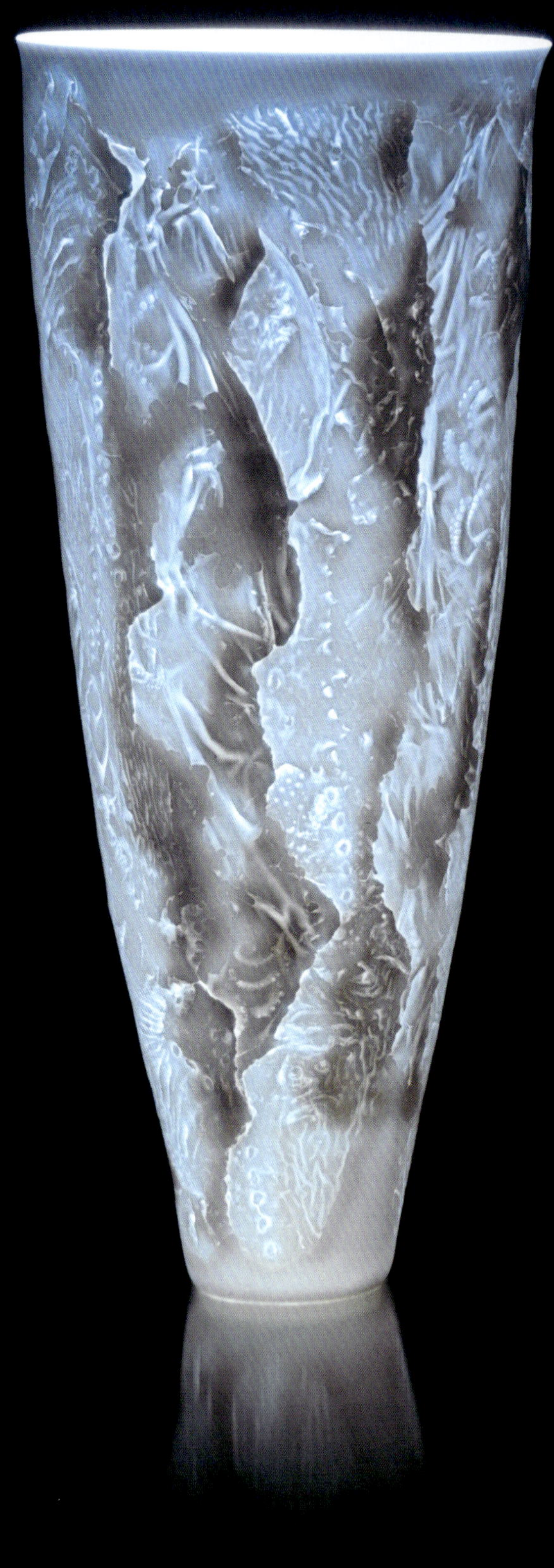

OPPOSITE: *Coastal Light* vessel, Angela Mellor, 2014. Bone china with paperclay inlay. 40 x 11 cm (h x w). *Photo: Stephen Bond.*

RIGHT: *Coastal Light* cylindrical vessel, Angela Mellor, 2012. Bone china paperclay. *Photo: Stephen Bond.*

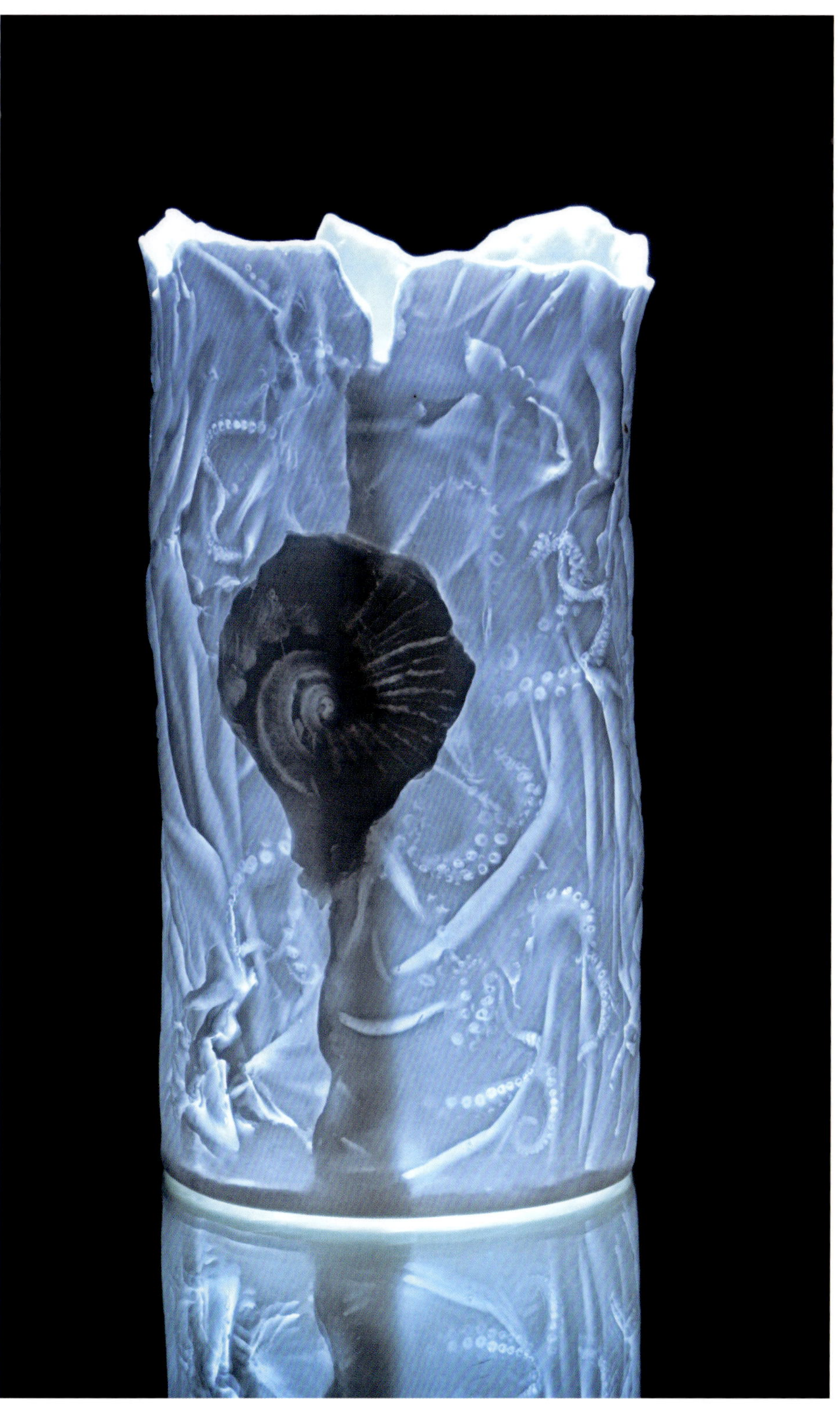

Attending workshops in other countries has been very rewarding, both in terms of learning and the interesting people I have met. In 2012, I participated in a paperclay workshop led by Rosette Gault and Jerry Bennett at the ICS, Kecskemét, Hungary, where I experimented with seaweed as a texture for the *Coastal Light* series. It was also a real privilege to work with Herend Porcelain, making our own paperclay and plaster moulds.

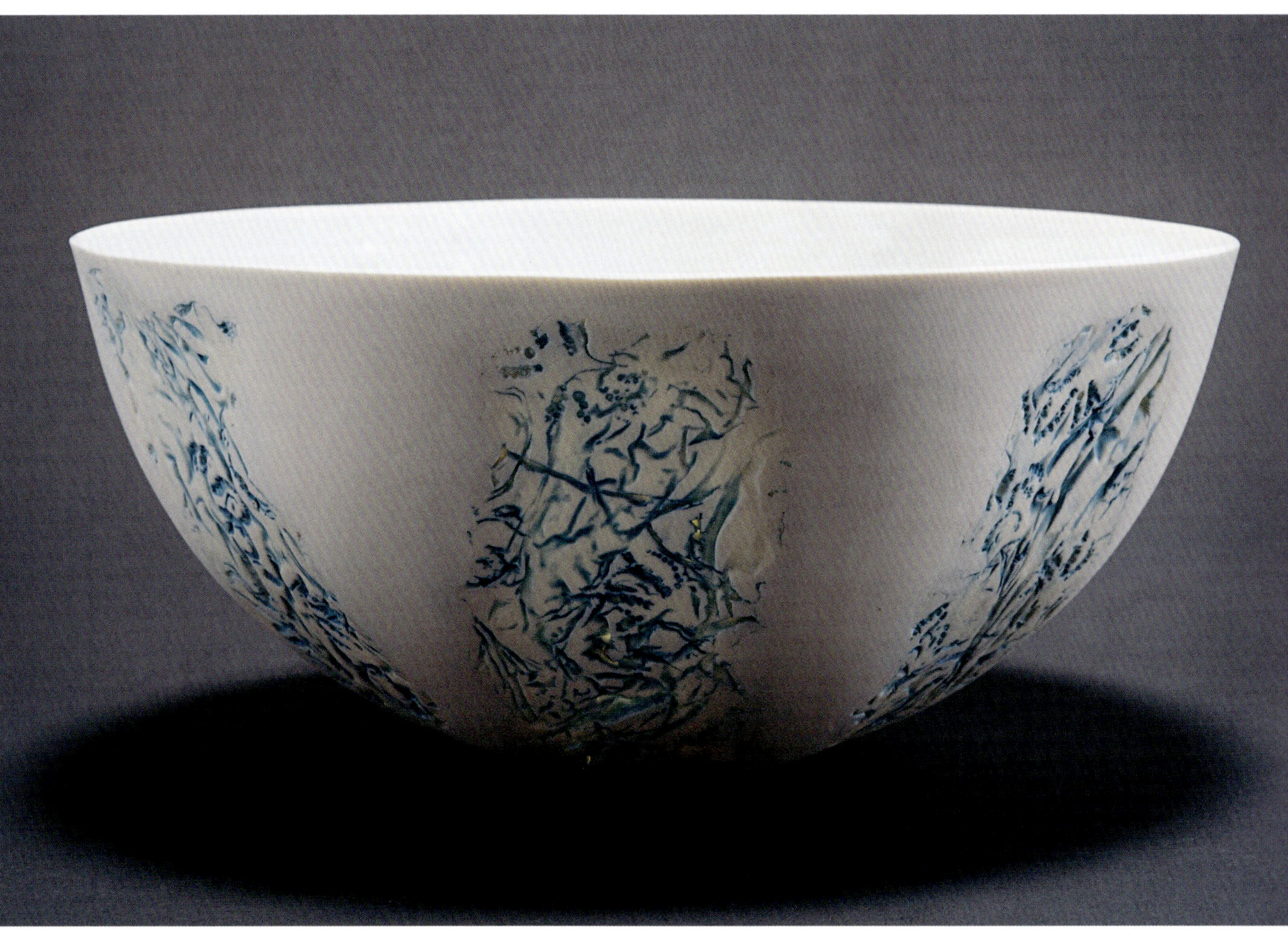

ABOVE: *Arctic Bowl*, Angela Mellor, 2012. Herend porcelain with paperclay inlay, painted with underglaze.

The *Spray* series

Often, while walking my dog around Thetford Forest in Norfolk, I was inspired by leaves and branches. I decided to incorporate these into my work, using textured paperclay. One of these finds, a large conifer branch, led to the *Spray* series.

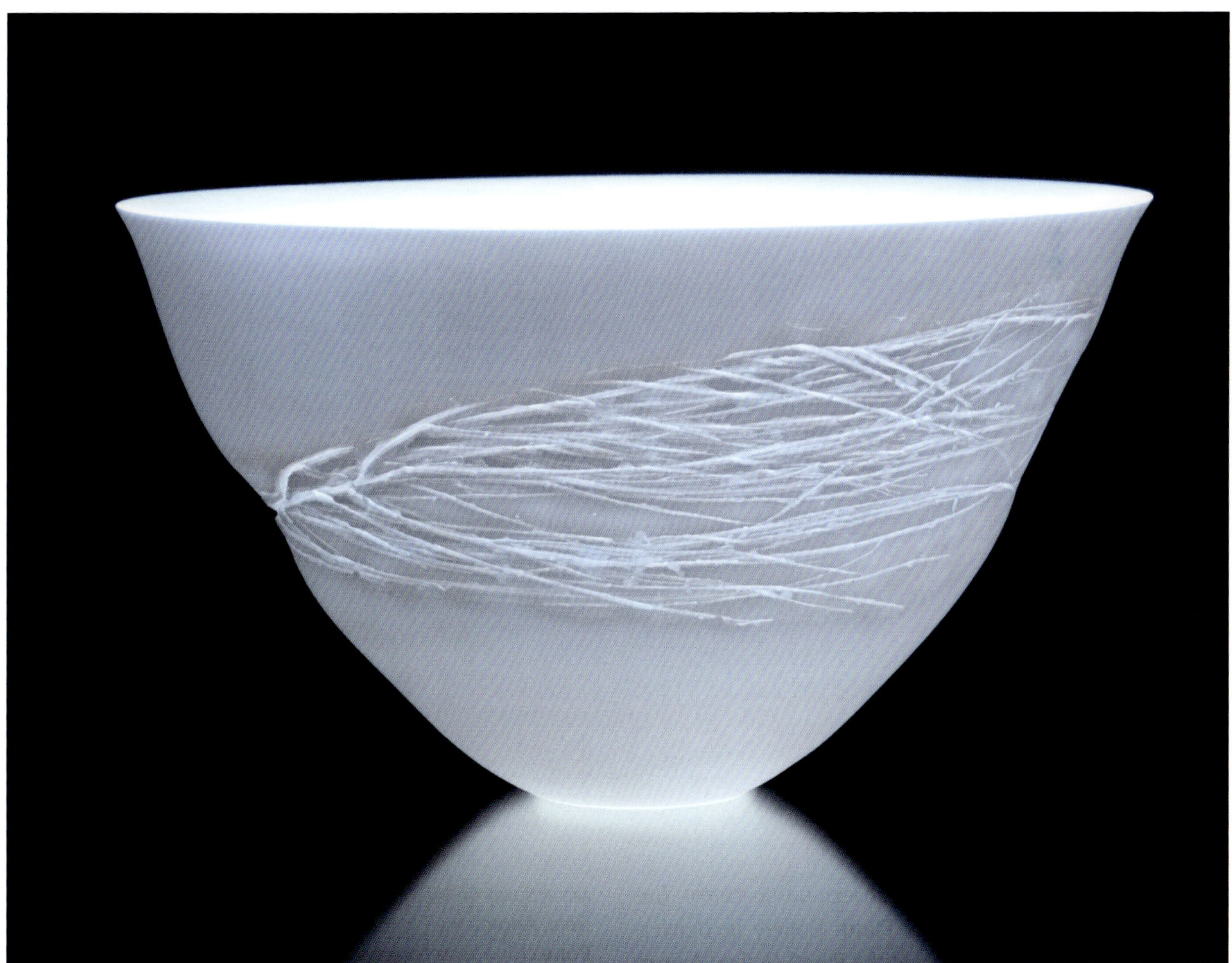

ABOVE: *Spray Bowl*, Angela Mellor, 2012. Bone china with paperclay inlay. 14 x 23 cm (h x w). *Photo: Stephen Bond.*

Another work from the *Spray* series, *Spray Bowl, Green and Gold* was exhibited at Ceramic Art London (CAL) 2014 at the Royal College of Art. The image of the bowl was used on the poster for the exhibition, which was displayed on the London Underground. The work was inspired by fleeting sunlight on water and was painted with green underglaze and applied gold leaf.

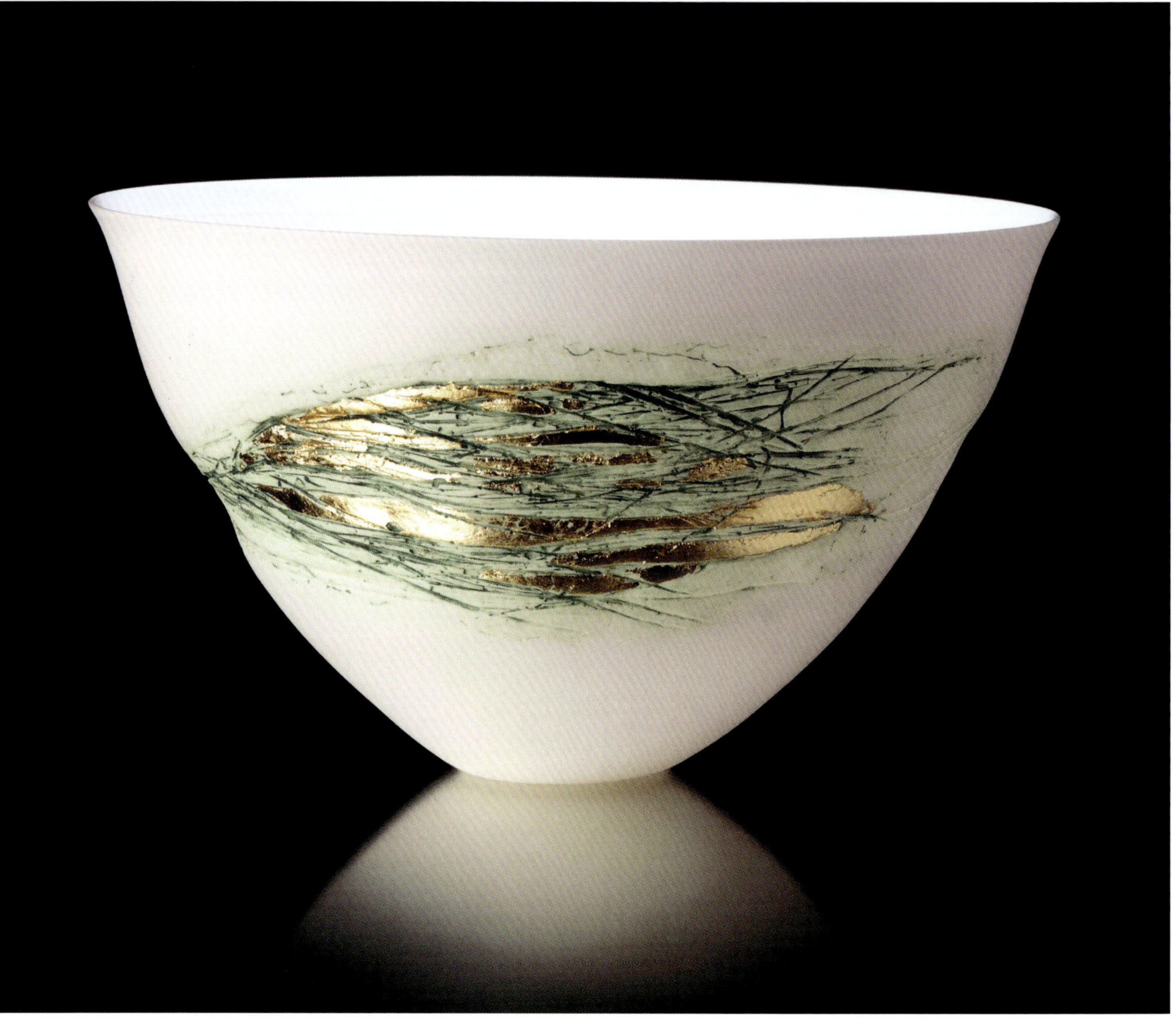

In 2015, I exhibited *Spindrift* in an exhibition called *The Slipcast Object Revisited* at the Plinth Gallery in Denver, Colorado, USA. The piece received the Plinth Gallery Award for Best Vessel. It was inspired by spray blown from the surface of the sea.

ABOVE: *Spray Bowl, Green and Gold*, Angela Mellor, 2013. Bone china with paperclay inlay, painted with underglaze and applied gold leaf. 13.5 x 23 cm (h x w). *Photo: Stephen Bond.*

OPPOSITE: *Spray Vessel, Green and Gold*, Angela Mellor, 2014. Bone china and paperclay inlay, painted with underglaze and applied gold leaf. 25 x 17 cm (h x w). *Photo: Stephen Bond.*

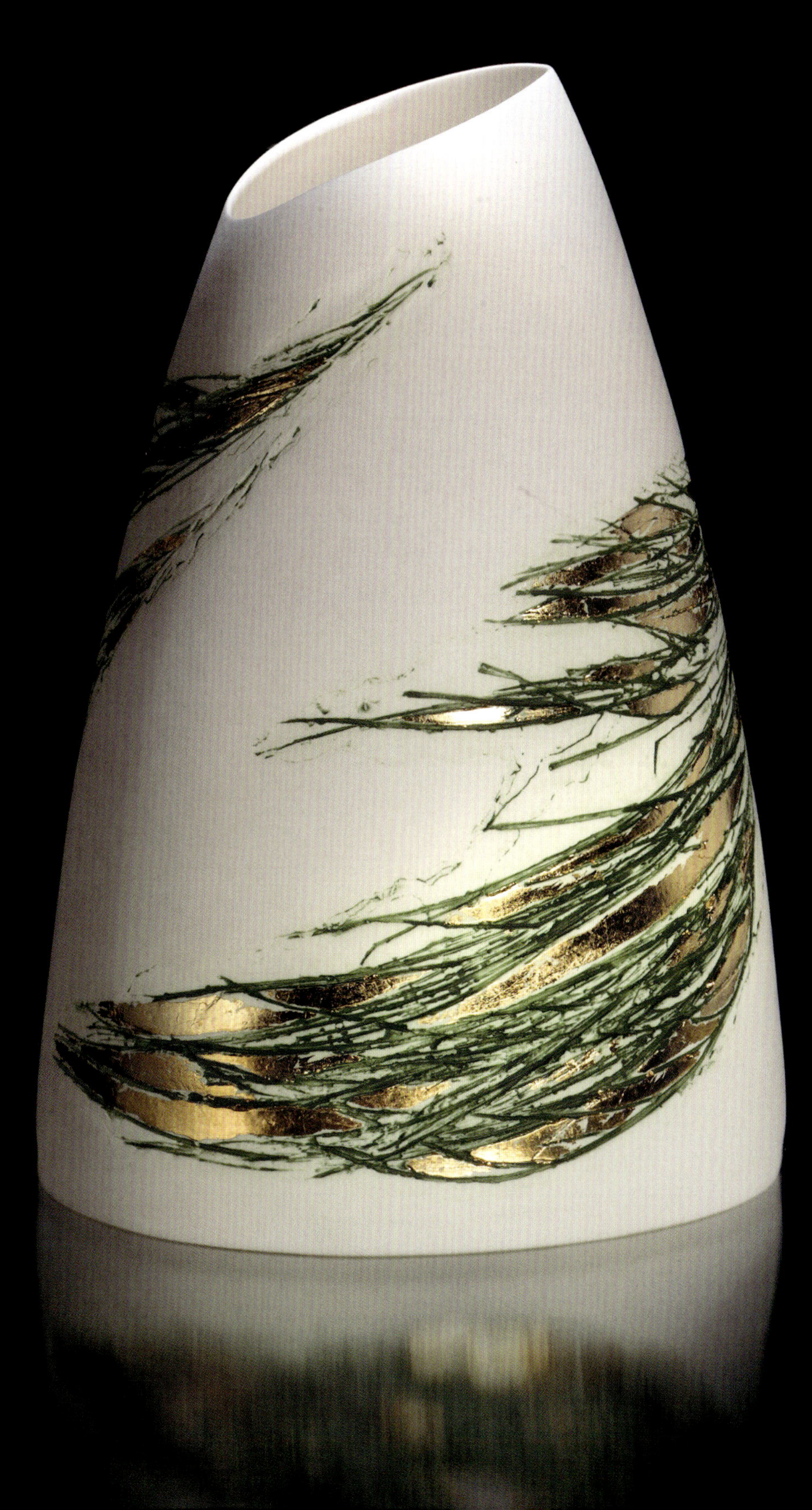

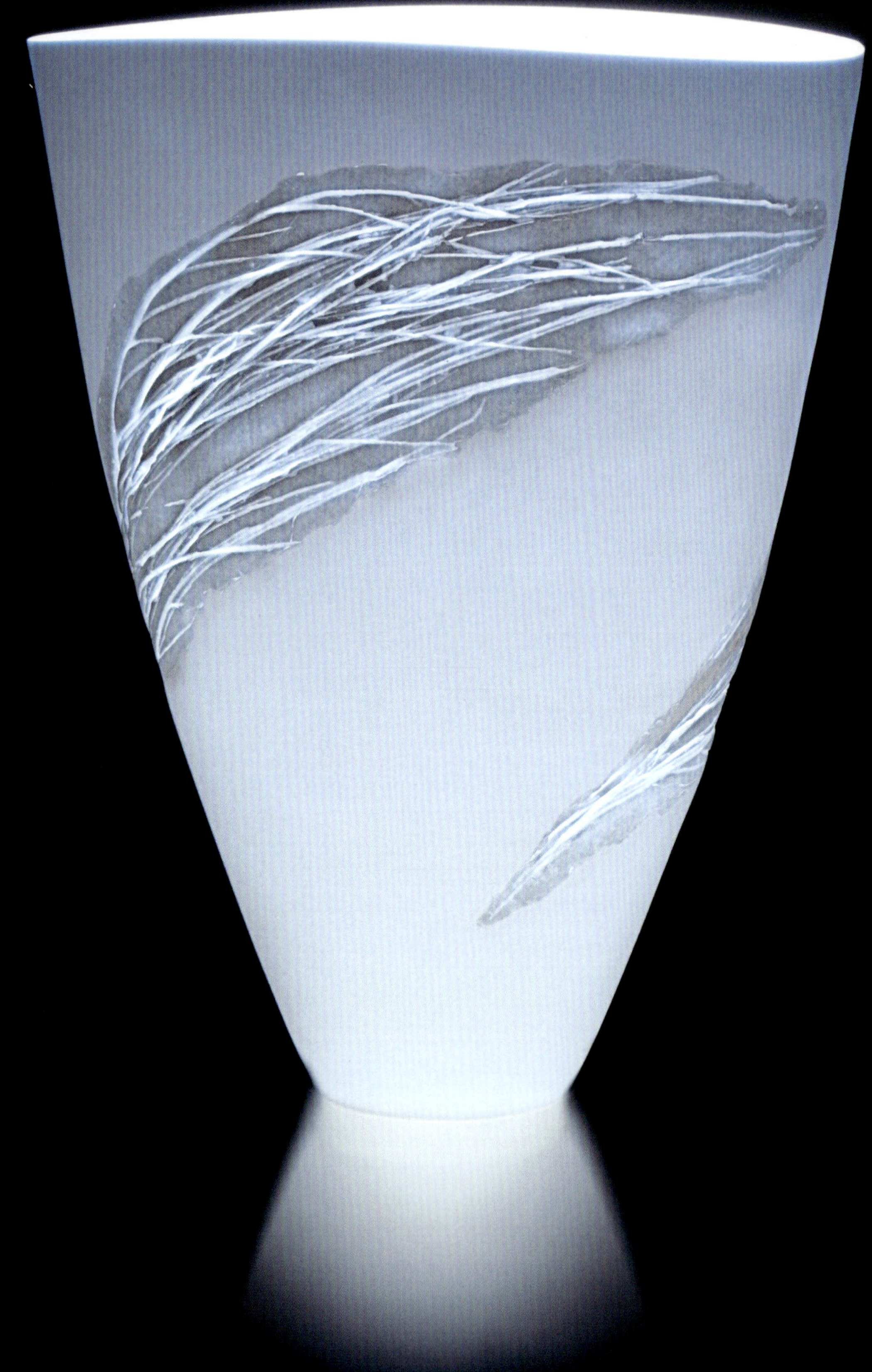

LEFT: *Spindrift*, Angela Mellor, 2015. Slipcast bone china with paperclay inlay. 25 x 20 cm (h x w). *Photo: Stephen Bond.*

BELOW: *Forest Light* bowl, Angela Mellor, 2014. Bone china with paperclay inlay. 8 x 10 cm (h x w). *Photo: Stephen Bond.*

The *Forest Light* series

I picked up a variety of leaves during further dog walks around Thetford Forest. This time, I was inspired by ferns. The *Forest Light* works were shown in the *Féerique* Christmas exhibition at Galerie du Don, France.

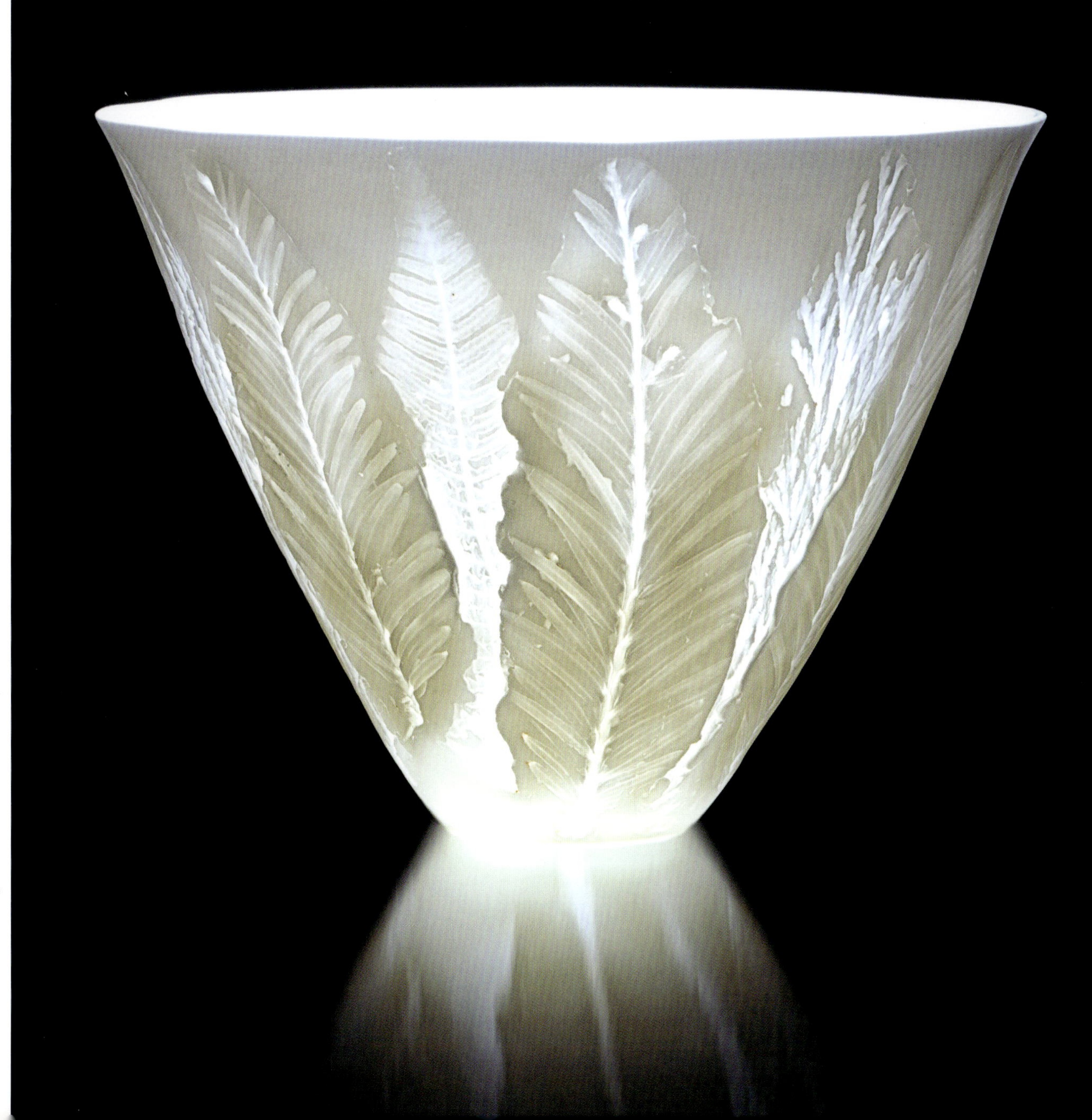

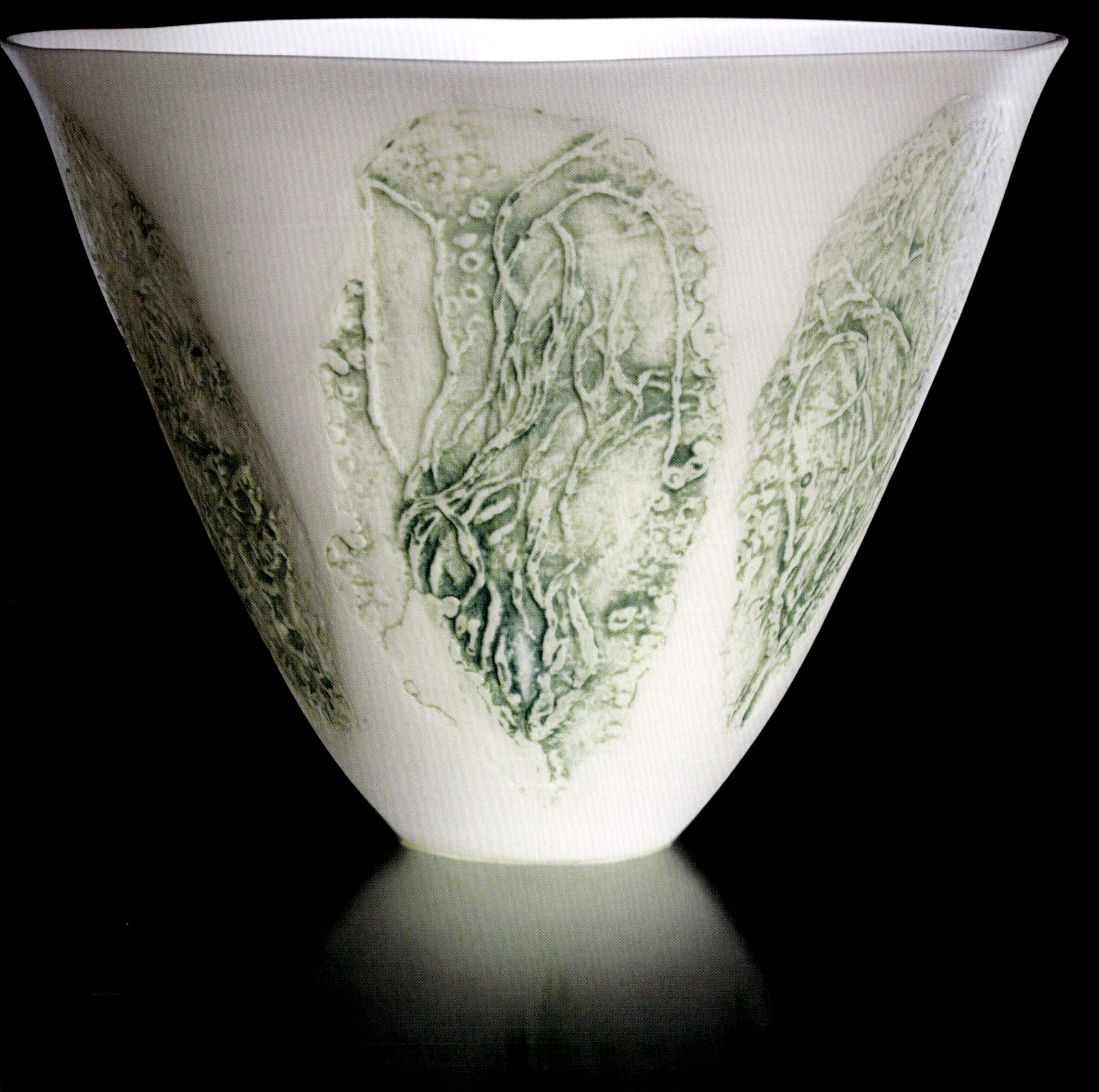

ABOVE: *Féerique*, Angela Mellor, 2014. Bone china with paperclay inlay, painted with underglaze colours. 10 x 13 cm (h x w). *Photo: Stephen Bond.*

The *Sculptured Light* series

In 2017, my work took a new direction. I began working on a larger scale, making more sculptural pieces entirely hand-built in textured paperclay.

Sculptured Light was a radical departure from previous series and marked an exciting development in manipulating bone china paperclay, a medium I had by then been pioneering for over 25 years.

These energetic forms grew out of a love of natural textures, the interplay of light and shade, and a desire to push paperclay to its absolute limits. No two are the same, as each one is painstakingly hand-built.

This sculptural work embraces bone china's tendency to warp and shift during the firing process – its very readiness to dance is what gives each piece its own character. Each of these sculptures captures a moment of movement, perhaps an unfurling bud, or tidal currents against the shoreline, and the corresponding interplay of light and shade. The rhythm of the inspiration soon gives way to something deeper during a hand-building process that pushes the medium of paperclay to its limits. Each new form begins to take on a life of its own, finally awakening in the magic of the kiln.

Blue Mountains was the first sculpture in the series. It was inspired by Australia's Blue Mountains, in homage to my late husband Manfred, who worked there at one time.

Kabuki Dancer embodied the graceful movement of a Japanese dancer.

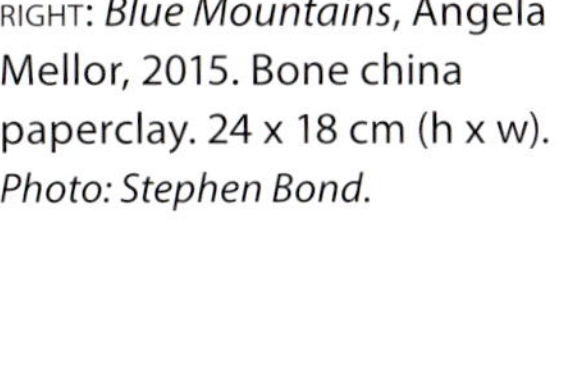

RIGHT: *Blue Mountains*, Angela Mellor, 2015. Bone china paperclay. 24 x 18 cm (h x w). *Photo: Stephen Bond.*

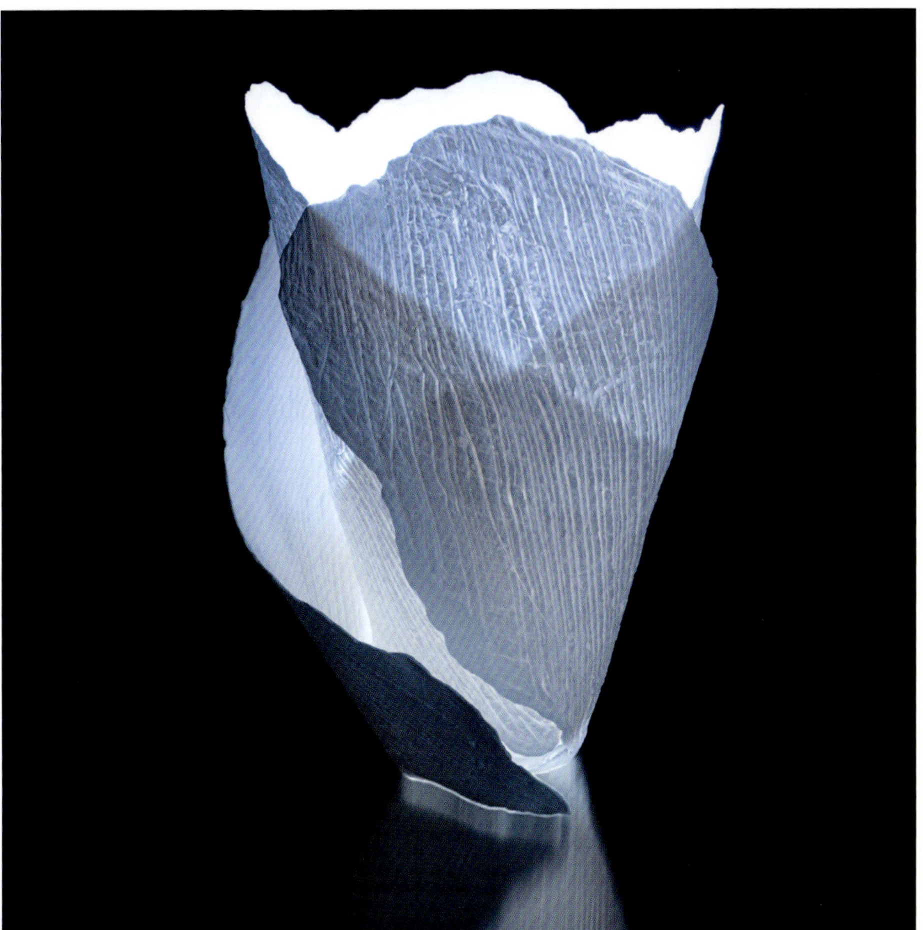

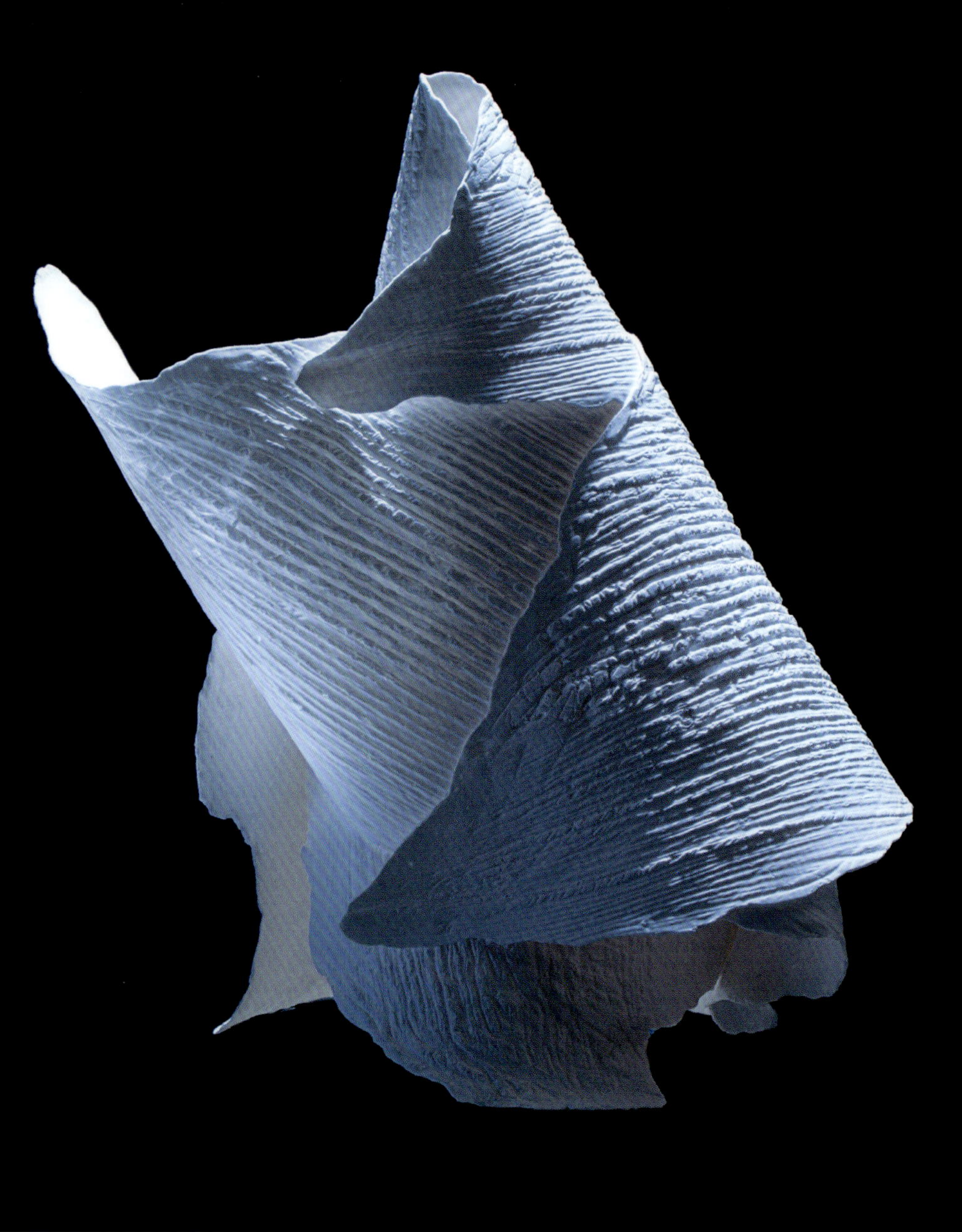

LEFT: *Kabuki Dancer*, Angela Mellor, 2015. Bone china paperclay. 26 x 23 cm (h x w). *Photo: Stephen Bond.*

OPPOSITE TOP LEFT: White orchid, showing the detail of the vein structure in sunlight.

OPPOSITE TOP RIGHT: *Unfurled Light I*, Angela Mellor, 2016. Bone china paperclay. 24 x 36 cm (h x w). *Photo: Stephen Bond.*

OPPOSITE BOTTOM: *Unfurled Light II*, Angela Mellor, 2016. Bone china paperclay. 31 x 48 cm (h x w). *Photo: Stephen Bond.*

The sculptures in this series formed part of an exhibition for Byard Art Gallery in Cambridge. Some of them were reminiscent of flower forms as petals unfurling. Other sculptures were inspired by waves breaking on the beach.

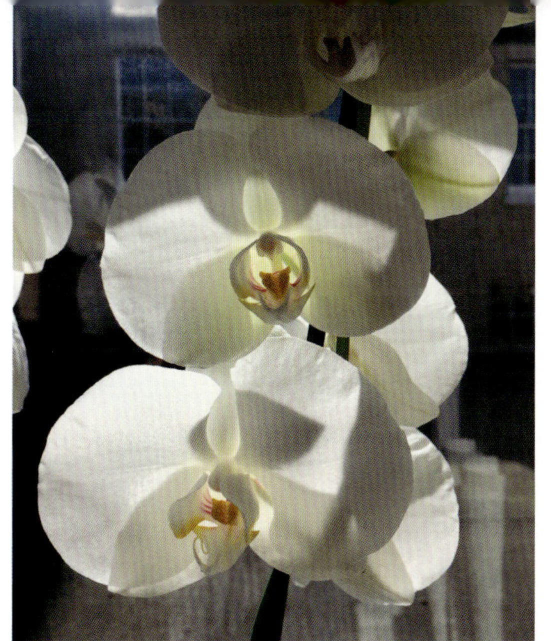

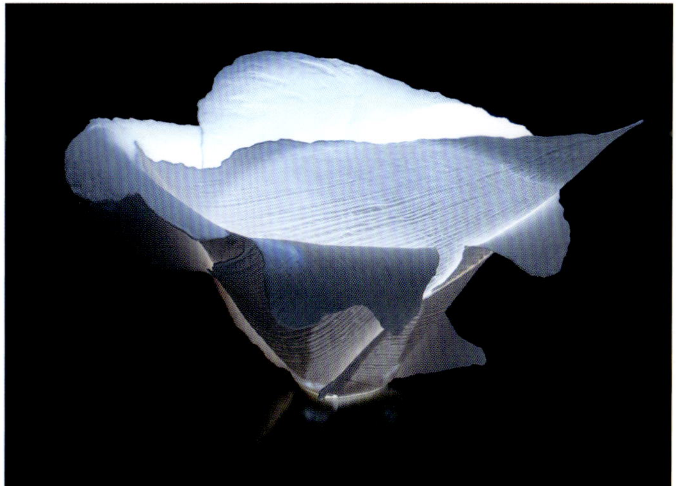

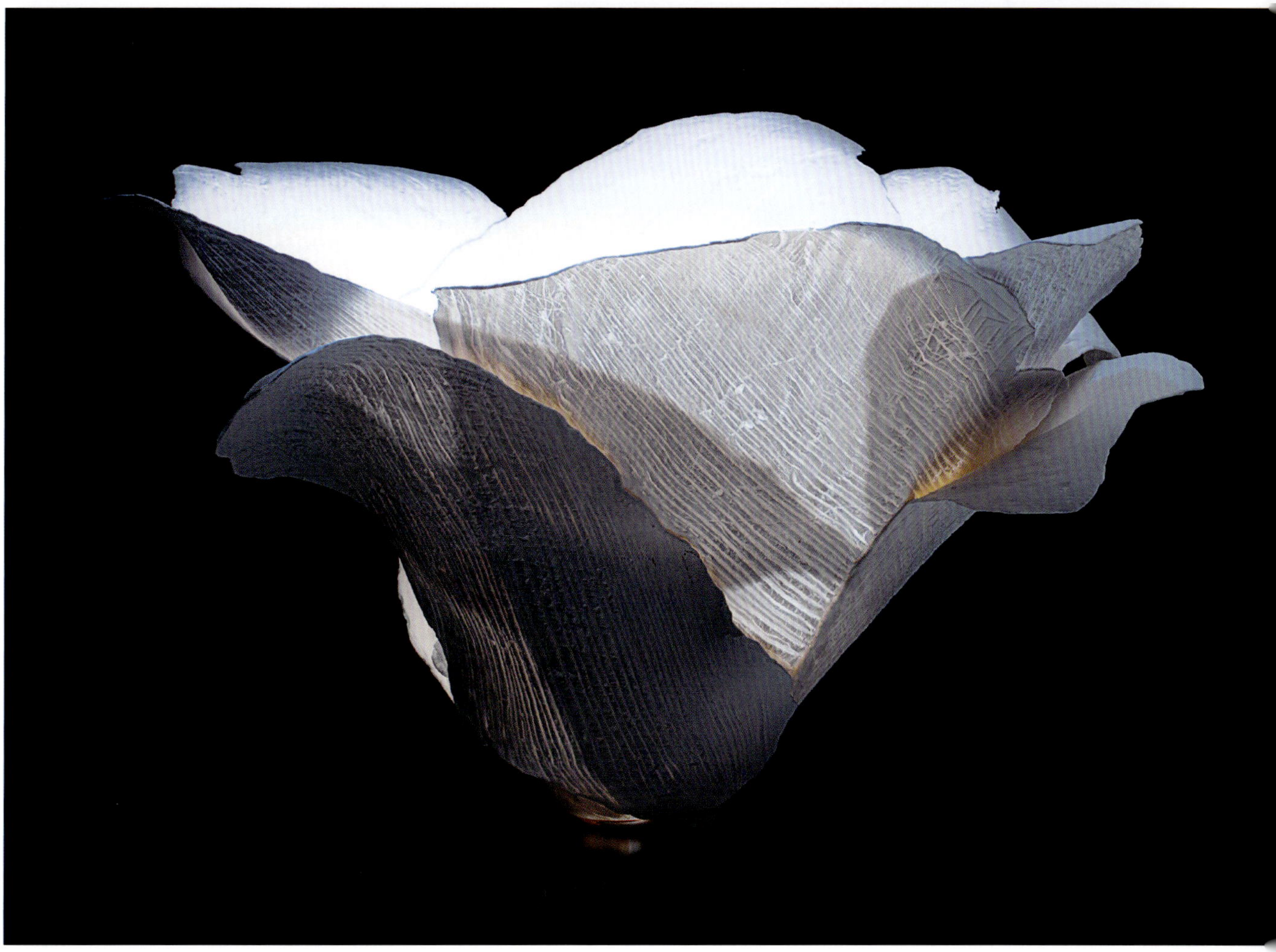

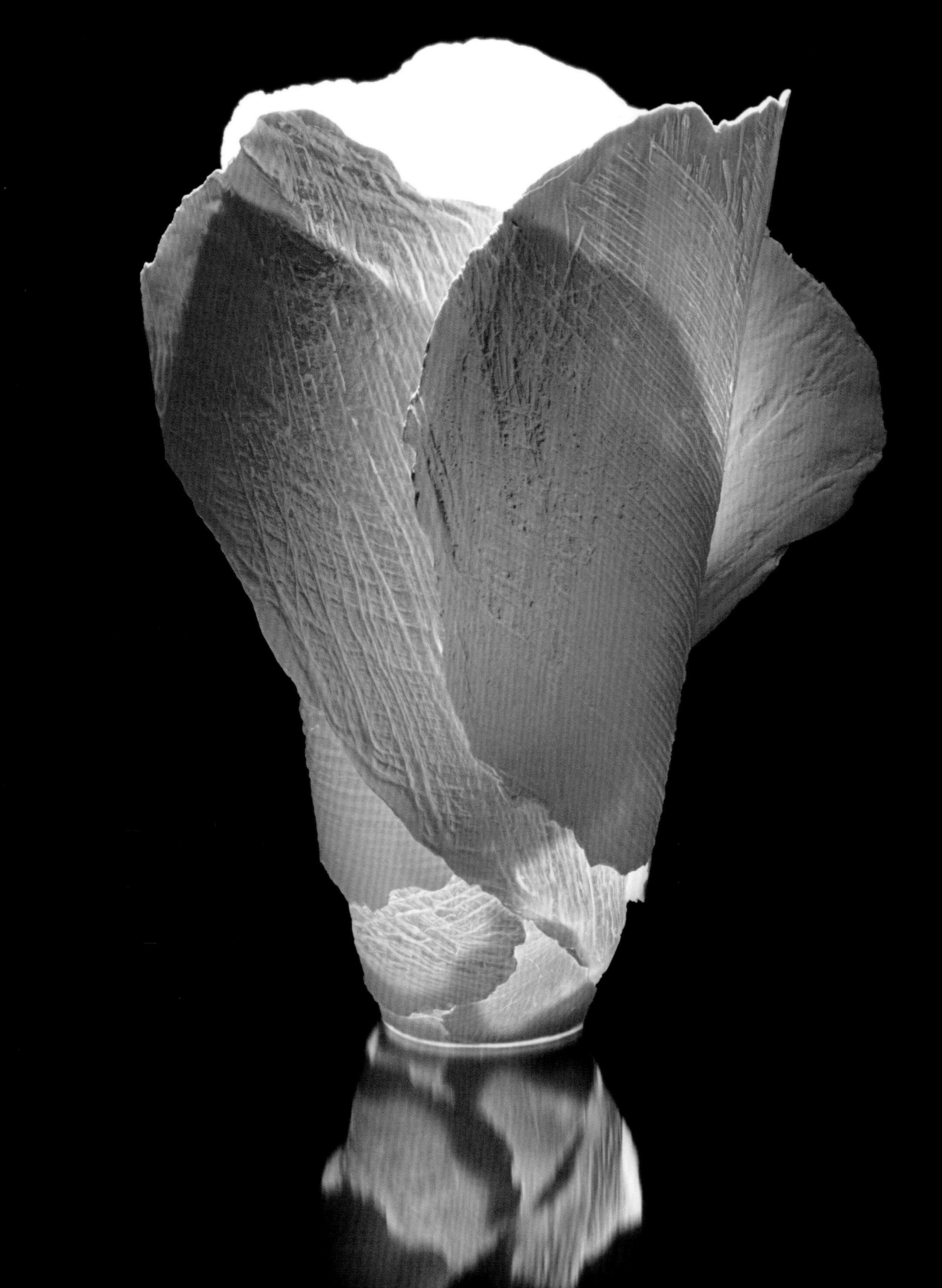

OPPOSITE: *Unfurling Light I*, Angela Mellor, 2016. Bone china paperclay. 24 x 18 cm (h x w). *Photo: Stephen Bond.*

RIGHT: *Unfurling Light II*, Angela Mellor, 2016. Bone china paperclay. 40 x 28 cm (h x w). *Photo: Stephen Bond.*

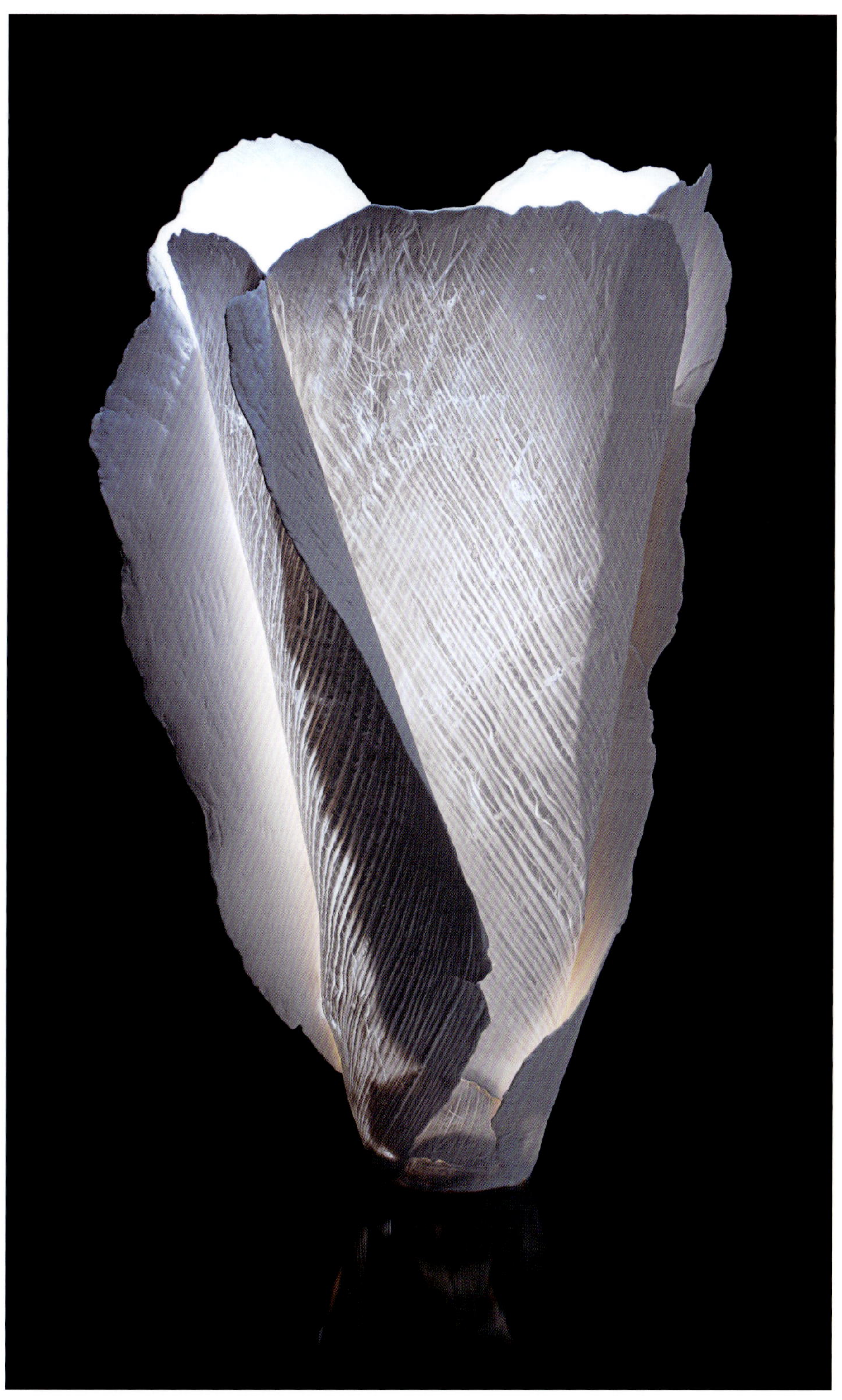

RIGHT: *Unfurled Light*, Angela Mellor, 2018. Bone china paperclay. 27 x 35 cm (h x w). *Photo: Stephen Bond.*

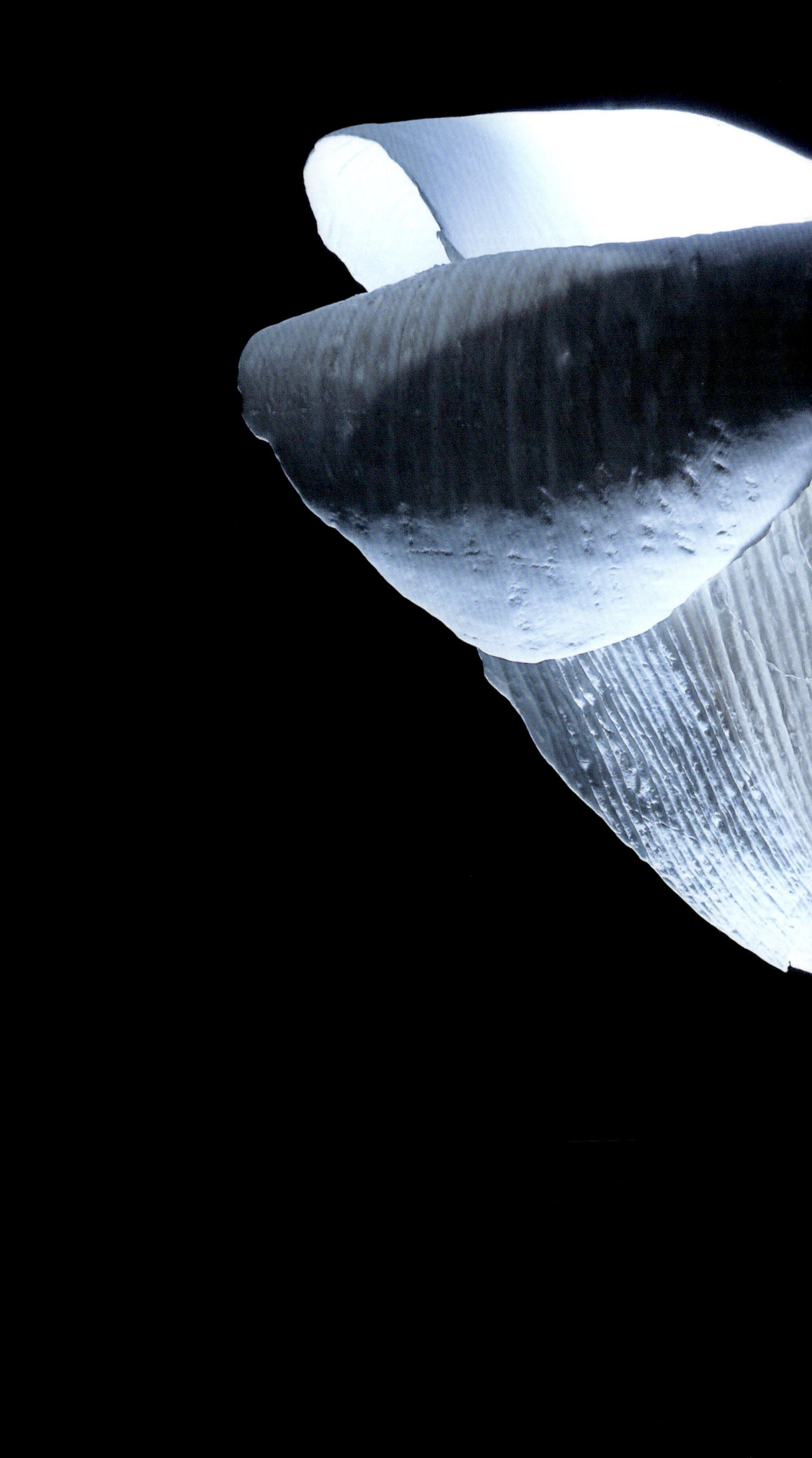

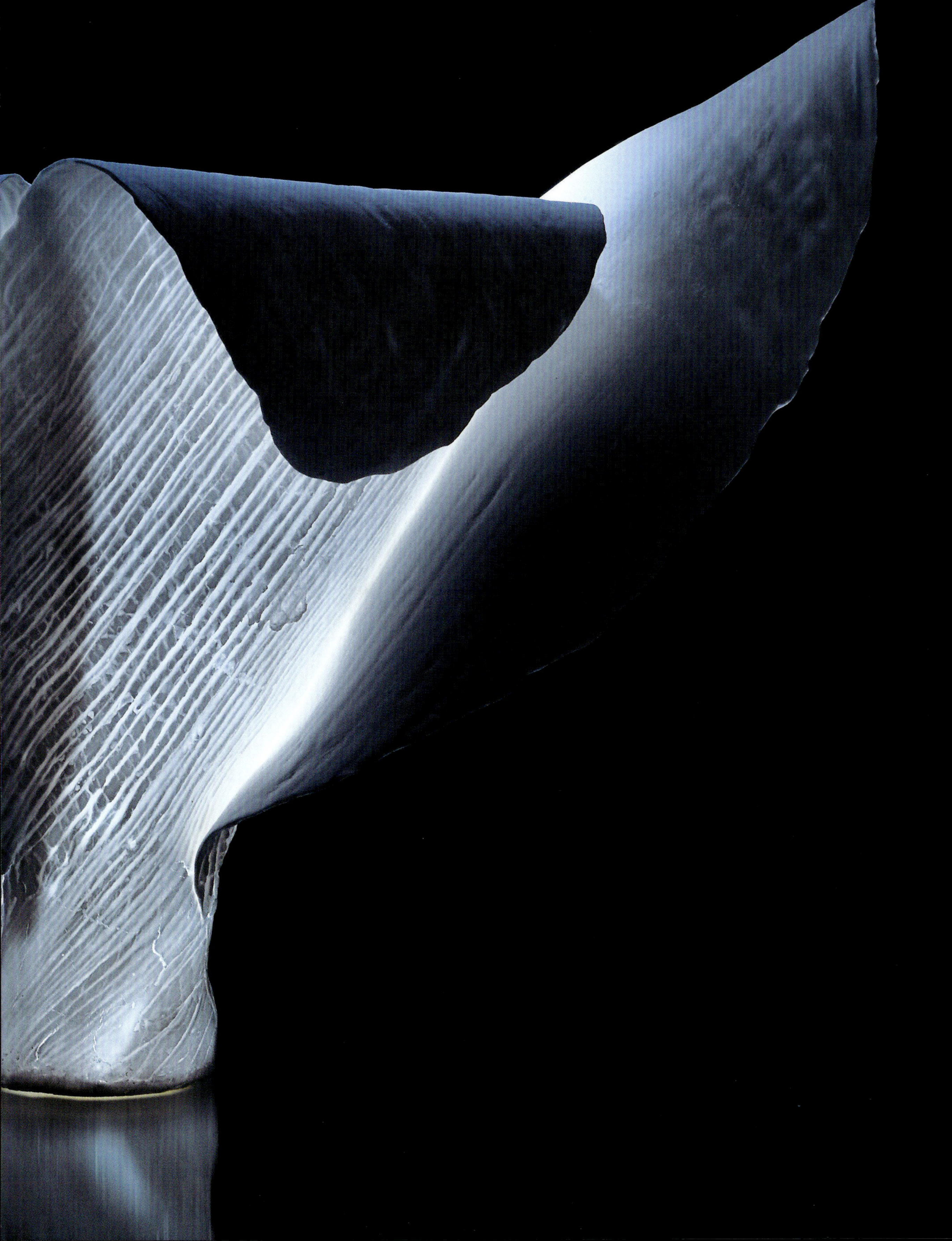

The *Breaking Waves* series

Breaking Waves is a further departure from my earlier work and the title relates not only to freezing in time the dramatic movement of breaking waves, but also to breaking new ground. They form part of my ongoing process of research and development in the use of bone china paperclay, which I have now been championing for over 30 years.

In 2018, Andrew Lamberty, a London gallerist, wrote of *Breaking Waves*:

> *Angela Mellor is an important contemporary ceramicist. She is widely recognised around the world as a special talent, particularly in Japan where the art of pottery is very admired. She has work in the prestigious Musée de Sèvres in Paris.*
>
> *There is a compelling fascination to observing waves. No two are the same, they have come from afar, and they speak of long experience reaching a destination. They have a mysterious story, each collapsing in a fascinating cascade of topography, tide and sky.*
>
> *We enjoy that they whisper to us of a fate.*
>
> *Angela's latest collection marks her evolution into ceramic sculpture, and her desire to capture a breaking moment and freeze it in time. Her œuvre of previous pots have all suggested an inclination to this arrival at pure sculptural form.*
>
> *These latest beautiful, delicate, diaphanous swirls of dancing light-lustred paper bone china waves were inspired by Angela's observations of the sea meeting the coast of Anglesey, North Wales. White horses.*
>
> *Acknowledging Sino-Japanese aesthetics and traditions of the Gongshi, the sculptures in this extraordinary collection are presented on slate from the beaches which inspired their creation.*
>
> *And in so doing, the artist has petrified translucent fluid in the moment of meeting land.*
>
> *Every work is of course unique, as such special, and cannot be re-captured, a moment in time.*
>
> *But actually, one can anticipate recapturing this rare ossified 'moment' of a prancing wave.*
>
> *Light it from above in a darkened room, or enjoy the interplay of light and shade over its textural surfaces in natural light.*
>
> *Once more you will re-experience its beauty, and revivify that rare delight that gives this collection its life.*[1]

[1] Review of *Breaking Waves* exhibition by Andrew Lamberty, Gallerist, *Art and Antiques*, London, December 2018.

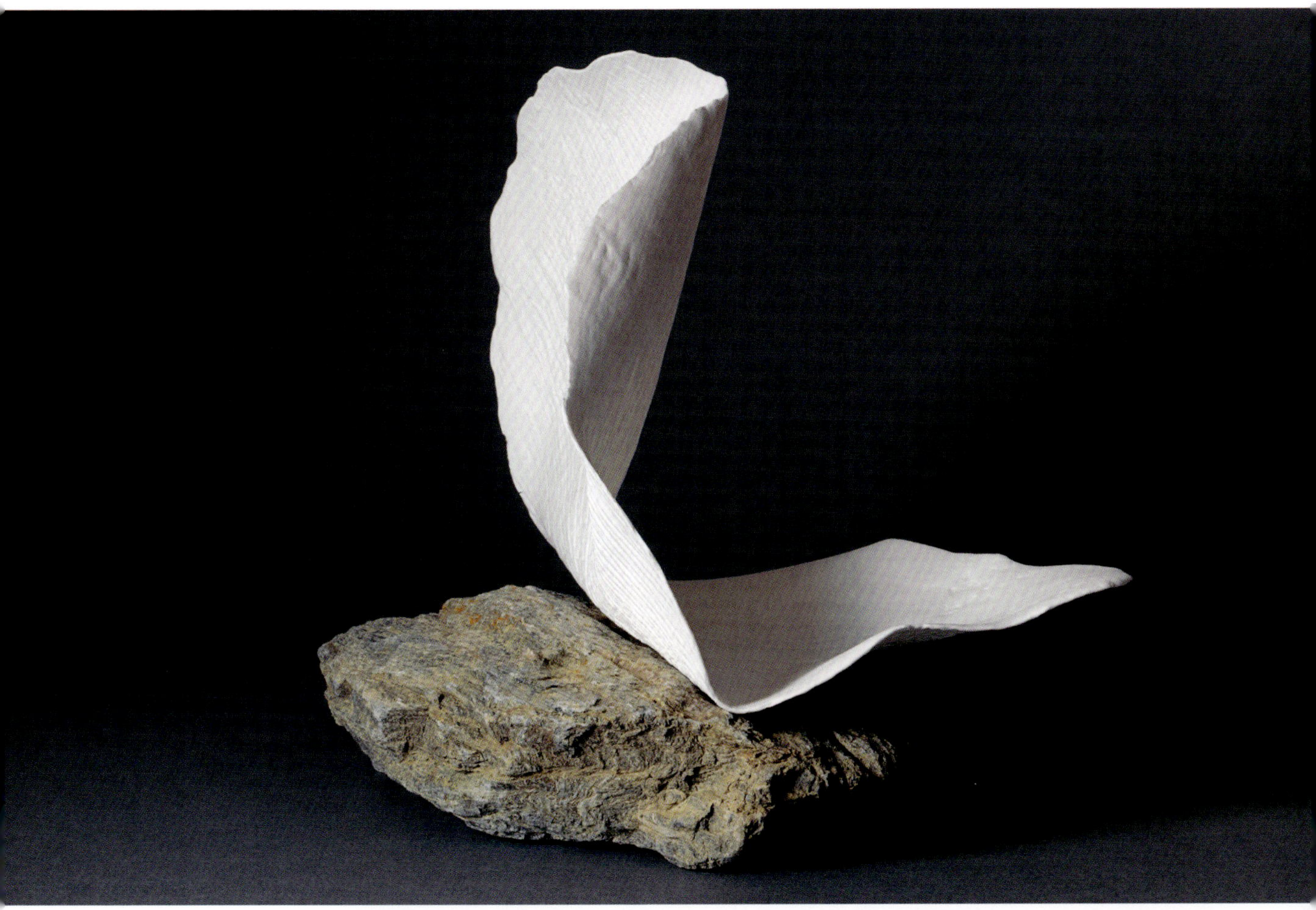

ABOVE: *Upsurge*, Angela Mellor, 2018. Bone china paperclay, slate. 24 x 32 cm (h x w). *Photo: Stephen Bond.*

RIGHT: *Ripcurl*, Angela Mellor, 2018. Bone china paperclay, slate. 17 x 32 cm (h x w). *Photo: Stephen Bond.*

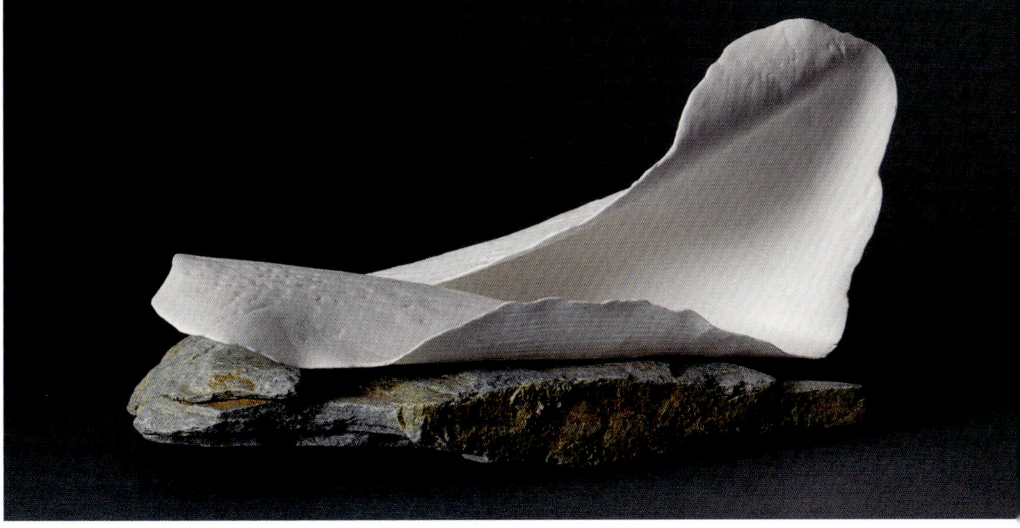

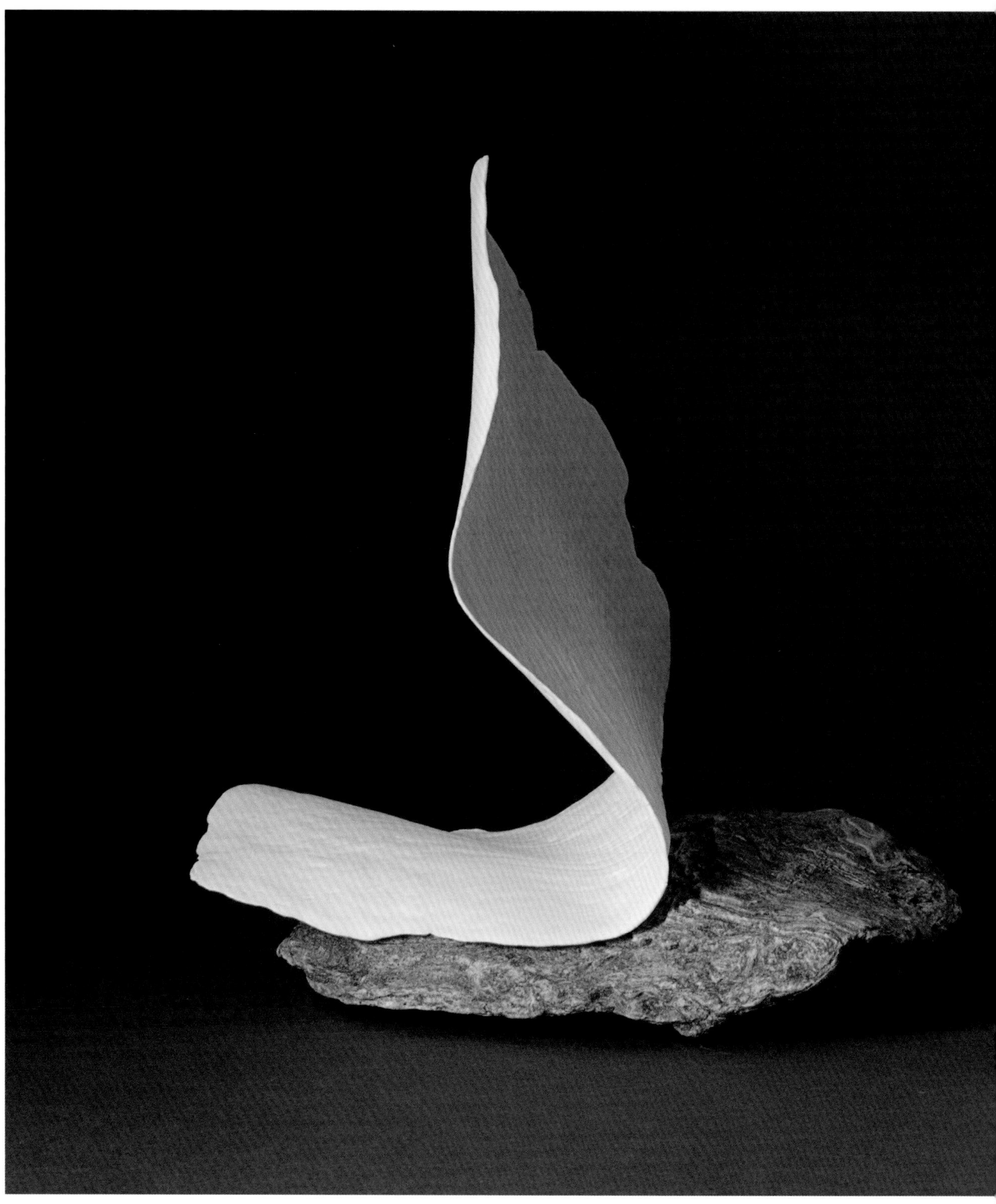

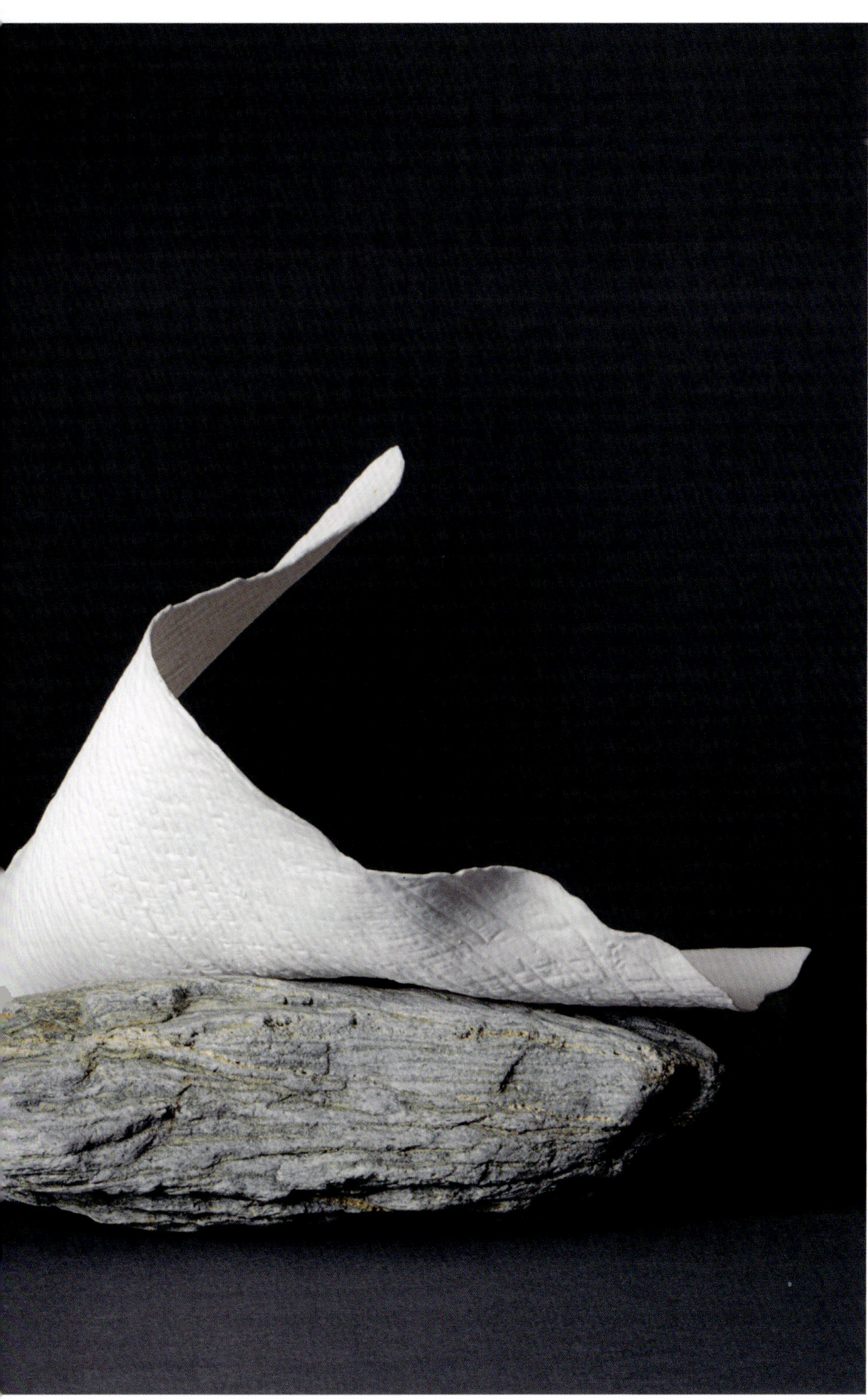

LEFT: *Dancing Waves*, Angela Mellor, 2018. Bone china paperclay, slate. 24 x 23 cm (h x w); 20 x 22 cm (h x w). *Photo: Stephen Bond.*

NEXT PAGE:
TOP LEFT: *Curling Waves*, Angela Mellor, 2018. Bone china paperclay, slate. 20 x 34 cm (h x w). *Photo: Stephen Bond.*

BOTTOM LEFT: *Rolling Waves*, Angela Mellor, 2018. Bone china paperclay, slate. 20 x 24 cm (h x w). *Photo: Stephen Bond.*

TOP RIGHT: *Crashing Waves*, Angela Mellor, 2018. Bone china paperclay (spotlit from above). 27 x 44 cm (h x w). *Photo: Stephen Bond.*

BOTTOM RIGHT: *Crashing Waves*, Angela Mellor, 2018. Bone china paperclay, slate. 27 x 44 cm (h x w). *Photo: Stephen Bond.*

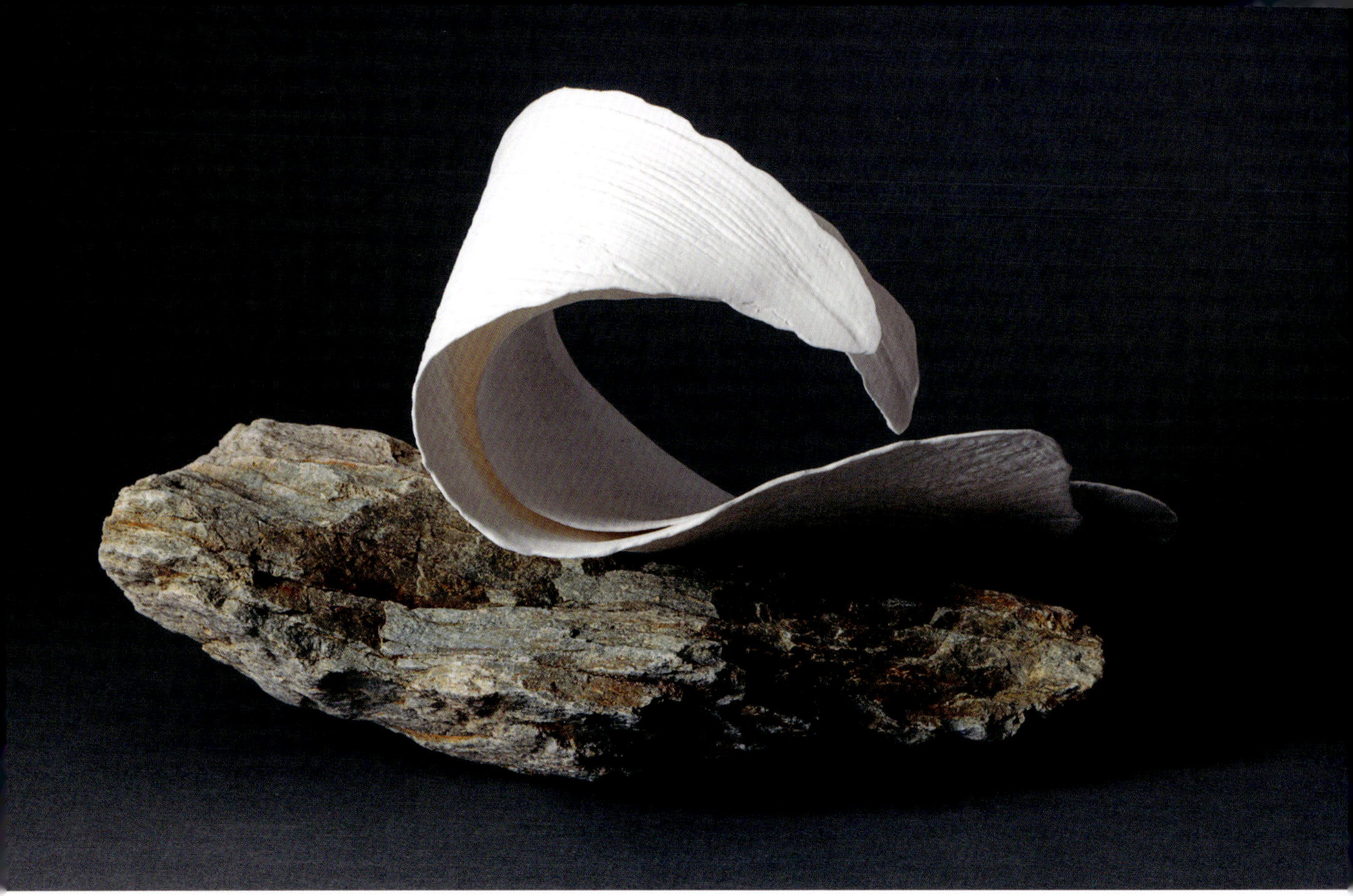

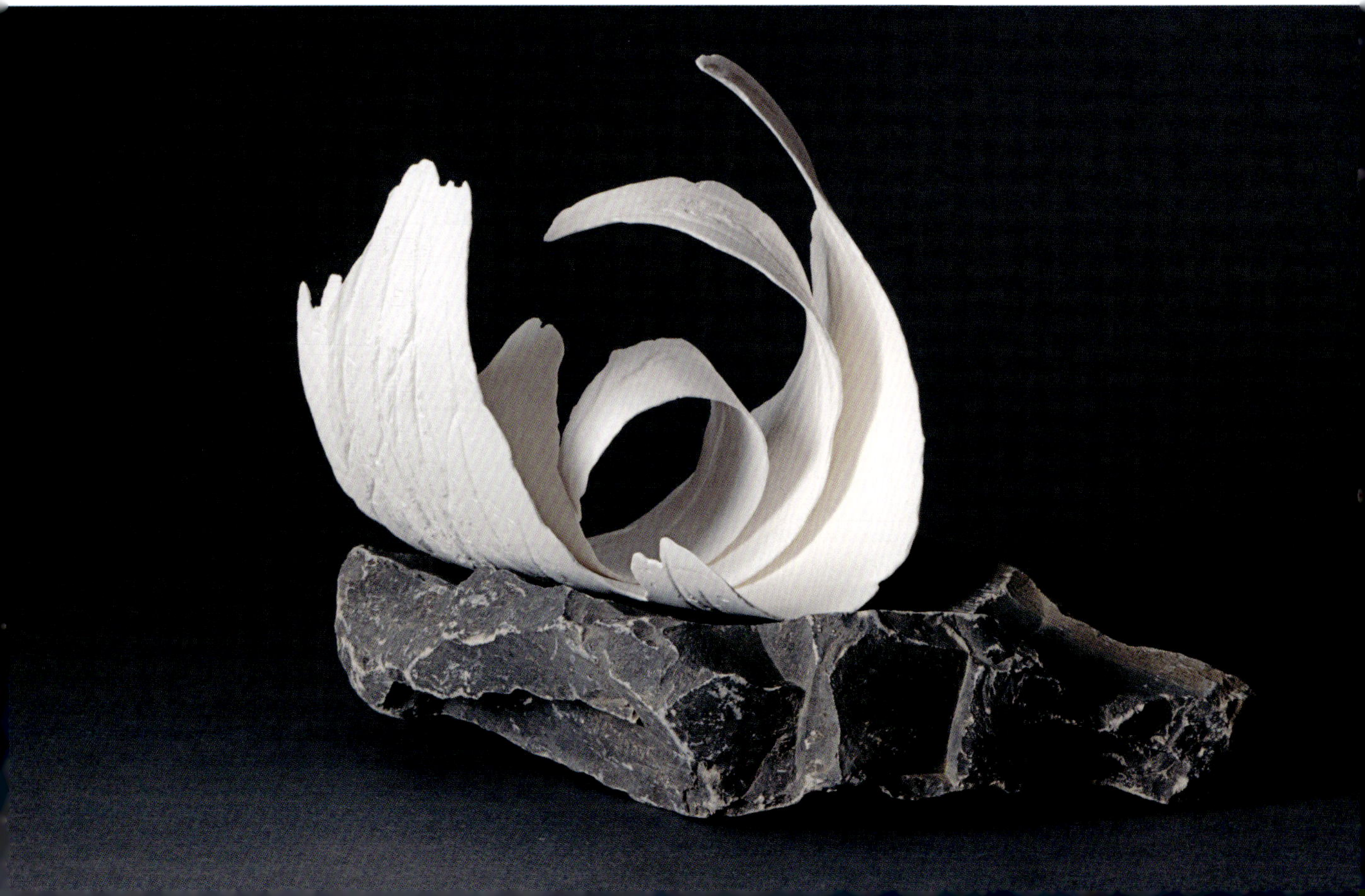

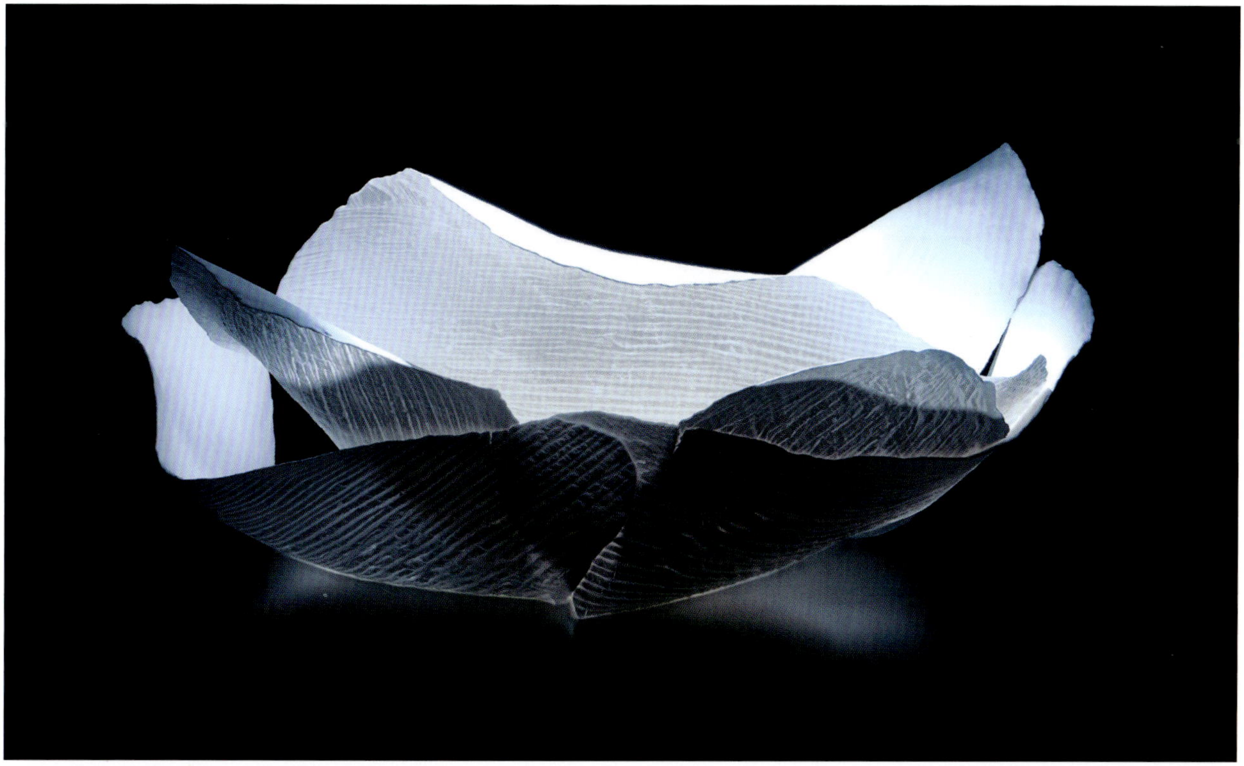

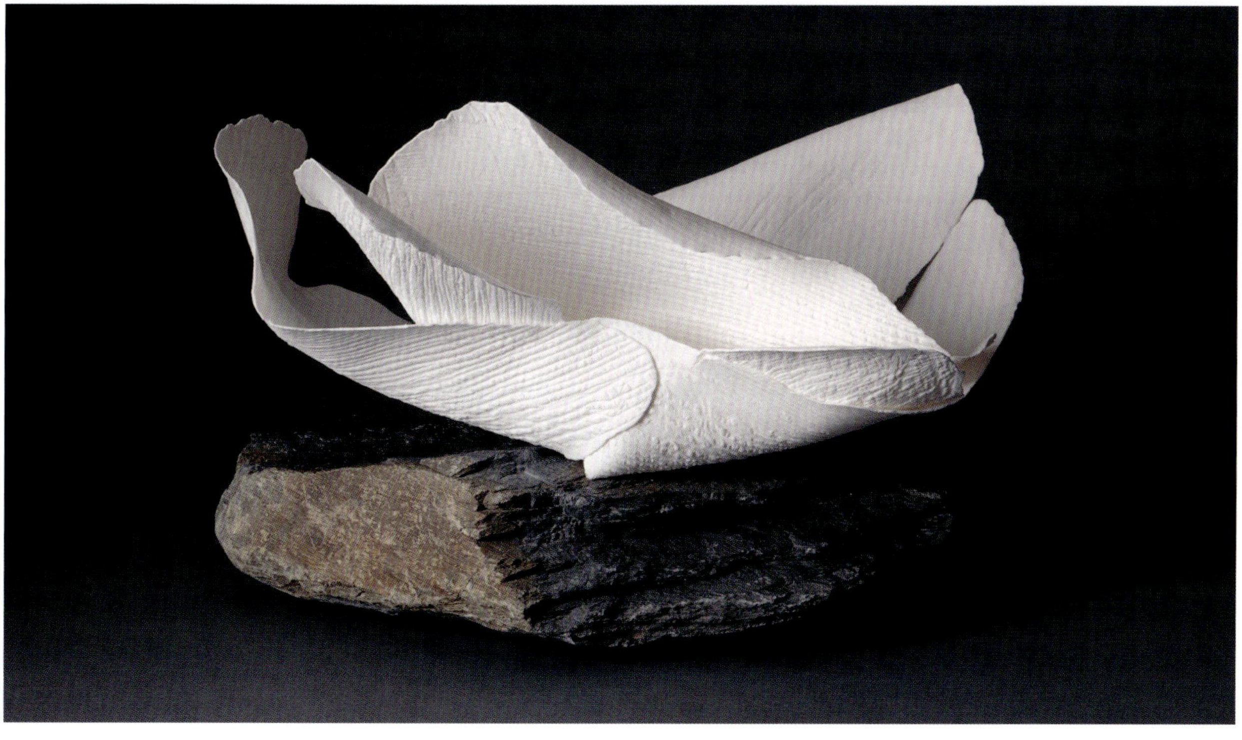

The *Unfurled Light* Series

The final development of *Sculptured Light*, and a final investigation of flower forms, was *Unfurled Light*. Made in 2018, the design was inspired by a cyclamen plant. When I took the work out of the kiln, the petals had unfurled slightly more than I anticipated, due to the idiosyncratic nature of bone china. After some thought I realised that it needed to be elevated, which posed a problem ... with what?

I felt that paperclay would not be suitable, as it would be too light once fired. I turned my thoughts to a more robust solution that would also complement the white paperclay, and decided on a turned aluminium base. A local company made this for me to my specifications. I was happy with the result and submitted this work for the *Particle & Wave: PaperClay Illuminated* exhibition as I had been invited to join the Advisory Committee on the basis of my work in paperclay over the years.

The exhibition toured many museums around the USA between 2019 and 2021, culminating at the Fuller Craft Museum in Boston, Massachusetts.

It was the brainchild of the exhibition's driving force, Lori Nelson, whom I had previously met at a paperclay workshop led by Rosette Gault and Jerry Bennett (both from the USA) at the International Ceramics Studio in Kecskemét, Hungary.

The team, comprising five early innovators in the use of fibre and clay, hired a curator, Peter Held, a studio ceramic artist and former curator of ceramics at the Arizona State University Art Museum. He reviewed over 200 potential artists and made his selections using his intuition and emotional responses, balanced by consideration of formal qualities of form, scale, and sustained visual engagement. His expertise, dedication, and patience took it to the next level and provided the energy and support to transform the exhibition from an idea to a reality. His selection of artists provided a richness of visual diversity, colour, and texture.

The result was a groundbreaking travelling international exhibition featuring 45 artists from across the world who were redefining the potential of the ceramic arts. It provided historical information about the evolution of paperclay as an artistic medium and showcased the diversity of form and expression that exists today amongst the growing worldwide community of artists.

Sincere thanks go to Lori for her vision and steadfastness in bringing the diversity of paperclay to a wide audience.

The mission of the exhibition was to:

> *... stimulate a conversation that explores innovation, creativity, and connectiveness by sharing the breadth of work being created internationally by artists who are incorporating cellulose-fibers in clay as part of their studio practice ... to inspire other artists to be limitless in their creative processes by showing the versatility of paperclay as a material enabling one to push the boundaries of traditional ceramics and to encourage experimentation with other new materials.*[2]

[2] Lorie Nelson, 'Paperclay Illuminated: Journey to an Exhibition,' *Studio Potter*, Vol. 47, No. 1, 2019, p.71.

BELOW: *Unfurled Light*, Angela Mellor, 2018. Bone china paperclay, aluminium base. 34 x 35 cm (h x w). *Photo: Stephen Bond.*

This was the last exhibition I took part in due to the COVID-19 outbreak in 2020. It coincided with a time when I was not in good health, and I realised it was the right moment to leave my beautiful gallery and studio in Ely and move to the country. I live on a quiet riverbank not far outside Ely, which I felt was the perfect spot to write about my work and fascination with translucent ceramics.

I hope you have enjoyed reading about my experience and perspective, and that you have taken something of value from my journey.

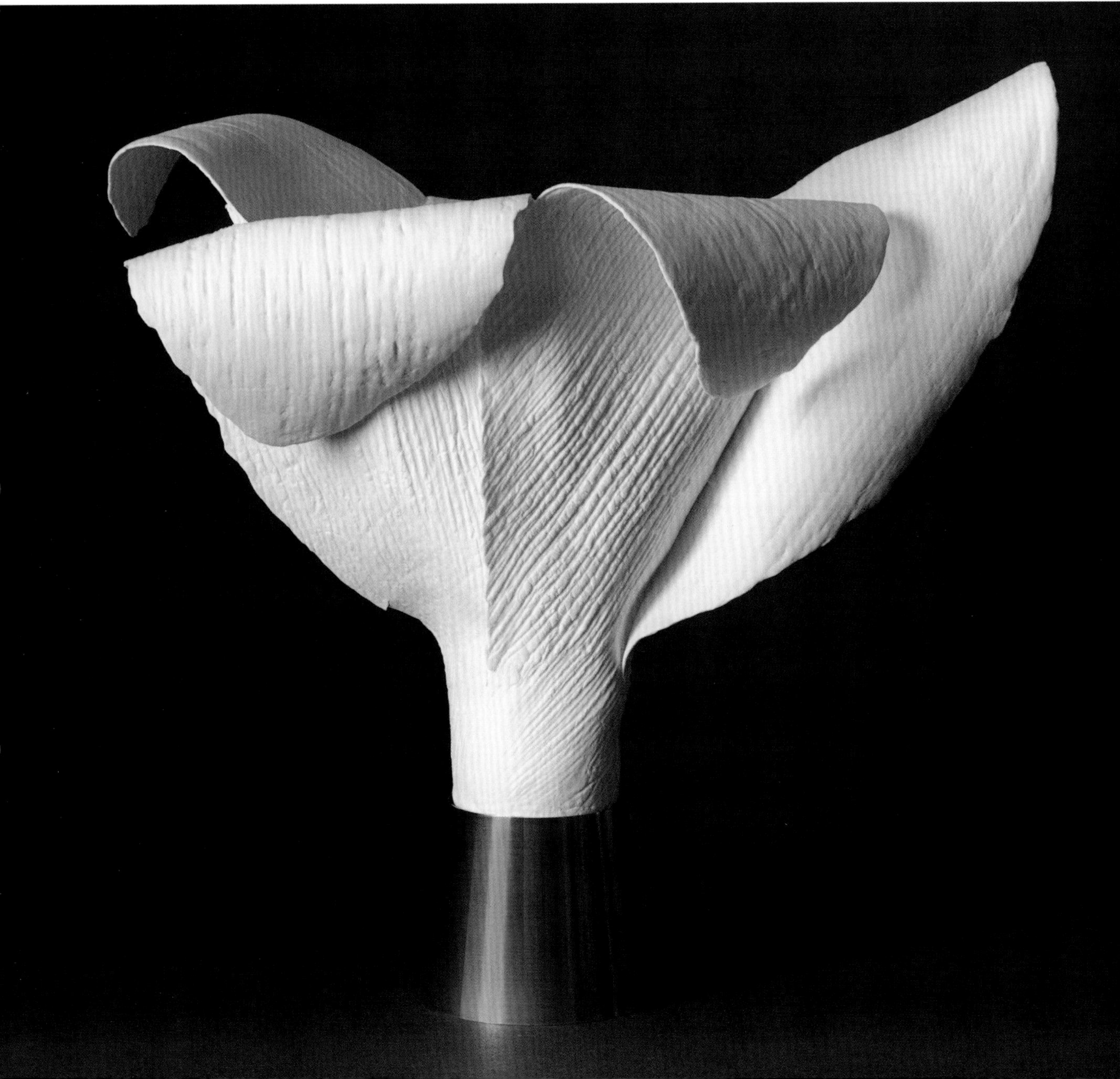

Bibliography

Books – Ceramics

Åse, Arne, *Water Colour on Porcelain: A Guide to the Use of Water Soluble Colourants*, Norwegian University Press, Oslo, 1989.

Atterbury, Paul, *The History of Porcelain*, Orbis, London, 1982.

Axel, Jan and Karen McCready, *Porcelain: Traditions and New Visions*, Watson-Guptill, New York, 1981.

Billington, Dora, *The Technique of Pottery*, Batsford, London, 1962.

Cowley, David, *Moulded and Slip Cast Pottery and Ceramics*, Batsford, London, 1978.

Danbolt, Gunnar, Susan Tyrrell, et al., *From Bowl to Art: Arne Åse and Modern Norwegian Ceramics*, Dreyer Bok, Stavanger, 1994.

Fleming, John and Hugh Honour, *The Penguin Dictionary of Decorative Arts*, Allen Lane, London, 1977.

Fournier, Robert, *Illustrated Dictionary of Practical Pottery*, A & C Black, London, 1992.

Frith, Donald E., *Mold Making for Ceramics*, A & C Black, London, 1992.

Gault, Rosette, *Paper Clay*, A & C Black, London, 1998.

–, *Paperclay: Art and Practice*, Bloomsbury, London, 2013.

–, *Paperclay for Ceramic Sculptors: A Studio Companion*, Art Seed Books, Seattle, WA, 2010.

Holmes, Jonathan, *Les Blakebrough: Ceramics*, Craftsman House, Fishermans Bend, Vic, 2005.

–, *Les Blakebrough, Potter*, Bay Books, Sydney, 1989.

Lane, Peter, *Ceramic Form: Design & Decoration*, A & C Black, London, 1998.

–, *Contemporary Porcelain: Materials, Techniques and Expressions*, A & C Black, London, 1995.

–, *Studio Porcelain*, Pitman House, 1980.

Lane, Peter and David Leach, *Contemporary Studio Porcelain*, A & C Black, London, 2003.

Leach, Bernard, *A Potter's Book*, Faber & Faber, London, 1945.

Lightwood, Anne, *Working with Paperclay and Other Additives*, Crowood, Marlborough, 2000.

Lynn, Martha Drexler, *The Clay Art of Adrian Saxe*, Thames & Hudson, New York, 1993.

Nigrosh, Leon I., *Claywork: Form and Idea in Ceramic Design*, Davis Publications, Worcester, Mass, 1986.

O'Rorke, Margaret, *Clay, Light & Water*, A & C Black, London, 2010.

Rogers, Mary, *On Pottery and Porcelain: A Handbuilder's Approach*, Alphabooks, Sherborne, 1979.

Wardell, Sasha, *Slipcasting*, A & C Black, London, 1997.

Wardell, Sasha and Sue Pryke, *Porcelain and Bone China*, Crowood Press, Marlborough, 2004.

Whyman, Caroline, *The Complete Potter: Porcelain*, Batsford, London, 1994.

Books – Glass

Arwas, Victor, *Glass: Art Nouveau to Art Deco*, Academy Editions, London, 1987.

Bloch-Dermant, Janine, *The Art of French Glass, 1860–1914*, Thames and Hudson, London, 1980.

Bray, Charles, *A Dictionary of Glass: Materials and Techniques*, A & C Black, London, 1995.

Chihuly, Dale and Sylvia A. Earle, *Chihuly Seaforms*, Portland Press, Seattle, WA, 1995.

Cummings, Keith, *Techniques of Kiln-formed Glass*, A & C Black, London, 1997.

Edwards, Geoffrey, *Art of Glass: Glass in the Collection of the National Gallery of Victoria*, National Gallery of Victoria, Melbourne, 1998.

Garner, Philippe, *Emile Gallé*, Academy Editions, London, 1990.

Ioannou, Noris, *Australian Studio Glass: The Movement, Its Makers and Their Art*, Craftsman House, Roseville East, NSW, 1995.

Klein, Dan and Ward Lloyd, *The History of Glass*, Tiger Books International, London, 1997.

Liefkes, Reino, *Glass*, V & A Publications, London, 1997.

Lundstrom, Boyce, *Glass Casting and Moldmaking: Glass Fusing Book Three*, Vitreous Publications, Camp Colton, OR, 1989.

Books – Miscellaneous

Greco, Carlo and Stefano Greco, *Piercing the Surface: X Rays of Nature*, Harry N. Abrams, New York, 1987.

Katoh, Amy Sylvester and Shin Kimura, *Japan: The Art of Living*, Tuttle, Rutland VT, 1990.

Opie, Jennifer Hawkins, *Scandinavia: Ceramics and Glass in the Twentieth Century*, V & A Publications, 1989.

Journal articles

Bergne, Suzanne, 'Colour and Light', *Ceramic Review*, Issue 86, Mar/Apr 1984, pp.6–8.

Blakebrough, Les and Ben Richardson, 'Southern Ice', *Ceramics TECHNICAL*, 1, 1995, pp.3–15.

Bromfield, David, 'Timeless Light', *Ceramics: Art and Perception*, 44, 2001, pp.6–10.

Franks, Tony, 'Bone China', *Ceramics TECHNICAL*, 4, 1997, pp.3–6.

Lewenstein, Eileen, 'Angela Verdon – Light Fantastic', *Ceramic Review*, 142, 1993, pp.27–30.

Nelson, Lorie, 'Paperclay Illuminated: Journey to an Exhibition', *Studio Potter*, Vol. 47, No. 1, 2019, pp.66–71

Nicholls, Andrew, 'Translucent Light. Ceramics and lighting combine in a collaboration between Angela Mellor and Mondo Luce.', *Object*, 42, 2003, pp.36–7.

Smith, Penny, 'Tripping the Light Fantastic: When Art and Science Converge', *Craft Arts International*, 60, 2004, pp.2–4.

Staffel, Rudolf, 'Light Gatherer', *Ceramics: Art and Perception*, 17, 1994, pp.34–8.

Wardell, Sasha, 'Living Light', *Ceramic Review*, No. 111, 1998, pp.15-9.

Web resources

Prika, Nemanja Nikolić, nemaniko.com/about (accessed 25 March 2025).

—, nemaniko.com/porcelainlightpanels/ (accessed 25 March 2025).

Romule, Ilona, www.ilonaromuleporcelain.com/artist-statement (accessed 25 March 2025).

Verdon, Angela, www.angela-verdon.com (accessed 25 March 2025).

Wight, Chris, cone8.co.uk/bio-statement (accessed 25 March 2025).

Workshop notes

Black, Sandra, *Porcelain and Bone China*, Workshop notes, 1997.

Holmesglen TAFE Institute, *Mould-making*, Workshop notes, 1997.

Rowbottom, Jessie M., *Slip-Casting – A Brief Practical Guide*, 1981.

Stokes, Helen, *Mould Making and Glass Casting*, Workshop notes, 1999.

Glossary

Alumina – A white, heat-resistant solid aluminium oxide.

Bat – A wooden board or sheet of plaster to support a model when mould-making. Can also be used to dry out clay.

Blunge, blunger – To blunge is to mix clay and water together to form a slip, usually done in a blunger, which is the container holding the mixture.

Bone ash – Calcium phosphate made from calcined cattle bones; a major constituent of bone china. It is also used to give a milky quality to certain glazes.

Bone china – A soft-paste porcelain developed by the British in the eighteenth century. Contains as much as 50 per cent bone ash, and has the translucency of 'true' hard-paste porcelain.

Calcined – Heating bone ash to a very high temperature, which causes loss of moisture and reduction or oxidation as well as decomposition of other compounds such as carbonates. Temperatures are controlled so that the ash does not reach melting point.

Casting – Pottery formed by pouring slip into a porous mould. When emptied, an even layer of slip coats the mould.

Cottle – A cylinder or band that surrounds an object to be moulded.

Deflocculant – A soluble material added to clay slip to increase its fluidity. It allows the slip to retain a high proportion of clay while remaining fluid enough to be poured out.

Feldspar – An alumino-silicate mineral, as clay is, but it has a proportionally higher content of fluxes.

Fettling – An industrial term for trimming excess clay away from the seams.

Flux – A substance which causes or promotes melting.

Frit – A material used in glazes and enamels which consists of a glass which has been melted, cooled, then ground to a fine powder.

Greenware – Ceramics that have been formed but not fired.

Hard-paste porcelain (*pâte dure*) – 'True' porcelain, fired above 1350°C.

Kaolin (china clay) – The purest and most refractory of clays. One of the major components of porcelain, containing very little iron and therefore white.

Leather hard – A clay body that has semi-dried and is stiff enough to work and decorate.

Natch – A small dimple and corresponding raised bump. Natches are used to align sections of plaster moulds.

Setter – A ceramic support to help reduce warping and cracking of the rims of high-fired clays. Setters are cast and fired at the same as the piece and discarded after firing. A pre-fired refractory ring can be used as a support during firing.

Shellac – a natural resin (lac), used as a hard coating.

Slake, slaking – Allowing the dry material (plaster or clay) to combine chemically with the water. Plaster must undergo a few minutes slaking before power mixing. Solid lumps of clay will take longer.

Slip – The suspension of clay in water.

Slipcasting – A process of forming whereby clay slip is poured into a hollow plaster mould, which absorbs water from the slip, leaving a dry, thin-walled object.

Soaking – Maintaining a particular firing temperature in the kiln to enhance translucency and fusion of clay and glaze or to achieve a particular effect.

Sodium silicate – Also known as 'water glass'; a common deflocculant.

Soft-paste porcelain (*pâte tendre*) – Not as hard or as translucent as 'true' porcelain; a product of Europe's attempts to imitate the Chinese clay body. It is fired to approximately 1100°C.

Spare – The extra or 'spare' space at the top of a mould, which acts as a pouring hole for the slip once the mould has been made. It also provides space for extra slip as the level drops during casting and allows for topping up.

Thixotropy – The ability of certain clay suspensions to thicken on standing; a characteristic of over-flocculated slips.

Turning – The process of paring down plaster to achieve a form while the plaster rotates.

Undercut – An area on a model that prevents removal from the mould.

Viscosity – The resistance of a liquid to flow; the opposite of fluidity.

Vitrification – 1: The furthest point to which the clay body can be fired without deformation. 2: The formation of a glassy, nonporous material in the clay body.

Wreathing – Small, uneven ridges or waves on the drained inside surface of a cast body.

Coastal Light bowl (detail), Angela Mellor, 2012. Bone china paperclay. 16 x 34 cm (h x w) *Photo: Stephen Bond.*

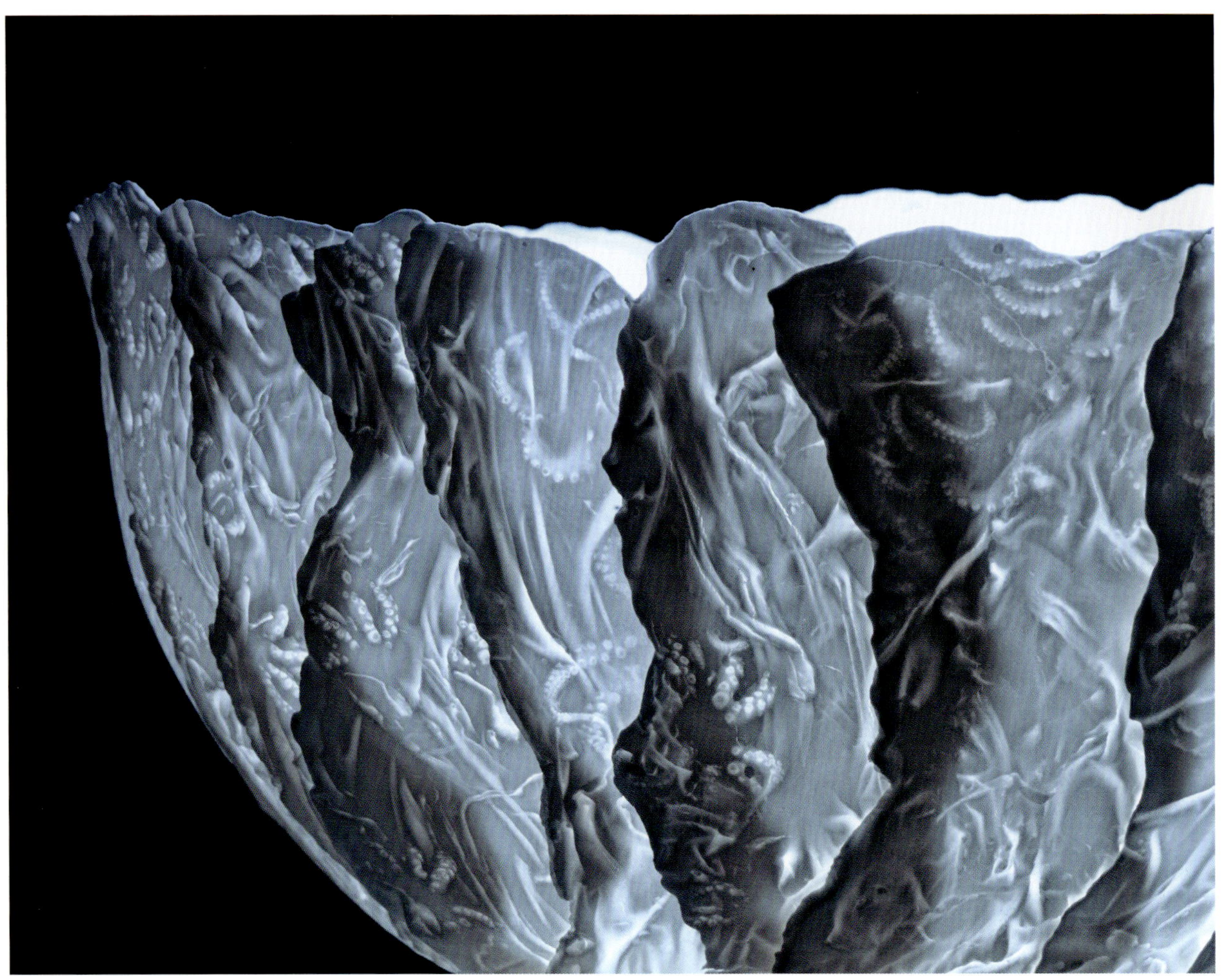

Acknowledgements

I wish to express gratitude to:

Peter Lane, teacher and author, for his valuable advice and encouragement.

My late mother Rose, who made great sacrifices to ensure I had the education to pursue my art ...

Michael and Nicholas, my sons, for their continuous support and trust in me.

Tony Linsell, for his consistent encouragement to write this book.

Louise Barr, for her devoted help in my UK studio and gallery for many years, and for her patience in helping to prepare this book.

Publishing Director, Jayne Parsons and Assistant Editor, Sara Simper for their guidance and support. Special thanks to the Designer, Laura Woussen for collating it all so beautifully.

Carolyn Postgate, past editor for *Anglian Potters* magazine, for her help and advice.

Photographers James Austin, Isamu Sawa, Uffe Schulze, Victor France, Stephen Bond, and Layton Thompson, for capturing the translucency in my work through their excellent photography.

The ceramic artists who have devoted their time to contribute to this book.

To family and friends for their unfailing support.

A special tribute to my late husband Manfred, for his unfailing support and encouragement throughout my early research.

Angela Mellor

Thank you to the contributors:

Arnold Annen, Switzerland
Ingrid Bakker, Holland
Maggie Barnes, UK
Antonella Cimatti, Italy
Dorothy Feibleman, UK/Japan
Margaret O'Rorke, UK
Mark Goudy, USA
Mette Maya Gregersen, Denmark
Françoise Joris, Belgium
Monika Patuszyńska, Poland
Nemanja Nikolić Prika, Serbia
Ilona Romule, Latvia
John Shirley, South Africa
Anne Türn, Estonia
Angela Verdon, UK
Sasha Wardell, UK
Chris Wight, UK
Patty Wouters, Belgium

Index